God's Clockmaker

1. Richard of Wallingford shown as abbot, in the act of dividing a circular instrument, probably meant to be his albion. Hanging in the alcove is a quadrant, while the books on the floor perhaps symbolise his many writings. British Library, MS Cotton Claud. E.iv, fol. 201r. (*British Library*)

God's Clockmaker

*Richard of Wallingford
and the Invention of Time*

John North

Hambledon and London

London and New York

Hambledon and London
102 Gloucester Avenue
London NW1 8HX

175 Fifth Avenue
New York, NY 10010

First Published 2005

ISBN 1 85285 451 0

© J. D. North, 2005

The moral rights of the author have been asserted

All rights reserved.
Without limiting the rights under copyrights
reserved above, no part of this publication may be
reproduced, stored in or introduced into a retrieval system,
or transmitted in any form or by any means (electronic, mechanical,
photocopying, recording or otherwise), without the prior
written permission of both the copyright owner and
the above publisher of this book.

A description of this book is available from
the British Library and from the Library of Congress.

Typeset by Van den Linden

Printed on acid-free paper and bound in
Great Britain by Cambridge University Press

Distributed in the United States and Canada
exclusively by Palgrave Macmillan,
a division of St Martin's Press

Contents

Illustrations ix
Preface xv

PART ONE : Foundations

1 Eclipse 3
2 The Black Monks 7
    The Order 8
    Federation 10
    The Monastery 12
3 Wallingford 17
    The Borough 17
    Son of the Smithy 18
    The Priory 22
4 Oxford 27
    The Beginnings of the University 28
    Grosseteste: The Forming of the University 30
    Theology and the Sciences 33
    Gloucester College 38
    Rival Institutions 42
    Nine Long Years and More 47
5 An Astronomer Among Theologians 51
    Cause for Regret 51
    Oxford Theologians Abroad 53
    'Mathematical Pursuits' 56
    Astrology and the Calendar 58
    New Instruments: Rectangulus and Albion 60
    The Astrolabe 64
6 The State of the Kingdom 69
    The Fall of Edward II 69
    Edward III and the Downfall of Isabella 72

PART TWO : An Abbot's Rule

7 A New Abbot	77
Goliath	78
Avignon	79
Why Avignon?	80
Pope John XXII (1316-34)	82
The Road	84
The Throne of Costly Grace	86
Fortune's Wheel	90
8 Reprove, Persuade, Rebuke	93
Discordant Notes	94
A Visitation	96
The Abbot's Dues	98
The Leper	99
A Good Shepherd?	102
Rights	104
9 The Visitor Visited	107
An Abbot in Parliament	107
Balancing the Books	110
Enemies and Friends in Adversity	111
10 The Litigious Abbot	115
Justice Within Whose Law?	118
The Mills of St Albans	119
Hand-Mills and Liberties	123
Morality and Bloodshed	125
Trials by Jury	127
The Men of Redbourn	129
Isabella's Mill	130
Mills, Malt, and Mast	131
Windmills	134
Unflagging Aspirations	135

PART THREE : Time and the Man

11 Builders and Clockmakers	139
The Builder	139
Roger and Laurence of Stoke	141
12 Horologe and History	145
Time and the Hour	145
Water Clocks	147

The First Cluster of Records	153
Perpetual Motion	154
Mechanisms and Motives	160
Astronomical Motives	160
Astronomical Motives Questioned	164
The Mechanical Escapement's First Application?	166

13 The St Albans Clock — 171

The Treatise	172
The Manuscripts	173
The Escapement	175
The Order of Invention	182
The St Albans Striking Mechanism	185
Developments in Italy	190
Richard of Wallingford as Engineer	194
The Building of the Clock	197

14 *Machina Mundi* — 201

The Clock as Instrument	202
Tides and Fortune	206
On Reading an Astrolabe Dial	208
The Sun's Variable Motion	211
The Moon and Dragon	216

15 Legacy — 219

Time the Controller	219
Time's Fell Hand	221
The Man	221
Dissolution and Survival	224

PART FOUR : The Springs of Western Science

16 The Migration of Ideas — 229

The Latin Tradition	230
From Cordoba to Western Monasteries	231
Al-Khwārizmī in England	233
The High Tide of Translation	235
Approaches to the Greek Aristotle	237
Jewish Contributions	240
Parallel Worlds: Theology as Censor	242
Provence and Profatius	243

17 A Primer in Aristotelian Natural Philosophy — 249

18 Natural Philosophy in Oxford — 259

A Metaphysics of Light	260

Grosseteste and Thirteenth-Century Optics	262
Aristotle and Geomety	266
Aristotle and Scientific System	267
Rationalists, Empiricists, and God	271
A New Dynamics	275
A New Kinematics: The Mertonians	278
The Rise and Fall of Aristotelian Science	280

19 The Astronomers — 283

Early Western Astronomy	284
The Renaissance of the Greek Tradition	287
Ptolemy's Almagest	288
A Painful Climb: Student Texts	292
Ptolemaic Planetary Theory	296
Astronomical Tables and Techniques	302
Natural Philosophy and the Astronomers	308
Heaven and the Heavens	312

20 The Astrologers — 319

Early History	320
Oxford Astrology	324
Exafrenon	327

21 Instruments of Thought — 333

Mathematics as Instrument	333
Material Instruments	340
The Rectangulus	345

22 Albion — 351

Early Equatoria	351
Tacit Geometry	358
New Ways with Old Theory	361
Sun, Moon, and Eclipse	367
Curves for Functional Relationships	372
The Fortunes of Albion	373

23 Epilogue — 381

Notes	385
Bibliography	411
Index	425

Illustrations

1	Richard of Wallingford inscribing an astronomical instrument (British Library, MS Cotton Claud. E.iv, fol. 201r)	ii
2	The region with which Richard of Wallingford was most closely associated	19
3	A smithy, as illustrated in Georgius Agricola, *De re metallica* (Basel, 1556)	21
4	The university church of St Mary the Virgin, Oxford	26
5	A bird's-eye view of the Gloucester College buildings, after Ralph Agas	40
6	Gloucester Hall, and surviving buildings of Gloucester College, engraved by David Loggan	41
7	Part of Wenceslaus Hollar's view of Oxford, 1643	44
8	Some points of reference in the Oxford known to Richard of Wallingford	45
9	Merton College looking south, engraved by David Loggan	54
10	The rectangulus	60
11	The albion	61
12	Exploded view of an astrolabe	65
13	A fourteenth-century astrolabe	67
14	The house of Plantagenet and the French royal house	68
15	Plan of the abbey church of St Alban	76
16	A corn-mill driven by an undershot wheel, after a drawing in the twelfth-century *Hortus deliciarum*	120
17	An overshot water wheel, from Georgius Agricola, *De re metallica* (Basel, 1556)	121
18	A simple hand-mill, or quern	122

19	A more ambitious fourteenth-century hand-mill	123
20	A post-mill, after the Luttrell Psalter	124
21	The Tower of the Winds, Athens	144
22	Schematic drawing of a water-clock (after a manuscript in the Archivo de la Corona de Aragón)	149
23	The water-powered drive of an anaphoric clock of Vitruvian type	150
24	The Grand (Vosges) fragment, as it might have been fitted into the anaphoric clock display	152
25	Villard de Honnecourt's design for perpetual motion	155
26	The Alfonsine mercury clock, as redrawn by Manuel Rico y Sinobas (Madrid, 1863)	157
27	Pierre de Maricourt's design for a magnetic motor to provide perpetual motion	159
28	One possible arrangement for the device described in the northern Italian *Fiat columna* text	164
29	Richard of Wallingford, with pastoral staff and abbot's mitre, pointing to his clock in the abbey (British Library, MS Cotton Nero D.7, fol. 20r)	170
30	The variable-velocity drive in the St Albans clock (Bodleian Library, MS Ashmole 1796, fol. 167v)	175
31	The astronomical trains of the St Albans clock (MS Ashmole 1796, fol. 176r)	176
32	The common form of 'verge and foliot' escapement	177
33	The earliest known drawing of the common verge and foliot (from Giovanni de' Dondi's treatise on the *astrarium* in MS 631, Biblioteca Civica, Padua, fol. 13r)	179
34	The main components of the St Albans strob escapement	180
35	The verge's oscillatory motion in the strob escapement	181

36	The strob mechanism, used not as an escapement but to ring a bell repeatedly	183
37	Drawings by Leonardo da Vinci of an escapement of the St Albans type	184
38	Richard of Wallingford's table for his hour-striking mechanism (MS Ashmole 1796, fol. 181r)	186
39	The fourteenth-century striking clock built by Henri de Vick for the French king	187
40	A rendering of the St Albans mechanism for hour-striking	188
41	A Leonardo da Vinci drawing of a mechanism for hour-striking (Biblioteca Nacional, MS 8937, known as Codex Madrid, I, fol. 12r)	189
42	A drawing by Leonardo da Vinci of a mechanism of the general St Albans type, here to produce a shuttling motion (Codex Madrid, I, fol. 7r)	190
43	Thirteenth-century ironwork in the south aisle of the St Albans presbytery	195
44	Typically crude wooden mill gearwork. Detail from Georgius Agricola, *De re metallica* (Basel, 1556)	196
45	Richard of Wallingford's spiral contrate wheel	197
46	A general view of the St Albans clock as recreated for the Time Museum, Rockford, Illinois	200
47	Table for cutting the teeth of planetary wheels in the St Albans clock (MS Ashmole 1796, fol. 116r)	205
48	Fortune turning her wheel (drawn from the twelfth-century text *Hortus deliciarum*, since destroyed)	206
49	Dial of the St Albans clock, as reconstructed for the Time Museum, Rockford, Illinois	209
50	The form of Richard of Wallingford's masterpiece of gear design, his oval contrate wheel of 331 teeth	212
51	A detail of the previous figure	213

52	Geometrical scheme for laying out the oval wheel	214
53	The Aristotelian cosmos. From Alessandro Piccolomini, *De la sfera del mondo: de le stelle fisse* (Venice, 1540)	256
54	Ptolemy depicted as a king, observing with a quadrant. From Gregorius Reisch, *Margarita philosophica* (Strasbourg, 1504)	289
55	Spherical triangle cut by a transversal, illustrating the theorem of Menelaus	291
56	Characteristic loop in the motion of a planet (Mercury passing through Leo and Virgo during August 2003)	297
57	The Greek model for the motion of the Sun	298
58	Simplified version of the Greek model for the apparent motions of Mercury	299
59	View of the paths of the Sun and Mercury, as would be seen from outer space	300
60	Specimen pages from the Alfonsine tables, from the St Albans manuscript closest to Richard of Wallingford (now Bodleian Library, MS Ashmole 1796)	302
61	The standard Ptolemaic model for Venus	304
62	The physical version of the Ptolemaic model for Venus, from Georg Peurbach's *Theoricae*	311
63	Summary diagram for the doctrine of planetary dignities	330
64	Marginal diagrams to assist with the theorem of Menelaus	337
65	Detail of a woodcut from Oronce Fine, *De solaribus horologiis et quadrantibus* (Paris, 1542)	341
66	Woodcut illustration (Strasbourg, 1539) of an 'old quadrant', or 'quadrant with cursor'	342
67	The construction of the new quadrant of Profatius	342
68	Sundial of the type known as the 'ship of Venice' (from the *Gentleman's Magazine* for January 1787)	343

69	Cylinder dial (detail from Hans Holbein, *The Ambassadors*, National Gallery)	344
70	Woodcut frontispiece to the *Computus manualis* printed for student use by Charles Kyrfoth (Oxford, 1519-20)	346
71	Armillary sphere described by Ptolemy in his *Almagest* (after the reconstruction by the engineer P. Rome)	347
72	Torquetum, from Petrus Apianus, *Introductio geographica* (Ingolstadt, 1533)	348
73	The form of Richard of Wallingford's rectangulus	349
74	The method of observing altitudes using the albion	352
75	The Ptolemaic model for the relatively simple case of the planet Venus	353
76	The main plate of the fourteenth-century equatorium in Merton College, Oxford	358
77	The probable form of the disc of epicycles, now missing from the Merton equatorium	359
78	The general arrangement of the discs of the albion	360
79	Construction advocated by Richard of Wallingford for the angle subtended by an epicycle at the Earth	362
80	Albion's so-called 'eccentric of the epicycle of Mercury', which takes into account the non-circular form of the Mercury deferent	365
81	A seventeenth-century example of the use of the universal astrolabe (*saphea*) projection as used on the albion (Museum of the History of Science, MS Radcliffe 74)	366
82	The general pattern of the lunar and solar eclipse instruments on the albion	368
83	Johannes Schöner's solar eclipse instrument; woodcut from his *Aequatorium astronomicum* (Nuremberg, 1521)	369
84	Johannes Schöner's lunar eclipse instrument; woodcut from his *Aequatorium astronomicum* (Nuremberg, 1521)	369

85	The lines for unequal (seasonal) hours as drawn on the plate of an ordinary astrolabe	370
86	The instrument for determining the true times of conjunctions of the Sun and Moon on the albion	371
87	Part of the brass albion now in the Astronomical Observatory of Monte Mario, Rome	374
88	One of the instruments in Petrus Apianus, *Astronomicum Caesareum* (Ingolstadt, 1540), following Richard of Wallingford's lead	376
89	Another of Apian's many instruments in the same work and tradition, here adopting non-circular contour lines	377

Acknowledgements

The publisher and author wish to thank the following institutions for their kind permission to reproduce illustrative material: Biblioteca Civica, Padua (Fig. 33); Biblioteca Nacional, Madrid (Figs 41, 42); Bodleian Library, University of Oxford (Figs 30, 31, 38, 47, 60); British Library, London (Figs 1, 29); Merton College, Oxford (Fig. 76); Museum of the History of Science, University of Oxford (Figs 13, 81); National Gallery, London (Fig. 69).

Preface

RICHARD OF WALLINGFORD was the most original English scientist of the later middle ages. His life began with few advantages and ended prematurely and in great misery, but it was lived with a burning intensity. He studied and taught mathematics and astronomy at Oxford, England's premier university, going on to become abbot of St Albans, England's premier monastery. At Oxford, he made important contributions to mathematics, and designed new astronomical instruments. At St Albans, in an environment that was anything but tranquil, he designed an extraordinary clock. It is the very earliest mechanical clock of which we have detailed knowledge, and in several respects was without equal in the following two centuries.

How should a person of such diverse talents be remembered? That he is not well known to history has much to do with the difficulty of his scientific thought. The biographer of a man born more than seven hundred years ago—even though he was an abbot whose life merited a lengthy account in his abbey's chronicles—does not have the luxury of personal documents to leaven the account, or to expose his subject's personal feelings. Richard's recorded actions as abbot give us a passable idea of his character, which in the ordinary sense of the word his scientific writings do not, but it is his writings which justify his place in the history of ideas. While they are not to the taste of every reader, any more than they were to that of the St Albans chroniclers, a man who made outstanding mathematical and astronomical advances deserves to have them not only listed but explained. Without his scholastic background he could not have designed his clock; without his scholarly reputation he would never have become abbot; and without the wealth of his abbey he would have remained just another scholar, dreaming of a machine that could never be brought into being.

The invention of the mechanical clock was one of the great turning points of history, an important junction on the tortuous road which eventually led to European economic and technological pre-eminence. Although the invention took place some years before Richard of Wallingford's birth, since his writings shed much light on its genesis I have asked how it came about, what motives lay behind it, what precise form it took, and where the breakthrough was made.

One of the difficulties of presenting medieval history to a wider public is that of conveying a feeling for what the lapse of seven centuries entails. Isaac Newton is nearer to us in time than Richard of Wallingford was to him. In the nineteenth century, on the rare occasions when English historians took note of pre-Copernican science, it was to dismiss it as a product either of ignorance and superstition or of hair-splitting logical vacuities. A few found signs of hope for the future of English pragmatism—for instance, in Roger Bacon's visions of marvellous inventions yet to come—but that was as far as the general level of scientific education allowed them to go. Many, on the other hand, considered themselves very close to the religious sentiment of the middle ages, notwithstanding the barrier presented by the Reformation. Today, people are likely to be more familiar with voodoo than with medieval Christianity, more familiar with Tolkien than with Dante. Historians who study the intellectual movements of the middle ages are of course in another category, but are not always free from prejudice. Some write as though there was no world outside the lecture room, paying deference wherever they detect depth and subtlety, but in an entirely bloodless way. Others, impatient of logic, natural philosophy, mathematics, and science in its more easily recognised forms, find all the social history they need—and an occasional scientific insight—in occult practices, and even in witchcraft. Richard of Wallingford does nothing to help his biographer in any of these respects. He was neither a white-coated experimenter nor a bloodless logician; he was no magician, and he did not know the secret of the Holy Grail. He was a creature of the age in which he lived, and should be judged by the manners, the religion, and the science of his own time. That is why I have thought it necessary to explain at some length what his lost world was like.

The final part of the book goes beyond the personal history of earlier chapters. It places Richard of Wallingford in the wider context of western science, which was rapidly gathering momentum in his day. While not intended as a comprehensive history of western scientific thought, this part carries a serious message about the vital importance of the exact sciences of the later middle ages to the scientific movement that followed in its wake. The thirteenth and fourteenth centuries witnessed a growing appreciation of scientific ideals which we now take for granted, in particular those relating to the mathematisation of science and the creative character of the act of formulating new concepts and theories. This part of the book still has a biographical purpose, but on a larger scale. It surveys the chief tributaries to the Oxford tradition in which Richard of Wallingford was nurtured, those originating in ancient Greece and medieval Islam; and it deals with some of the streams which issued forth from Oxford after his time. It is not intended as a history of Greek scientific genius or Islamic virtuosity, except in so far as they touch on the life of the subject of this book.

Central to the sciences discussed here are natural philosophy, cosmology, mathematics, astronomy, and astrology too—where it followed similar norms. The high point of Richard of Wallingford's science was reached with his use of certain mathematical techniques, which passed from him to later generations. They were exploited, to be sure, by only a discriminating few. Then as now, the imagination was caught far more easily by an expensive clockwork artefact than by an esoteric mathematical theorem—but the design of his clock also rested heavily on mathematics. To do even rough justice to the technicalities of Richard's work it is necessary to include a certain amount of formal detail, but I hope I have included signposts enough for those who wish to pass it by.

One purpose of this book is to make materials I first assembled between 1964 and 1971 more easily accessible. They were published as *Richard of Wallingford* (Oxford, 1976), in three volumes, which those who require further detail will need to consult. In the preface to that edition I thanked the many friends who helped me with it at the time. I have dedicated this volume to the memory of three of them. Francis Maddison is a fourth whom I must mention, for it was he who first made me aware of the many unanswered questions surrounding the origins of mechanical timekeeping, and who led me into a medieval maze from which there was to be no escape. It was my wife Marion, however, who—having learned to live with the edition all those years ago—suggested that the time had come to introduce Richard of Wallingford to a more general readership. In doing so, it has been my good fortune to work with Martin Sheppard, a devoted editor of an almost extinct species. I thank him, and all at Hambledon and London, for their timely support.

March 2004   Oxford

By the Same Author

The Measure of the Universe

Richard of Wallingford (3 vols)

Horoscopes and History

Stars, Minds and Fate

The Universal Frame

Chaucer's Universe

Fontana History of Astronomy and Cosmology

Stonehenge

The Ambassadors' Secret

In Memoriam

Willy Hartner   1905-1981
Alistair Crombie   1915-1996
Olaf Pedersen   1920-1997

who shared the belief that medieval science did
not begin or end with the middle ages

Part One

Foundations

1

Eclipse

WHEN JOHN LELAND stood at the crossing of the nave in the abbey church of St Albans in 1534, and looked towards the great window in the southern transept, he saw something he thought to be a marvel without equal in the whole of Europe. What he saw was already two centuries old. It had been built for his monastery by Richard of Wallingford, abbot of St Albans, with great labour and at enormous expense. The abbot was a man whose office was a symbol of the wealth and power of the church, but what he had offered to his community was meant as a proof of his skill, his great learning, and his piety. The modern mind is so sated with ingenious contrivances that it is no longer easy to understand the awe, even veneration, shown towards a clockwork mechanism, however complex, but those sentiments were real enough.

We may call it a clock, although a word closer to Leland's Latin would be 'horologe'—an instrument for telling the hour. Neither word really explains its purpose. As Leland looked at the colossal dial of the clock, high on a gallery below the great south window, he might have seen the places of the Sun, Moon and stars as they were at that moment in the heavens. He might have seen that the Moon showed the same phase on the dial as it was showing currently in the sky, and it might have been explained to him that the Moon's eclipses in the heavens would likewise be correctly displayed at the appropriate day and hour. In looking south at this colossus he was looking in the direction in which the true Sun, the Moon, and most of the visible stars, reached their highest points in their daily motions around the sky. Some of those occurrences might have been visible through the window above the clock, but there were things shown on the dial that were not to be seen at all in the St Albans sky: the ebb and flow of the tide at London Bridge; a moving image of changing human fortune; and numerous lines and geometrical figures that in all probability Leland did not understand, but that only served to enhance his admiration.

Leland tells us something more, namely that Abbot Richard the clockmaker had composed a set of rules concerning the clock, lest it deteriorate through the fault of the monks, or cease to function because they were ignorant of its structure. That book, which to all intents and purposes was lost for more than four centuries, has now been found. As a result, most of the workings of the machine—it was as much a celestial theatre as a timepiece—are now known to us. We know nothing whatsoever of the detailed workings of any earlier mechanical clock. The fact that this is the earliest known mechanism of its type, and yet was in many ways more complex than any other from the two centuries following Richard's death in 1336, makes it doubly remarkable. Leland's instincts were right. He was right to describe the abbot as 'easily the first in mathematics and astronomy in his day'.[1] It is true that to justify those claims he had only the written testimony of others, and a few of the abbot's own writings that he could not readily understand, but no one knew the manuscript remains of the English monasteries better than Leland. He was far more than a royal chaplain on a sightseeing tour. As the king's librarian and antiquary, he was expressly commissioned to search for antiquities of all kinds, in all the cathedrals, religious houses, and colleges of England. The clock caught his attention twice over, partly for its own sake, and partly because it related to an important group of manuscripts.

The circumstances of John Leland's appointment are notorious. Henry VIII's wish to divorce Catherine of Aragon and marry Anne Boleyn had brought to a head many a long-running question of discontent at the idea of English subservience to the pope, and widespread corruption among the clergy. The idea that divine law was on a higher plane than human law offered comfort to those who thought they would benefit more from the former than the latter, and those who identified more closely with the state usually accepted—however grudgingly—that church and state should stand apart. The break with Rome spelt the end of monastic wealth and privilege, but also of monastic learning. The new English church, with the king at its head, almost inevitably became an arm of the state: it was too rich and powerful to be allowed an independent role. When Henry's personal, political, and diplomatic plans began to outrun his finances, he was easily persuaded that he had the right to seize the great wealth of the monasteries. Most church institutions eventually bowed to royal authority, but excuses were easily found. The open allegiance of some of the monasteries to the bishop of Rome was considered a threat to the state, and Catholic uprisings in 1536 and 1537 seemed to confirm that they offered a serious threat to the new order. Their wholesale dissolution was therefore pursued with great vigour. Within three years the English monasteries had virtually disappeared. Their wealth had been enormous: when their income was transferred to the crown it virtually doubled all previous revenues. In time, there came another kind of dissolution, when Henry and his descendants sold off the land to pay for wars and adventures in foreign policy. Reminders of the great achievements of medieval monasticism were lost to the public consciousness, and with them the memory of Richard of Wallingford, his clock, and his learning.

For all this, Leland of course carried none of the blame. While he approved of the church's reformation, he recognised that there was much more to be rescued than gold, silver, buildings and land. He had impeccable qualifications for the commission given to him in 1533. Touring the country between 1534 and 1542, he listed the contents of thousands of monastic manuscripts. He tells us that at St Albans he was shown their 'parchment treasure' by a 'duly erudite monk'. The man in question was an Oxford scholar, Thomas Kyngesburye, and the irony of the situation is that it was he who was later called upon to sign the parchment transferring the monastery to the crown.

The break with Rome marks not only a break in history but one in historical knowledge, and we owe much to Leland for making this break less serious. In 1546 he printed what amounted to an advance notice of a projected bibliography of British writers, as a gift for the king. It was the prototype of a succession of similar works. Through them we can begin to build up a picture, not only of the intellectual and monastic life of England in the middle ages, but also of the medieval universities, and of such scholars as Richard of Wallingford who belonged to both worlds. Leland's work was re-edited by John Bale in 1549. Within a year, Leland was certified insane; and by 1552 he was dead, with the great bulk of his papers unedited and left to others to put into print. What he had provided on his 'laborious journey'—Bale's very apt description—was a window into a past that most of his fellow-countrymen seemed at the time to be happy to forget.

Among the manuscripts preserved from the monastery of St Albans there was one that mattered more to the history of the place than any other, and Leland made much use of it for his own rescarches. It was the *Deeds of the Abbots of the St Albans Monastery*—in Latin, the *Gesta abbatum monasterii Sancti Albani*—and it is now in the British Library as manuscript Cotton Claudius E.iv. Compiled at various times over a long period, it includes an account, now known to be faulty in several details, of the abbots of this rich and important monastery. It opens with the reputed refoundation of the monastery by Offa, the powerful king of Mercia, who died in the year 796. The chief compiler of the *Gesta abbatum* was the St Albans monk Thomas Walsingham, writing in or around 1440. For information before 1308,

Walsingham simply took over the writings of Matthew Paris, the renowned thirteenth-century artist and historian of England, and William Rishanger. Both of them were monks of the abbey. Matthew Paris had inherited much of his material from another, his predecessor Roger of Wendover, who died in 1236. (Roger's writings are an important source of early English history, but they are often quite fanciful, as when he tells of how the whereabouts of St Alban's remains were revealed by an angel to Offa during a visit to Bath; and of how the king then journeyed to Rome to get the papal blessing for his grand project.) For the period from 1308, Walsingham relied on notes and remarks made by other monks, and on his own experience. We owe our first debt to him, as we try to piece together the life of Richard of Wallingford. There are perhaps three significant sources of information to be detected in his account, apart from something more extensive and infinitely more valuable: Richard's own writings. For the most part, these were simply beyond the understanding of Thomas Walsingham and most of the monks of St Albans.

2

The Black Monks

THE ABBEY OF ST ALBANS was in one respect unexceptional. It was a monastery under the jurisdiction of an abbot, in which monks and lay brothers followed the rule of life of St Benedict, who had lived more than eight centuries before Richard of Wallingford's time. It stood in a tradition with a long and rich history which had begun in Benedict's Italy, but had eventually spread to Gaul and the rest of Europe. The order was not strongly centralised. In principle, each Benedictine monastery had a large measure of autonomy, cohesion being provided simply by adherence to a shared rule. This left plenty of scope for individualism. Like all Christians, those who followed the Benedictine rule were taught to shun pride—the first of the seven deadly sins, and considered to be the fountain of all others—and yet the monks of the St Albans cloister knew that their abbey was in many respects out of the ordinary. Their abbey was well endowed. It had greater privileges, and greater revenues, than almost any other in England, and with the exception of Canterbury, for most of its history it supported more monks than any rival house.

The abbots of St Albans claimed precedence over all other English abbots, in view of the fact that their monastery was founded to honour Alban, Britain's first Christian martyr. The story of that martyrdom, which very probably took place in the middle of the third century, contributed much to the *esprit de corps* of the foundation. Alban had been a high-born native of Verulamium—the Roman name of the town that would eventually be named after him. He held Roman citizenship and probably military rank. Tradition has it that, although he was a pagan, he sheltered a persecuted British Christian priest, whose piety made such an impression on him that he was himself converted to the Christian faith. When the Roman authorities tracked down the priest, Alban gave him his own cloak, so allowing him to escape. Arrested, Alban refused to make a pagan sacrifice and was condemned to death. The story had it that he converted one executioner, but was beheaded by a second—whose eyes were said to have dropped out as a well-deserved punishment. In time, the story of the disguise led to the name Amphibalus—from the Greek word for cloak—being assigned to the priest whose life Alban saved.[2]

Whatever the date, a church was eventually built to commemorate the event, with a shrine to which the sick were taken to be cured. There was some sort of monastic settlement by the fourth century, and in the year 429 Germanus, bishop of Auxerre, visited the place and spoke highly of the community. According to tradition, the church was eventually allowed to decay and was forgotten, only to be newly discovered by divine revelation in the time of Offa, king of Mercia. He was credited with founding a monastery on the site in the year 793. It was said to have been for a hundred monks under the Benedictine rule—and a hundred was still more or less the number of monks when Richard of Wallingford was elected abbot in 1327. The old story was challenged at an early date, and there was a long dispute with the monks of Ely as to who held the true relics of Alban—relics, of course, produced revenue from pilgrims. The case for the St Albans monastery was reinforced by the supposed rediscovery of their patron saint's original grave in 1257—not the last archaeological counterfeit in the neighbourhood, if that is what it was. In 1439, at the request of the abbot of the day, the legend of the saint was put into verse by the English poet John Lydgate. Alban's cult had spread to France at a very early date and Lydgate was able to draw heavily on a French poem. The last of the abbots of St Albans had Lydgate's account printed at a press in the town, shortly before John Leland's visit, unwittingly marking the end of a chapter of history covering at least twelve centuries.

The Order

Pride in the monastery was coupled with pride in the Benedictine order itself, which had a thousand years of history behind it by the time the English monasteries were eventually dissolved by Henry VIII. Strictly speaking, Benedict had never been a priest, but by the time he was driven out of his native Umbria around the year 525 he had already founded a group of twelve religious houses, with ten monks in each. He finally settled in Monte Cassino, half way between Rome and Naples, and there established not only a monastery but a monastic rule which would in time provide the guiding principles for almost all monastic life in Europe, not only that of the Benedictines.

The phenomenal success of the rule owed much to the fact that several talented Benedictine monks found favour at the courts of the Holy Roman Empire and with other influential potentates. One of the great strengths of the rule was its flexibility, but at its inflexible core was its insistence on prayer, the reading of the scriptures, and manual work. As a young man, Benedict had been so appalled at the behaviour of his fellow students that he had gone into retreat, away from all learning—'knowingly unknowing, and wisely untaught', to use the words of his biographer Pope Gregory. In short, Benedict's did not begin as an intellectual movement, and yet the combination of prayer, biblical study and labour fitted his order for a role that he had not foreseen: to educate, first the monks, and later, others, who were destined to live and work in the world.

Through their missionary work among people of all ranks in society, the black monks—so called because of their dress—spread rapidly across Europe. Their rule was introduced into England by St Augustine of Canterbury, as early as 597. The monks became a civilising influence, partly by their teaching but also by their example—in agriculture, arts and crafts, and even in the ways of organising daily life efficiently. These men were far from being entirely otherworldly, but in any case the earliest monasteries consisted largely of laymen, with a relatively small number of priests among them. That strong lay element never disappeared, and it provided the order with much of its strength.

Benedictine writers constantly remind us that their communities were primarily meant to study virtue, rather than learning for its own sake; but then they proceed to point out how so many of the great members of the order in the past, and the libraries of the houses in which they lived, testify to the honour in which learning was held. Benedictine historians often observe with pride that their rule has been issued in nearly a thousand editions in the course of the fifteen centuries of its history; but rule books are a better index of obedience than of erudition. More convincing early witnesses were two Englishmen who seem to typify the ideal of the scholar-monk. The first was Bede, whose writings on grammar and the calendar of the church, chronology and history, music and poetry, scripture and the lives of the saints, all tell us much about the extraordinary flowering of Northumbrian culture before its suppression at the hands of Viking invaders. Bede's works were fortunately copied in England and on the continent before this happened, and they had earned for him the title of 'Venerable' at least as early as Alcuin, the second of our great English Benedictine scholars. Born about 735, the year of Bede's death, Alcuin was the most able of those whom the emperor Charlemagne gathered around him. 'The schoolmaster of Europe' had been trained in the school of York, of which he had become head by the time he met Charlemagne on a journey to Rome in 781. By his writing and teaching, Alcuin inspired and guided a new intellectual movement, not only in theology and biblical studies, but in the philosophy and secular learning of antiquity which was then being rediscovered.

Learning apart, there were many regional variations in the ways the Benedictines conducted themselves. In Burgundy, the order became highly centralised under the autocratic leadership of the abbey of Cluny. A movement for spiritual reform, reaching a climax in the eleventh century, spread from there to other parts of Europe, including England. There is no doubt that St Benedict intended his monks to do their own manual work, but as time passed, they did less and less. The Cluniac understanding of the rule placed much greater emphasis on the time to be given to prayer, which eventually came to occupy a large part of the typical monk's day. In due course—in the eleventh and twelfth centuries especially—new orders of monks were established with new identities, accepting Benedict's rule but supplementing it in ways that led to greater austerity and a channelling of worldly into spiritual energy.

This change brought a number of problems in its wake. In England, the Cistercian order was the most successful of those new orders. The Cistercians were at first of a more pragmatic cast than most, and in the twelfth century they attempted to revive the ideal of self-sufficiency in manual labour. Even there, the attempt was short-lived. In the course of the following century, leaders of monastic communities of all persuasions were growing increasingly concerned with the need to fill the time on the hands of their monks. Prayer had its limits. By order of a general chapter of black monks in 1277-79, abbots were to find administrative duties for more of their monks, while others were to be put to copying, illuminating, and correcting manuscripts. (Such pursuits were all easily justified by a number of instructions laid down by Benedict himself.) There were monasteries in plenty where the monks were either too few or without the necessary level of literacy for this to be an effective solution to the fundamental problem, but the monastery of St Albans was certainly not one of them. It had long been a hive of intellectual labour, and was well equipped to adopt the changes ordained—to the benefit of Benedictine scholarship more generally.

Federation

The English Benedictine monasteries were to be numbered in hundreds, great and small. For many centuries they remained true to their ancient tradition of independence. There was a slight tendency for clusters of houses to form, following the lead of Cluny, but it was not until the year 1215 that confederation was raised to the status of a general principle, throughout the church as a whole. In that year the fourth and greatest of the five Lateran councils—so named because they were held in the Lateran Palace in Rome—met to consider church reform. Among many other weighty matters, it was decided that the monasteries within each province of the church should be federated for the sake of strength and discipline. (The English provinces were Canterbury and York.)[3] It was decided that the heads of the monasteries, their abbots, should meet every three years, in their so-called 'provincial chapters', to decide on important matters. Not only should they make laws that were binding on their communities, they should appoint visitors who would report back to the chapters on the state of affairs, spiritual and material, in their abbeys.

The loose federation allowed for competition, but within tolerable bounds, and this helped to stimulate resourcefulness in economic, social, and political affairs. These had not been Benedict's own priorities, but times had changed, and the church with them. The new formula for federal government often failed miserably in Benedictine houses on the Continent, but in England it worked reasonably well, despite an abundance of jealousy and personal animosity between abbots. Then as now, there was one easy way of avoiding a clash of human wills in the upper echelons of administration: members of committees simply stayed away from meetings, so weakening the federal ideal. Interference by visitors from outside the order was something they found harder to bear than enmities within it. There were to be Benedictine visitors, but augmented by bishops appointed from outside, bishops who of course did not belong to the monastic system. The abbeys became accustomed to visitation by bishops, although they both resented and feared the experience. There is one story of a monk who broke out and enlisted in another monastic order altogether at the thought of an impending visitation by the tireless scholar-bishop Robert Grosseteste. (It is true that Grosseteste had an unusually fearsome reputation, for in his first wave of visitations, held during the first six months of his episcopate, he deposed no fewer than seven abbots and four priors.) The bishops, on the other hand, resented sharing their powers of visitation with insiders to the system; and then again, some abbots greatly resented the thought that rival abbots would sit in judgement on them, and preferred to subject themselves to episcopal visitations.

The two English provinces, with their federal organisation, were on the whole well regulated, and the larger houses especially flourished. By the end of the thirteenth century, the organisation controlled about three hundred Benedictine houses, many of women, but more of men. These included some of the country's greatest religious institutions. Especially noteworthy for their wealth, privileges, and numbers were the abbeys at St Albans, Canterbury, Westminster, Bury St Edmonds, Peterborough, York, and Durham. To be a black monk implied renunciation of the world, but to be a member of such an organisation as theirs in the thirteenth and fourteenth centuries was to be conscious of being member of an elite. To be an abbot within such an organisation meant something more. St Benedict had planned that all within the cloister should be equal, and that those holding high office should be elected by their fellows. By the later middle ages the forms were still observed, but the powers of the elected abbot had grown so great that humility was all too often stretched to breaking point.

The Monastery

Apart from pride in his order, and pride in the fact that his abbey was first in order of foundation in England, the St Albans monk would have taken pride in its architecture. The abbey church—now the cathedral—was set high on one slope of the green valley of the River Ver, across from the site of the Roman town of Verulamium. During the middle ages that city was the first of any importance north of London, about twenty-five miles distant. A Roman road, called Watling Street by the Saxons, ran from Dover through London to Wroxeter near the Welsh borders. It had originally gone through the middle of the town of St Albans, but at the end of the eighth century or thereabouts it was diverted—surely Britain's oldest town by-pass. By the tenth century, the abbots were beginning to dismantle the ruins of Roman buildings for materials. Roman Verulamium had covered about 200 acres, and the store the abbots assembled was so massive that most of it remained unused for centuries. What the Saxons built was modest, even so, and was treated with some disdain after the Norman Conquest, when Paul of Caen was made abbot in 1077.

Paul was well connected—some even said that he was the natural son of Lanfranc, whom William the Conqueror made archbishop of Canterbury. True or not, it was Lanfranc who appointed Paul to St Albans. Both had previously been at the Benedictine abbey of St Stephen (St Etienne) at Caen, Lanfranc as abbot. When they began to build their English churches they modelled them on that church at Caen, but with one great difference: whereas at Canterbury the church was a copy of the church of St Etienne in plan and measurements alike, the building at St Albans was more elaborate than either, and on a more ambitious scale. Abbot Paul—or rather his great architect Robert the Mason—was there able to make use of a massive supply of well-seasoned timber that his predecessors had laid by. Whatever they thought of the elegance of the other materials there assembled, they were not above making use of them. Among other properties of the main church, this explains the broad Roman tiles that are still to be seen in large numbers in its walls. Paul of Caen considered his Saxon predecessors coarse and uneducated (*rudes et idiotas*), and he is even said to have destroyed their tombs. The chronicler Matthew Paris found many of Paul's actions hard to forgive, and the arrogance towards native English traditions which gave rise to them. It is a historian's foible to protest at the loss of the past, added to which it was not Paul's own past he was destroying. Memories are short, however, and the monumental scale of his new buildings at St Albans was something of which the monks could be proud.

The massive crossing tower, originally capped by a pyramidal roof, was itself an architectural masterpiece. No other great church of the eleventh century still has its crossing tower still standing: those at Winchester, Lincoln, Wells, York, and Chichester all suffered disasters of one sort or another, the first two shortly after completion. The ambitious outline of the Norman church at St Albans was not very different from that which survives today, although the structure was often altered in its details. When the church was eventually extended at the eastern end, the form of the apse of Abbot Paul's church was lost. Even some of the extensive twelfth-century work that was added to the Norman original has been lost. At the very end of that century, John de Cella (abbot from 1195 to 1214) pulled down the west front and parts of the aisles, and began to replace them with something more to the taste of his own time. In 1250 there was an earthquake, and by 1257 the opening of cracks in the fabric made it necessary to demolish and rebuild much of the eastern end of the Norman church. For the rest of the century the resources of the abbey were spent on rich embellishments—a fine painted timber vault over the rebuilt presbytery, a Lady chapel, a sanctuary and an ante-chapel, for example.

When Richard of Wallingford first set foot in the abbey as a young monk, at the beginning of the fourteenth century, its church would have been a source of great admiration, even to someone who is likely to have visited the great London churches. Under Abbot Hugh of Eversdone, the new Lady chapel at the eastern end of the church had just been roofed and glazed, after long delays. The interior of the church had grown to about 515 feet long (157 m). Where church builders in Normandy vied with one another for the height of their vaults, in England they seem to have valued length. For those who measured architectural glory with a rod, the St Albans church was only marginally shorter than the cathedral at Winchester—although both were admittedly shorter than St Paul's in London. A contest based on the length of the nave alone, however, gave the palm to St Albans. At about 300 feet long (91.4 m), it had the longest nave in Christendom.[4]

The tower too was larger than most, the tallest surviving Norman tower in England. Apart from its broached spire, it still stands as it stood when built. Despite the presence of Early English and Decorated styles of architecture in the church, the Norman presence is still felt. The only Saxon remains are some substantial lathe-turned baluster shafts in the arcading of the transepts, where they had been reused by the Normans. Some say that they came from Offa's church, but they are more probably from the tenth century.[5] In overall form, the church Richard of Wallingford would have seen was much the same as that we see today. One striking difference would have been in the use of colour, which the English Reformation swept away. A fine series of paintings on the piers—scenes from the life of the Virgin, crucifixion scenes, and others, all in tempera—was uncovered in the course of restorations in 1862. They are a reminder of changed attitudes to what a church should be. Much of the painted ceiling fell victim to the Victorian renovators. That at the eastern end of the church is from the fifteenth century, overlaying thirteenth-century painting that Richard of Wallingford would have known.[6] Perhaps the most striking of all the early painted murals, detected with great difficulty in modern times, would have been one showing 'Christ in Majesty'. With his right hand raised in blessing and his left holding a chalice, Christ looked down on the altar. He was seated on a double rainbow within a great lobed mandorla, against a ground with diaper pattern. Flanking the whole stood the apostles Peter and Paul, each in a pinnacled tabernacle, Peter with a key and Paul with a sword.

Then as now, a person entering at the western end of the church would have been unable to appreciate all of its complexities at first sight, since its great length is broken up into spaces appropriate to different uses. First is the main part of the nave, in which the general congregation gathered. It ends in an altar to the Virgin, and a fine carved stone screen from 1350, replacing the wooden screen that Richard of Wallingford would have known. Beyond, and hidden by it, is the choir, where the monks worshipped at appropriate hours throughout the day. Beyond that is the presbytery, but first there is an interruption at the crossing, with the interior of the great tower and its fine lantern above us, and transepts to left and right. Today in the south transept there is a lonely thirteenth-century angel with outspread wings, retaining some of its early paint. In the north transept there are still a few medieval tiles, and examples of mid-fourteenth-century glass.

Eastwards from the crossing is the presbytery, which had a thirteenth-century timber vault and floral bosses above it, again much more colourful then than now. The eyes were and are drawn in the first place, however, to the high altar at the far end. The present richly carved stone screen behind it is one of the finest of its kind in England, but it dates only from the late fifteenth century. In Richard's day it would have been of wood and much simpler. Here in the presbytery and its aisles there are tombstones in the floor, many of them of past abbots. Richard of Wallingford's is now among them. Like most of the others it has lost its brasses, although we do at least know in his case what the inscription was. In Norman French, it promised indulgence to the passer-by who said a prayer for his soul.[7]

Beyond the great screen, and hidden by it, is the chapel of Saint Alban, with the shrine to which medieval pilgrims flocked in large numbers—an attraction carefully restored around 1300. It is richly carved with scenes of the saint's history; and again we can still make out traces of paint on the stone. A magnificent grille of Sussex wrought iron, blacksmith's work, dating from around 1275, protected the shrine from the press of pilgrims (Fig. 43 below). In the fifteenth century it was thought necessary to build a 'watching chamber' to keep guard over the shrine and the gifts left by pilgrims. Its frieze of oak shows scenes of contemporary life and the martyrdom of St Alban. Finally, beyond the shrine of St Amphibalus, the priest in the Alban story, the church ends at the Lady chapel.

Such a church as this was a wonder, scarcely to be matched in all of England, but it was only one element in a greater complex of monastery buildings. Most of the others are no longer standing, but we can still get an idea of their great scale from the traces of their foundations in the fields and lawns surrounding the abbey church. There is still an imposing gatehouse, built as the main entrance to the abbey court, although it is one which replaced the gatehouse known to Richard of Wallingford, destroyed in a hurricane nearly thirty years after his death.[8] There are a few remains that give us an impression of other buildings—of the vanished cloister quadrangle, of a smaller cloister, kitchens, chapter house (now replaced), dormitories, and guest houses. The church has kept much of the old character provided by its massive proportions and the texture of its brick, flint, and stone. The fabric of the vanished buildings would not have been very different. To envisage its appearance in Richard of Wallingford's day it is necessary to screen out the nineteenth-century additions, which speak loudly for what Lord Grimthorpe and his fellow restorers thought the middle ages should have been.[9] The west front is all Grimthorpe's work, but the tower and main transept walls, have been little altered since Paul of Caen's time. The transept windows are Grimthorpe's, but those he replaced were not as old as the fourteenth century.

Buildings on such a grand scale were not immune from occasional misfortune. When Richard of Wallingford first entered the abbey church as a young novice it was much as we have described it. When he became abbot, in 1327, it was in a less happy state. Two great columns on the southern side of the main church had collapsed on 10 October 1323, during the celebration of mass. Within an hour, the wooden roof they supported, and the aisle on the south side of the church, followed suit. 'Only two monks and a boy were killed.' The catastrophe—due to poor foundations—was followed by another, when temporary supporting timbers fell with more masonry and much of the cloister. Those events led to an eventual rebuilding of the Norman cloisters and five bays of the southern side of the church. (It is easy to distinguish the Decorated style of architecture on the south side from the rest.) The glories of medieval building which most of us now admire are those which have stood the test of time. Very many did not. The architectural miseries continued later in the same year, when the greater part of a stone wall behind the dormitory also collapsed. Elsewhere in the church a wooden beam fell on the shrine of St Amphibalus and broke the thigh of a mason working there. The beam demolished the marble shafts supporting the shrine, but miraculously left the wooden shrine itself intact.

The saint was duly thanked for his intervention on this point, but speaking more generally, the calamities of 1323 put a great strain on the monastery's resources. To Richard of Wallingford, when he became abbot, this was a serious obstacle standing in the way of a project much nearer to his heart than the repair of stonework. He needed large sums of money to build a great clock for his abbey, the like of which no other institution could rival. Throughout the nine years of life left to him, he struggled with the problem of abbey finance. This struggle helped to dictate his behaviour in other respects, but his character was moulded by more powerful forces of a very different kind.

3

Wallingford

RICHARD OF WALLINGFORD was not born into privilege, nor in the ordinary way of things would his circumstances have led him into a scholarly or religious life. He was the son of a certain Isabella and her husband William, a blacksmith of the town of Wallingford in Berkshire. The couple, as the St Albans chronicler reports, were 'prosperous with respect to the poor, and moderately so in the eyes of the rich', but 'they lived frugally and without complaining'. This sounds like a literary conceit, but no doubt harks back to Richard's own reminiscences. The year of his birth was most probably 1292. Edward I was on the throne of England, although he was then rather less concerned with home affairs than with events in Scotland and the risks of awarding its crown to John Balliol. Continuing Anglo-Scottish hostility was something to which the abbots of St Albans could not be indifferent: not only were they expected to provide troops and support for the war, but by virtue of their location they were often expected to house the king's army on its journeys from London northwards.

The Borough

Wallingford is on the River Thames, fifty miles or so to the west of London and twelve short of Oxford. It owed its early importance, indeed its existence, to the fact that the Thames was easily forded there, although by the thirteenth century the ford was supplemented by an impressively long bridge. Wallingford was no mere backwater of history. It had been an important Saxon stronghold that had grown to become the largest defended town in the kingdom of Wessex. In the year 1006 it was almost obliterated by a Danish raid under Swein Forkbeard, the father of King Canute, and the memory of that event no doubt explains why its inhabitants knew better than to oppose the army of William the Conqueror, sixty years later. In the Domesday survey, Wallingford was still by far the largest borough in Berkshire.[10]

A royal castle in Wallingford gave the town another small niche in English history, when the besieged Empress Matilda fled there following her famous escape from Oxford Castle over the ice in 1142. It was at Wallingford in 1153 that a compromise was struck between her and King Stephen: he was to reign until his death, after which the crown should pass to Matilda's son, the future Henry II. Henry held his first parliament in Wallingford Castle, and it was he who presented the town with its most important charter, although the borough was then already entering a period of decline. By the time Richard was born, well over a century later, it could be said that the town was like his parents—prosperous in the eyes of poorer places but less so in the eyes of richer. It was a town from which an ambitious boy would gladly escape.

Son of the Smithy

When Richard was barely ten years old, his father died. A boy of ten is not able to take over the heavy work of a blacksmith, but at that age the son of a smith is old enough to have learned much about the trade. He would not have known that more than two thousand years of hard experience lay behind the craft, but he would have known the workings of the smithy and the various stages in the production of iron implements, on which the whole of society was heavily dependent.

Finding iron ore in reasonable quantity had never been especially difficult, but smelting it was not easy, and was often left to specialised smiths. There were many good English ores, but by this time traders from Spain, France, Sweden, and Germany were bringing into England raw iron rods and ingots of high quality for sale. Henry of Eastry, for instance, bought large quantities of Spanish iron for work at Canterbury Cathedral in 1308-9, and similar purchases from sources in Normandy are recorded at Westminster, beginning in 1294. A Wallingford smith is more likely to have bought his iron from a bloomery nearer at hand, for instance from one of several in the north of Oxfordshire, where the ore was good enough for steel.

In the bloomery, the ore was first crushed and washed and roasted, using green timber, then quenched with water to get rid of sulphur and other impurities. This preliminary work was often done by women. Charcoal—this often costing much more than the ore itself—was then mixed with the ore, and the mixture fired in a smelting furnace, the 'bloomery fire'. Bellows were the commonest device for enhancing a natural draught, without which high temperatures were impossible—and without high temperatures nothing more than a useless cindery mass of slag with embedded iron globules was obtained. The ore having been smelted in this way, a spongy mass of iron remained, namely the 'bloom'. Its quality de-

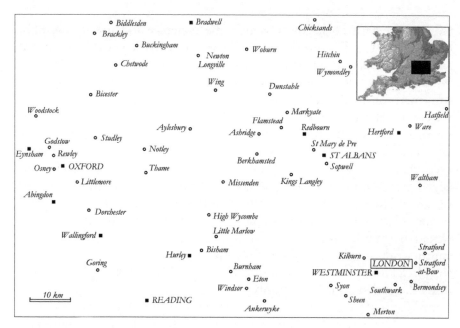

2. The region with which Richard of Wallingford was most closely associated, including the Thames valley (Oxford, Abingdon, Wallingford, and on to London) and the St Albans area. The small black squares mark places which at some period in the middle ages had religious houses of Black Monks. Circles mark small towns and villages which almost without exception had religious houses of other sorts. Woodstock (north of Oxford) was a royal seat. London was already too large for its monasteries to be mapped at this scale.

pended heavily on the efficiency of the bloomery. The slag remaining in it after the first smelting was beaten out by heating and hammering the bloom repeatedly, leaving a decent mass of wrought iron as the end product. This is what the Wallingford smith would have bought, but it is quite possible that the boy and his mother helped to improve the bloom further in the same way, according to the quality needed. It was not unusual for women to help their husbands in the smithy. In a characteristically medieval piece of misogyny—in a northern Passion play—it was said that a woman smith made the nails for Christ's crucifixion after her husband, feigning a sprained hand, had refused.

Turning wrought iron bars into implements of hardened steel, by repeated hammering and heating in contact with charcoal, was the next stage in the smith's labours. Even that technique had been mastered in various parts of the world in antiquity, so that the hammer and anvil and tongs that we associate with a blacksmith were in the ancient world already a symbol of his craft. The Roman legions had swords of hardened and tempered steel, and scarcely any part of the Roman empire was without its mines. New furnace designs were already bringing iron working to a fine art in some parts of the Roman empire. Techniques developed in Roman Spain were particularly successful, and were later exploited by the Muslim conquerors, although the rest of medieval Europe had no access to their products, and different procedures were evolved. A few medieval centres of excellence became renowned for the production of fine weapons and edged tools—northern Italy, parts of the Rhine valley, and Burgundy, for instance. There is no reason to think that Richard's father, or other smiths in the Wallingford district, would have had any special expertise. His would most probably have been the humdrum work of the ordinary blacksmith, work of a traditional sort that was needed everywhere. It was one of the blacksmith's boasts that, unsociable though his trade might be, every other depended on it for its tools.

As a Wallingford smith, it is more likely that William wrought his iron in a furnace blown with hand bellows than in one blown by water power; and more likely that in his forge he used charcoal bought from the charcoal burners in the local forests than coal brought from a distance. Again, in both respects, the young Richard no doubt often lent a hand—the wages of paid blowers could be a costly item for the ordinary smith. Water-mills to provide powered help with the endless hammering were as yet uncommon. It is unlikely, but not impossible, that the smith had a mill-wheel, even on the slow-moving Thames. There was a revolution in iron production then beginning in England to which he might just possibly have been party. Higher furnace temperatures, together with iron of a higher carbon content, were at last making it possible for cast iron to be produced in sizeable pieces. The new process made it easier to manufacture cannons, for example, but it was in its infancy, and the military revolution to which it gave rise was still some decades in the future at William's death.

While he is likely to have known little of such innovations, William would very probably have been done work for the soldiery of the royal castle. He could not have competed with distant specialists, skilled armourers in London and the larger towns, with their strong guild organisation, but he would have made horseshoes and arrow heads for the garrison. He is more likely to have repaired than to have made their best swords, chain mail or plate armour. Much of his time would have been spent making and repairing implements for other artisans: hammers and nails, tongs and pincers, saws and files, sickles and scythes, bill-hooks and axes, adzes and chisels, braces and drill-bits. He would have made hinges and scroll-work for the doors of the richer townsmen and church buildings, and also glazing bars for their windows. It is just conceivable that he had been called upon to forge the wrought iron bars that were needed for the frame of a clock, or to forge and cut its shafts and wheels, but if so his experience would have been rare indeed. His son, however, would one day be in an informed position to organise his servants to do such things.

3. A smithy, as illustrated in Georgius Agricola, *De re metallica* (Basel, 1556). The scene would have been little different in Richard of Wallingford's day, although the bellows here (B) are rather grand, and like the trip hammer (D) water-powered, something unlikely in the Wallingford smithy. The forge (A), the tongs and anvil (C) and the quenching water (E) would have been familiar, as would the boys, with their leather aprons.

The smoke and noise from smithies, especially at night, were a common source of complaint by townsmen, and a cause for legislation, throughout the middle ages. However mixed Richard's own feelings, he would have understood those of a poet of a later date whose thoughts are famously recorded in a manuscript now in the British Library. The poet's words bring the medieval smithy to life in a way that ours cannot. While the following modern rendering of them does not catch the wonderful alliteration of the original—'Swarte smekyd smethes smateryd with smoke', and so forth—it will at least be easily understood:[11]

> Black smutted smiths, besmirched with smoke,
> Drive me to death with the din of their strokes;
> You never did hear such noises at night,
> How the lads shout, what a clatter their knocks!
> Those crooked dwarfs, they shout Coal! Coal!
> And blow their bellows till all their brains burst.
> Huf! Puf! says one, Haf! Paf! says the other.
> They spit and they sprawl and spin many a tale,
> They grate and they grind and they grumble together,
> Kept all hot with their hard hammering,
> Their leather aprons are hides of the bull,
> Their legs are wrapped against fiery sparks.
> Heavy hammers they have, and hold them tight,
> Strong strokes they strike on an anvil of steel,
> Lus! bus! las! das! they snort in turn—
> Let the devil get rid of so doleful a tune!
> The master lengthens a little, lashes on a less,
> Twists both together and tacks on a third.
> Tik! tak! hic! hac! tiket! taket! tyk! tyk!
> Lus! bus! lus! das! This is the life
> Of these clot-heads all. Christ make them suffer!
> Can a man have no sleep for the hiss of the quenching?

Richard knew how hard and unglamorous was the life of a smith. When his father died, he was perhaps relieved to think that he was not yet strong enough to take over the work of the smithy himself.

The Priory

The boy was ten and without a father, but within a year or two he found himself adopted as a son by the Benedictine prior of Wallingford, William of Kirkeby. This was Richard's salvation. Had it not been for the presence of a foundation of black monks in the town, his destiny might have followed a very different course. In a roundabout way, his fate was another outcome of the Norman Conquest. In the redistribution of wealth that had followed the change of dynasty, the Wallingford church of the Holy Trinity had been handed over to Paul of Caen, the abbot of St Albans whose building plans transformed the great abbey. Paul soon afterwards built a house for a convent of black monks next to the Wallingford church, making it a cell of St Albans—a dependent monastery under a prior. The foundation was modest. It is not easy to say precisely how large it was, for the simple reason that, long afterwards, Cardinal Wolsey foreshadowed the great monastic dissolution to come with a lesser version of his own: in 1528 he appropriated the revenues of Wallingford Priory and a few other places to help him found a new college at Oxford. There are now no priory buildings remaining.

In Richard's lifetime, Wallingford Priory was one of eight cells subservient to St Albans, none of them large. The dependent priories at Hatfield Peverel, Hertford, Redbourn, and Wymondham were all closer to the parent abbey, while others at Belvoir (Lincolnshire) and Tynemouth (Northumberland) were much more remote. There was also a dependent hermitage at Markyate. Each cell supported between five and a dozen monks, and each would have had a similar number of servants, and perhaps a few lay brethren.

The Wallingford priory had been first colonised by the monks sent by Abbot Paul of Caen, at the end of the eleventh century. It was never a large institution, but it launched Richard on his Benedictine career. He would certainly not have been expected to take the vows of a monk immediately: we are told that he was simply adopted as a son by Prior William on account of his 'loneliness and aptitude and great promise'. That such a judgement was possible might indicate that the priory had already given tuition to Richard, and perhaps to other boys in the borough. His father had very probably done smithing work for the priory. It is of course conceivable that William and his wife Isabella had at some stage promised the boy to the church. In the early years of the Benedictine order it had been customary for parents to offer their sons to monasteries at seven or eight years of age, but by the end of the thirteenth century this habit of 'oblation' had officially disappeared, and monks had to be at least eighteen or nineteen, and know their own minds, before committing themselves. (The St Albans chronicler Matthew Paris, however, at one point puts the age at fifteen.)

How might Richard have lived as the 'adopted son' of the prior, a man whom he later described as 'gentle and much loved'? Monasteries and their subsidiary houses were communities of four main types of person in addition to the monks. (The analogy with nunneries is straightforward and complete.)

The *novices* were those who lived in the monastery, awaiting the day when they would take their vows and commit themselves totally to the monastic life. They were few in number. The novices were not to be admitted before giving evidence that they had a true calling, but this they were normally expected to do within a matter of months of joining the community.

After the novices there were the *lay brothers*. They did manual work of many sorts, and inhabited what was almost a monastery within a monastery, having their own rooms and their own part of the church during worship. They were of the highest importance to the economic well-being of the monastery proper, and almost always outnumbered the monks by two or three to one.

These very qualities often made them seem threatening to the ordinary monks. As a result, with the passage of time, the lay brethren tended to be replaced by *paid servants,* who were more easily controlled—porters, bakers, brewers, cooks, tailors, and so forth. In some cases the servants in men's convents were women, a fact that could give rise to scandal, or at least innuendo.

The fourth class was one of people who simply *lodged* in the monastery. They were often retired abbots or priors who had been allocated a pension, wealthy laymen who had purchased the privilege, or even complete families. The provision for maintenance was called a 'corrody'. The sale of corrodies could be a useful source of monastic income, while granting them liberally as gifts could be a burden to an abbey, as Richard would one day discover from the behaviour of his predecessor as abbot. Corrodies were sometimes traded for a flat sum of money, and there were many instances where the monastery began to regret a recipient's longevity. There is a known case of the sale of a corrody to a Jew, and another (at Dunstable) of a corrody sold on behalf of two boys, one of them at school. The king reserved the right to a permanent corrody at St Albans, to which he could nominate anyone he chose.

Residents in all four categories might be found in dependent cells, just as in the larger abbeys, although of course the numbers in the cells were much smaller. It is quite possible that Richard was resident in the priory in the fourth category, without formal payment, schooled by the prior or another of the monks. Whatever evidence William of Kirkeby might have had for the boy's promise when he adopted him, the prior had enough confidence to send Richard off in due course to study at Oxford. This he did at his own—which is to say the priory's—expense. Nothing is known of the prior's own scholarship. He is not recorded as a member of Oxford University, but St Albans had its own school, and he might have been educated there. Monks as a whole, and priors and heads of houses in particular, were expected to be reasonably well educated. Only occasionally did it happen that senior monks would appoint an ill-educated or otherwise weak man, in the hope that they would find him easy to manipulate.

Richard remained under the prior's eye between the ages of about twelve and sixteen—that is, between 1304 and 1308, or thereabouts. His later achievements suggest that he made good use of this time to prepare for his future entrance to the university. The Oxford courses began at an elementary enough level, with the 'trivium' of grammar, rhetoric, and logic. By the beginning of the fourteenth century, however, the universities were beginning to leave large parts of these 'trivial' subjects, especially the first two, to the lesser schools that were beginning to spring up, especially in the university towns of Paris, Oxford, and Cambridge. The rate at which students fell by the wayside in Oxford was high, and a poor grasp of Latin had much to answer for. The main purpose of the universities was to produce an elite, capable of serving the church and the state, and yet it was a constant complaint of bishops that some their clergy did not even understand the Latin they spoke, which they had learned only by rote. The problem was no less serious in the monasteries: from 1290 onwards, the Benedictine provincial chapters made frequent reference to breaches of the rule enjoining conversation in Latin. Records of visitations give ample evidence that questions put to the monks in Latin were often answered in French—the language of upper-class daily life—or English. That Richard of Wallingford had risen above the ranks of the illiterate was something he surely owed to William of Kirkeby as much as to his native intelligence. He doubtless owed his mechanical bent to his father, and experience of the smithy, but it was in Wallingford Priory that the foundations of his learning and his spiritual vocation were laid.

4. The church of St Mary the Virgin, seen from the north, the very heart of the university in Richard of Wallingford's time. The tower (*c.* 1280), spire (1310-20), north chapel (to the west of the tower, 1320s), and the university's first Congregation House (to the east of the tower, 1320-27), were all known to him. The first university library, above the Congregation House, was begun in his day, but not completed until 1411. The style of the windows was changed somewhat in the sixteenth century.

4

Oxford

OXFORD WAS ONLY about twelve miles from Wallingford, an easy morning's walk. Most of the inhabitants of the Berkshire market town would have been aware of Oxford's many imposing buildings, of the life on its streets, and even of its fame beyond the Thames valley. Oxford, a county town and borough, was by modern reckoning a small place. It had numbered rather more than a thousand houses at Domesday, when it was the sixth largest town in the kingdom after London, York, Norwich, Lincoln, and Winchester, but soon after that it began to decline in wealth and importance. Its fortunes were turned with the advent of the university. England's premier university was in Richard's time well over a century old, vigorous, filled with optimism, and already of repute throughout the Christian world.

Throughout the kingdom, those who were recognised as scholars were 'clerks' (clerics, clergy), and were in religious orders of one sort or another. The church made a distinction between those in holy orders proper (subdeacon, deacon, priest, and bishop) and those in minor orders, that is, persons not ordained (porter, lector, exorcist, and acolyte). Since few outside the church, however, were capable of notarial or secretarial work, the word 'clerk' came to be used of scholars in general; and with the rise of the universities it was inevitably applied to all students there, whatever their place in the church hierarchy. There was another important division, however, cutting across the first. The 'religious' clergy were those who were members of the monastic and religious orders. (They are also called 'regulars', because they followed a rule, *regula*.) The 'seculars' were those who worked in the world, whether for a parish, for a diocese, for a king, or other wealthy lay patron. Even when offering their scholarly expertise to lay patrons, they too were subject to a measure of clerical discipline. No matter what his precise circumstances in his native town, as a student in Oxford Richard of Wallingford was a clerk, with a style of dress that marked him out as such, dress that in its detail might have given the onlooker some further clue as to the Oxford institution to which he was affiliated. We are not told which this was, but that he studied grammar and philosophy for about six years, after which he took his bachelor's degree—or 'determined in arts', to use the parlance of the time. The St Albans chronicler, Thomas Walsingham, goes on to say that he left Oxford in his twenty-third year, and that not until then did he make the vows necessary to becoming a monk at St Albans.

The phrase combining 'grammar' with 'philosophy' is slightly odd in this context: in mentioning two widely separated parts of the curriculum it omits most of what the young student would have been obliged to study. Yet the six-year period rings true. The subjects which Richard later made his own are in any case mentioned shortly afterwards in the chronicle. We shall eventually try to fill in a few of the gaps of which the writer was seemingly ignorant.

The Beginnings of the University

Anglo-Saxon England, in the tenth and eleventh centuries, had no great centres of learning outside the monasteries and cathedrals. English students seeking a higher education gravitated towards the leading continental schools, such as those in Laon, Liège, Montpellier, Orléans, Salerno, Bologna, and Paris. Increasingly they patronised Bologna and Paris, but as numbers grew, so did the chances of survival of an English institution. Oxford, which was neither the capital of a kingdom nor the see of a bishop, seems an unlikely place in which to find an alternative seat of learning, but its university was not a deliberate creation, resulting from any single decree. It evolved gradually and fitfully, out of schools in the town: the first recorded Oxford schoolmaster began teaching as early as 1095. Searching for origins, we are prone to exaggerate the importance of such modest references as this, but lectures were certainly being offered at a higher level in the second decade of the twelfth century. When Gerald of Wales, in 1187, recorded the existence of several different faculties in the town, Oxford was not the only such centre offering itself. Even at the end of the twelfth century, Northampton had a better reputation for learning. The tide gradually turned in Oxford's favour for two main reasons. One was the strategic importance of the place. In 1133 Oxford was granted a royal charter, after the king had built a new palace there, and it became a centre of royal government on a small scale. It was also at a principal point of crossing the Thames, and therefore served as a useful centre of trade and communications. An even more important reason was that several religious communities had settled in or near the town, making it an influential clerical centre.[12] Quite apart from such religious houses it had more than its share of parish churches; it was occasionally used as a venue for important ecclesiastical councils; and it was ever more frequently used as a meeting place for church courts.

This last fact seems to have been the catalyst that turned the early Oxford schools into a university. If the urge to study any one advanced subject brought students to Oxford in the twelfth century, it was jurisprudence—canon law, the law of the church, and Roman civil law, which underpinned it. English law had always placed a high value on custom and case law, even when interpreting canon and Roman law, and this pragmatic tradition made it especially useful to have courts near those schools in which law was taught. It was not unknown for foreign students to come to Oxford to study law, even before the end of the twelfth century, and there are many more signs of vigorous native activity in the subject. In arts and theology, however, Oxford could not be considered a serious alternative to Paris until external forces came into play—in particular the increasing seriousness of the wars between England and France, which made it difficult for students and masters alike to travel between the two countries. It is difficult to quantify such effects, but the increasing tempo of study in Oxford during the last decade of the twelfth century can be roughly gauged from the quality of the surviving writings of a handful of noteworthy Oxford masters. One in particular should interest us, an Englishman from St Albans who had studied in Paris. Alexander Neckham is known to have been giving regular lectures in biblical theology in Oxford at that time, and the fact is especially interesting since his concerns were not typical of continental scholars. He showed a strong predilection for the natural sciences, and his writings helped to create a characteristic Oxford scientific style which was valuable and long-lasting.[13]

Neckham's view of the natural sciences was in no way typical of a person we should regard as a scientist, ancient or modern. He accepted simple forms of scientific doctrines already developed by others, and applied them to theology—for example, to the theory of the human soul, and to the account of the six days of creation found in the first chapter of the Book of Genesis. In Paris he had learned the value of the writings of the Roman consul and philosopher Boethius, and of Aristotle—several of whose works Boethius had translated from Greek into Latin. The recovery of the remaining works of the Greek philosopher, and the painstaking study of them, characterised the university science of the next two centuries. It is impossible to overestimate the importance of Aristotle to later medieval learning His doctrines of space and time, potentiality and actuality, continuous and discontinuous, different kinds of cause, matter and void, quantity and quality, relation and analogy, substance and essence—to mention only a few of his most pervasive concepts—are at the very heart of even our own basic scientific language. Aristotle's account of scientific method in general is likewise at the very root of ours. It is for such reasons that we should value the Oxford scientific movement of the thirteenth and fourteenth centuries.

Grosseteste and the Forming of the University

There was another side to this movement, on which Aristotle's writings had only an indirect bearing, and that concerned mathematics and astronomy. It was to these subjects that most of Richard of Wallingford's scientific work belonged, although by his time the university was more than a century old. During that period the university had grown rapidly in size and in international standing, helped by the endeavours of a man who combined great intellectual powers with enviable administrative skills. Robert Grosseteste was the most influential Oxford theologian of the thirteenth century. Like Neckham he applied his scientific knowledge to theological questions, but—unlike Neckham—he had a very original scientific mind. He had much astronomical and optical knowledge; and, without having a very profound knowledge of mathematics, he appreciated its importance to the physical sciences. There was nothing especially new in this, although it was a principle that had been largely overlooked in the West. It did no harm to have the principle proclaimed repeatedly by Grosseteste's leading advocate after his death, the Franciscan Roger Bacon, lecturer in both Oxford and Paris.

The chronology of Grosseteste's early life is not known with any certainty, but it is likely that he first taught at Oxford before the most disturbing event in the university's early history, when the students and masters were forced to disperse for almost five years, between 1209 and 1214. The event almost put an end to the infant university, and yet from this ordeal the institution finally emerged stronger than ever. The dispersion was in part due to the papal interdict by which King John and his subjects were excommunicated—an event which led many Oxford men to move to the spiritual safety of Paris. In part it was a consequence of violent quarrels between town and gown, in which two innocent students were hanged by the townsmen as a reprisal for the murder by a third student of his mistress. After the papal interdict was withdrawn in 1214, the papal legate took it upon himself to bring masters and students back to Oxford. (Some of them had settled in Cambridge and some in Reading.) This he did by limiting the rents of Oxford lodgings, by fining the townsmen annually in perpetuity for the support of poor students, and by making the town provide a dinner for a hundred of them on every St Nicholas's Day thereafter. In this way, the legate created a sense of insult in the town which some would say has persisted to the present day.[14] His purpose was of course quite different: it was to assert the church's jurisdiction over both parties. The task was not one to be easily achieved, in view of the remoteness of the bishop in whose diocese—Lincoln—Oxford lay. The church's hold on the place was certainly not weakened when Robert Grosseteste was himself made bishop of Lincoln in 1235. He died in that office in 1253.[15]

At some time between 1214 and 1241 Grosseteste had held the title of *magister scholarum*, master of the schools, an office which—like the chancellorship of a cathedral—carried with it the right to grant to clerks on behalf of the bishop the licence to teach as masters. It also gave him jurisdiction over all masters and scholars within the diocese, and marked an important advance in the university's status.[16] The resident university chancellor assumed—or tried to assume—the powers that had formerly been those of the chancellor of Lincoln. He was head of all schools in the town, including the grammar schools, and every scholar and master in the place was subject to his jurisdiction. The Oxford chancellor was henceforth elected by the Oxford regent masters as 'first among equals'. His status was quite different from that of the chancellor in Paris, who was primarily a functionary of the cathedral, constantly defending its rights and privileges from an increasingly assertive university. The university of Oxford had become a corporation, entitled to own property in common, to use a common seal, to obtain professional advantages for its members, to protect them in many different ways, and to make its own statutes—the last privilege being especially important in the coordination of teaching and of planning the curriculum.[17] Later in the century, certain non-regents obtained privileges in university government. There were minor problems here, and others in connection with the peculiar status of the friars, but from a broader perspective the town of Oxford would henceforth be split into two rival bodies, the civic and the academic. For centuries to come, the university would have the whip hand.

Oxford was now well and truly under the control of the church. Indeed, from the early thirteenth century onwards, almost all schools and universities of the Christian middle ages north of the Alps formed an arm of the church, the interests of which remained their first priority. They could be expected to take the side of the church in its conflicts with secular rulers, and there was a strong theological bias to all subjects studied in them. The theological atmosphere in Oxford was never as stifling as that in Paris, but later medieval chancellors, men with considerable power, were almost invariably regents in the faculty of theology. Despite the university's corporate status, church courts had ultimate jurisdiction over it. The chancellor's court was supreme within the university, the ultimate court of appeal in temporal matters being the king's court and in spiritual matters the pope.

There are many similarities between the ways in which the universities of Oxford and Paris developed, for the simple reason that many of Oxford's most senior masters had first studied in Paris. In Paris, scholars of the university were divided into 'nations', according to their country of origin. Oxford divided its scholars into southerners and northerners, *australes* and *boreales*. The former included the Irish and all Englishmen from south of the River Trent, while the latter included the Scots and the English from north of the Trent. The curriculum was much the same in Oxford and Paris. Oxford's first extant statute, dating from 1252, required that no one be licensed in theology without first graduating in arts: this made the university more useful to society at large than a purely theological institution would have been, since most students failed to go beyond an arts degree, and many failed even to complete that. Both universities grew steadily in national importance. Each provided the secular church with most of its senior members, but in Oxford, far from the centre of the diocese in which it was situated, there was always a greater measure of intellectual freedom.

As the university produced ever more graduates, the chances naturally grew stronger that any external control of the university would be exerted by Oxford men, familiar with the workings and prejudices of the place. It was not that the university always obtained the freedom for which it was hoping. The relationship with the bishops of Lincoln in particular remained problematical. When Grosseteste was himself elected to the see of Lincoln, in 1235, this created a curious situation, for he was by now a man in his sixties, used to positions of power, and in ecclesiastical polity he was conservative and even puritanical. His attitudes can be found even in his writings on cosmology, Aristotelian and Platonic, and on optics—a subject he thought to be at the foundation of all physics. He believed it right that certain hierarchies which he thought to be present in the cosmos should be mirrored in the structure of the church, which he wanted to be centralised and hierarchical. He had known life at the foot of the pyramid. Although never a Franciscan, already by 1232 he had given up all secular positions to live something approximating to the Franciscan life, while from about 1229 to 1235 he had been lecturer in theology to the Oxford Franciscans. He would in due course bequeath his fine library to them. Now, however, he had been made a bishop in the hierarchy. He felt he could attack monastic irregularities and worldly clerics, even Pope Innocent IV.[18] He certainly found it natural enough to take a paternalistic stand in his relations with his old university. When he had acted as Oxford chancellor, he had conferred the licence to teach on Oxford graduates. Now, as bishop, he was determined not only to keep that right—which had always formally belonged to the bishop of Lincoln—but to supervise teaching arrangements in Oxford, at least in theology.

What followed shows how the forces of conservatism may work in such cases. In 1246 Grosseteste obtained a papal decree confirming his rights, and he forced the chancellor, Ralph of Sempringham, to refrain from using the common seal of the university. The university had to wait until after Grosseteste's death in 1253, when at last it obtained papal recognition of its corporate status and statutes. In this way, it came one step nearer to its eventual status as a property of the English church as a whole, rather than of the one bishop.

From the point of view of later Oxford excellence in the natural sciences and the study of Aristotle, Grosseteste's influence was very great. He, more than any other person, turned the interests of the English Franciscans to those studies, with excellent results. Richard of Wallingford was no Franciscan, and he was born forty years after Grosseteste died, but he could not escape the influence of the man. Grosseteste, however, fought another rearguard action which, if it had succeeded, would have adversely affected the course of Oxford scholarship in the later middle ages. He resisted a change to a new method of teaching—one that was already in use in Paris—whereby theological debate was organised around a work known simply as the *Sentences*. This was a biblical summary that had been prepared by Peter Lombard, a pupil of Abelard, who had died as bishop of Paris in 1160. Grosseteste, who was far more erudite and widely read, differed fundamentally from Peter Lombard in outlook and in his choice of sources. He was not to know how important those commentaries on the *Sentences* would eventually become, not on account of any special virtue in Lombard's rather pedestrian work, but because it offered a standard list of topics for theological, philosophical, and scientific discussion. At all events, in this case the Oxford Dominicans made a successful appeal to Rome against Grosseteste's decision, and lecturing and writing on the *Sentences* became standard Oxford practice.[19]

Theology and the Sciences

Increasingly, after Grosseteste, the university contrived to hold in check the powers of successive bishops of Lincoln. In 1281, for example, Bishop

Sutton's visitation was resisted by the chancellor of the day, Henry of Staunton, on which occasion an appeal was made to John Pecham, archbishop of Canterbury. An uneasy compromise was reached: Pecham decided that where the bishop found fault with Oxford scholars, the chancellor should correct it. The bishops of Lincoln continued trying to turn back the clock, but their attempts became progressively less spirited. By the end of the thirteenth century, Oxford had self-confidence enough to present itself as a more or less independent *studium generale*, a school with as high a standing as that of any other in the Christian world. Its graduates were considered to have a right to teach in any other school of the same status, without further examination. Petitions were made to a succession of popes—in 1296, 1303-4, and 1317—in the hope that this double claim would be given formal recognition. While the university waited in vain for an answer, the claim was recognised *de facto* throughout Latin Christendom.[20]

Oxford's standing was enhanced in other ways. The second half of the thirteenth century saw the foundation of its first colleges, corporate bodies which assisted students preparing for degrees. Secular students usually lived in academic halls, often governed by one or more masters. Such halls occasionally went out of existence, while others took their place, and occasionally a defunct hall would be refounded. Colleges, on the other hand, enjoyed a continuous existence, by dint of their endowment. University College (funded by a bequest in 1249) and Balliol College (founded as part of a penance, no later than 1266) were the first two foundations, but Merton College was far more influential than either. Its founder, Walter de Merton, gave it a rich endowment, to support male members of his family, to honour his political friends and patrons, and to serve church and state. He first intended the student body he created in Oxford to be administered by Merton Priory in Surrey; by 1264 he had drawn up statutes transferring endowments and responsibility to a 'house of the scholars of Merton' in Malden, Surrey; but by 1274, largely as a result of political circumstance, he decided to move the Malden administration to Oxford.[21] Almost by accident, the resulting institution acquired an unusual measure of independence—far greater, for example, than that of the eleven Parisian colleges then in existence. The founder's third set of statutes, presented in 1274, have survived to the present day with relatively few changes. Even now we know the names of 136 members of the college who were there before the end of his own century. Several of them had earned international recognition, and there was a tendency developing for the fellows to interest themselves especially in the sciences. The numerous merits of the Merton statutes soon became apparent, and the college became a model for others in both Oxford and Cambridge, although during the first century of its existence none could match it in endowment or reputation. Not the least of Merton College's strengths was its relative freedom from outside interference. It was supervised externally only by its visitor, the archbishop of Canterbury. The first Oxford visitation was by Archbishop Robert Kilwardby, a scholar with a deep interest in the natural sciences, and strong views on their place in the curriculum.

Oxford, by the end of the thirteenth century, could hold its own on the European stage. It shared in the broad institutional patterns of all other universities, and the security provided by the new colleges to a few leading scholars gave Oxford an added advantage over many. More important, from a European point of view, was the large measure of doctrinal agreement with Paris. The shared curriculum and the many shared texts could not entirely guarantee this state of affairs, which perhaps owed more to the large number of scholars who had studied or taught in both places. The cut and thrust of intellectual debate, in which members of religious orders with different loyalties and priorities took part, inevitably gave rise to opinions that were open to the charge of heterodoxy, or even heresy. Universities so beholden to the church could not escape occasional censorship. One episode in particular affected both Paris and Oxford late in the century. It touched directly on questions of natural science and philosophy, and it shows us how Oxford could be constrained to follow the lead of the senior institution.

Throughout the thirteenth century, Aristotelian philosophy had been gaining a foothold in both universities, and it seemed to many a conservative theologian that the natural sciences were beginning to pose a serious threat to the Christian faith. As early as 1210, soon after Aristotle's natural philosophy had become available in Latin, the provincial synod of Sens forbade Paris masters to read them in private or public, on pain of excommunication. The ban was repeated in 1215, and in a modified form was made the subject of a papal bull issued by Pope Gregory IX in 1231. At about that time, John of Garland—a young scholar who had attended John of London's lectures on natural philosophy at Oxford—wrote an open letter from the new university of Toulouse. He was one of the first to be appointed to teach there, and wrote advertising the fact that the Toulouse students were free to read the books forbidden in Paris. That special situation was not brought to an end until 1245, when Pope Innocent IV extended the ban to Toulouse.

Within a few years of Innocent's edict, the ban seems to have been ignored with impunity everywhere. By the mid-1250s, all of Aristotle was being read in Paris, as in practice had been the case in Oxford as far as the availability of texts allowed. In the 1260s, however, the campaign against the new learning flared up once more. Bonaventure—minister general of the Franciscan order, and future cardinal bishop of Albano—became ever louder in his insistence that it had become a threat to the true faith. A group of like-minded churchmen now invoked the help of Etienne Tempier, bishop of Paris. In 1270 Tempier condemned thirteen propositions as heretical, and in 1277 he went further: helped by a committee of theologians, he condemned no fewer than 219 theses. That second list of propositions was ill-written, repetitive, and full of inconsistencies, but the penalty for holding or teaching any item on the list was excommunication. Many items were included to counter the teaching of the Dominican Thomas Aquinas—it was no accident that the list was issued on the third anniversary of his death. Not until his canonization in 1325 were those particular examples annulled by the then bishop of Paris.[22]

After Tempier's edict, Oxford was at last drawn into the mire. Only eleven days after the Paris condemnation of 1277, a shorter list was issued by Archbishop Robert Kilwardby, forbidding the teaching at Oxford of thirty articles, all closely related to Tempier's. Kilwardby's main concern was to counter Thomist teaching on the unity of form and the nature of the soul, but there was a distinct anti-scientific flavour to his edict, and his only concession was to threaten milder penalties for ignoring it. In 1284 his successor John Pecham—who had previously attacked Kilwardby on a theological question in the bitterest of language, and who was no mean scientist himself—made a visitation of the university and renewed some of Kilwardby's prohibitions. Both were Oxford men who had taught in Paris. Pecham's stance is easier to comprehend than Kilwardby's. Like Aquinas, Kilwardby had been a Dominican—he had been provincial prior of the Dominicans in England before becoming archbishop. Pecham, however, had been provincial minister of the Franciscans in England, and was regarded with much hostility by the Oxford Dominicans, who forced him to defend himself against the charge of having maliciously stirred up enmity between the two orders. It is natural enough to infer that what brought the two English archbishops to take the same action, in this matter, was their wish to follow the Parisian lead. Recent writers have inclined to the view that the closeness in time of the Kilwardby and Tempier prohibitions was a mere coincidence, but even if this is right, the two pontiffs must have felt they had a common cause.[23]

There has been much modern disagreement as to the impact of Tempier's prohibitions on the development of the natural sciences. One of the main purposes of the condemnations was to defend God's absolute and infinite creative and causal power against the inroads of the philosophers, whose system seemed to threaten it. Instead, the effect was to arouse philosophical curiosity and to stimulate greater interest in Thomas Aquinas and his hero Aristotle. In the wake of the condemnation, philosophers began to examine the Aristotelian texts even more carefully than before, and then to consider many other conceptual possibilities than those strictly sanctioned by Aristotle.[24] More than twenty of the forbidden theses concerned time—the eternity of the heavens, of the world, of matter, of the soul, and of time more generally. The new interest in Aristotle produced invaluable discussions of the nature of motion and space, not only touching on motion through space (kinematics) but on the possibility of empty space, void, a possibility that Aristotle had denied. Many of the resulting discussions would no longer be considered to fall within the realm of philosophy or natural science. This is true, for example, of discussions by such leading fourteenth-century scholars as Thomas Bradwardine, Jean de Ripa, and Nicole Oresme, when they investigated the idea that the void might in some sense be filled by an omnipresent deity.[25] Such discussions, however, reveal a refreshing freedom of thought.

Those who so avidly debated such questions, those who discussed the omnipresence of angels, or their placelessness, their motion, or their way of acting, arouse little real sympathy today, even among the few who read them, unless it is for their logical consistency or sheer ingenuity. It is a simple fact, however, that some of their debates contained the seeds of fruitful science to come. The language of the sciences encouraged an appropriate frame of mind. Just as Neckham and Grosseteste had used their knowledge of astronomy when writing commentary on the Book of Genesis, so now the Paris and Canterbury condemnations encouraged the tendency to include ever more scientific commentary on Peter Lombard's ostensibly theological *Sentences*. The language of university debate in theology gradually became increasingly imbued with natural philosophy and logic, mathematics and science. When scholars at the apex of the university hierarchy used such language, those in the lower ranks could not be expected to refrain. In the Oxford which Richard of Wallingford entered as a young man, theology was queen of the sciences. He owed it great respect, and learned its language, but it is not unlikely that in his lost theological commentaries his preferred dialects would have been those of mathematics, natural philosophy, and the cosmic sciences.

Gloucester College

Some decades after his death, when it became fashionable for Oxford colleges to lay claim to as many great names as possible, Richard of Wallingford was claimed by Merton College, which was still Oxford's leading academic institution. Since he was the protégé of a Benedictine prior it is extremely unlikely that he was ever there. After he became a black monk he would certainly not have been allowed there, either by the rules of his order or by those of the college, which had been established to educate the secular clergy. He might at first have done what most students did, and have lived in private lodgings or in the more regulated environment of an academic hall—of which there were then more than a hundred—but there is a far more likely alternative. He could have lived within the walls of a Benedictine foundation. After he became a monk he would have been obliged to live in such an environment, and as a southerner this would have meant the place generally known as Gloucester College.

Gloucester College no longer exists, and in the common sense of 'college' it never did. 'Now then', asked Archbishop Courtenay of a monk from St Albans in 1389, when he planned to make a visitation of the Oxford institution, 'don't you have a prior who can arrange a meeting of the chapter? Someone who lives in the place with you?' 'No', answered the monk, 'We're not a college. Those who live here have no common seal and no endowments, besides which they lack all the other things that a college needs.' William Courtenay was chancellor of the university, but at the time was much more concerned with violent affairs of state than of scholarship. He decided to abandon his visitation, having been warned in a diplomatic letter from Abbot de la Mare of St Albans that a visitation would be disruptive. More than half a century had passed since Richard of Wallingford's death, but the episode throws light on the status of the place at which he most probably studied. There was perhaps a touch of envy of the colleges in the monk's answer, but no doubt the speaker had some inkling of fruitless struggles that had taken place in his order a century earlier, to create something resembling the institution that the archbishop was expecting to find.

In the end, there were five monastic colleges at Oxford, but they did not survive the dissolution of the monasteries that had given them life, and they are now largely forgotten. Durham College already existed in Richard's day, as a house of black monks from the northern province. It was refounded, after the dissolution, as Trinity College. The colleges known as Canterbury, St Bernard's, St Mary's, and St Swithun's were all founded after Richard left Oxford. Their names have gone, at least as college names, although in one way or another they were absorbed by colleges that still survive. Gloucester College—if that name is to be used—was refounded after the dissolution as Gloucester Hall, the buildings of which were later made a part of Worcester College. A few of the old buildings are still there, but nothing that can give an accurate impression of what greeted Richard when he entered Oxford.

The place—not the name—owed its existence to orders made by general chapters of the southern Benedictines, in particular one held in 1277, that a house for study be set up to serve the monks in their province. It was to be financed by an annual levy on all their monasteries. The year 1277 was that in which the bishop of Paris, Etienne Tempier, attacked what he considered to be heretical teaching in Paris. Perhaps there was a feeling that Benedictine learning should be insulated from dangerous unorthodoxy. Whatever the initial stimulus to action, the typical Benedictine abbot of the southern English province had a more material outlook.

There had long been rumblings of discontent with the centralised form of Benedictine organisation imposed following the Lateran decision of 1215. For many abbots, the proposed tax to support an Oxford foundation was the last straw. Some refused to pay. Some refused to attend chapters to discuss the case further, and then refused to pay the fine for non-attendance. Some of the bishops took their side, and absolved those who would not pay the Oxford levy. So heated did hostility to the proposed Oxford house of studies become that the matter was referred to the papal court itself.[26] Nothing happened for six years. In 1283 St Peter's abbey, Gloucester, set up an Oxford cell; but another fifteen years went by before it was entirely supplanted by an institution shared by all the Benedictine monasteries in the province of Canterbury—about sixty in all. The first monk to take his degree was William de Brock of Gloucester Abbey, in 1298. To members of the order, the occasion was one of great symbolic importance. For centuries their monasteries had been centres of intellectual life. Now, however, the intrusive friars had captured the high ground, and were swarming like flies over Oxford. They could not be allowed to steal the limelight. A retinue of a hundred horse went to Oxford from Gloucester to celebrate, while most of the southern abbots made gifts to the new bachelor. The southern Benedictines were for the time being filled with good intentions.

Their new Oxford foundation had a peculiar status. In Richard's day it was neither cell nor college, and the responsibility for administering it had passed from Gloucester to Malmesbury. The southern abbeys went on gently resisting all requests for contributions towards the cost of buildings. An agreement to give financial support was eventually reached at a general

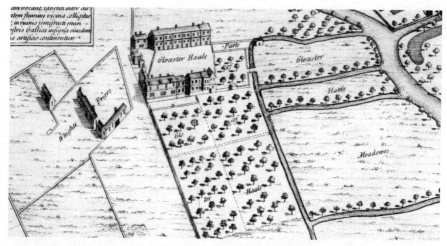

5. A bird's-eye view of the Gloucester College buildings surviving in the late sixteenth century, and those formerly occupied by the Carmelites across the street, where Beaumont Palace had previously stood. The view is taken from the re-engraving (1728) of the oldest known complete map of Oxford, that prepared by Ralph Agas between 1578 and 1588. By Agas's time, the old chapel and hall had been demolished and the remaining buildings had been acquired by St John's College, as a hall ('Gloucester Hall') for its students. The range of 'mansions' on the far side of the quadrangle still survives as part of Worcester College, an eighteenth-century foundation.

chapter of the black monks held in 1315 at Northampton, but even those abbeys who sent monks to Oxford to study were reluctant to contribute much. When Richard was in Oxford there were still abbots under threat of excommunication for their tardiness. In 1321, however, the presidents of the young foundation had a stroke of luck. Their site was next to a house of white friars, Carmelite friars, whose prior happened to be at King Edward II's side when he was forced to flee before the Scots at Bannockburn in 1314. The king made a vow that in return for deliverance he would grant the Oxford Carmelites a home in Beaumont Palace, and in 1318 they were finally presented with some of the palace buildings. These, however, were separated from their existing site by Stockwell Street. The Carmelites had hoped to connect the two sites with a tunnel, but in the end they decided instead to sell off their old site to the Benedictines. In this way, Gloucester College accrued more land and ready-made buildings. The abbots protested as loudly as ever at the cost of the enlargement, but it added something to the gravitas of the institution.[27]

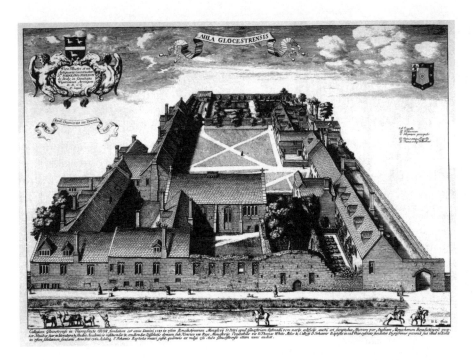

6. Gloucester Hall, with many of the surviving buildings of Gloucester College. Engraved by David Loggan for his *Oxonia Illustrata* (Oxford, 1675).

The precise arrangement of the early buildings is uncertain. They included a series of houses ('mansions'), each comprising one or more rooms belonging to a different monastery (Figs 5 and 6). In Richard of Wallingford's day, the house of St Albans monks was probably a modest affair, wooden framed. (Even the buildings of the endowed colleges were as yet spartan, with clay floors, unglazed windows, and unplastered walls.) Each house was supposedly governed by the general chapter, but the abbots were so proud of their rights and so accustomed to their independence that the institution never acquired a corporate spirit in the way that the colleges and friaries did. The site itself was still under the control of the abbot of Malmesbury, who allocated an area to each abbey on which it could build for itself. There was a common hall, kitchen, bakery, and chapel, with a sort of college head, a 'prior for students'. At some uncertain date, the custom was instituted of having the students elect their prior, and the resulting rowdiness caused much concern to a general chapter of the province. While we know little of the daily life of the Gloucester College students, we can at least see that they did not always behave as though they were in their home monasteries.

In one sense, this Benedictine institution was a microcosm of the Oxford university of later centuries, a consortium of independent houses of learning. The students of the college were never all monks: some were still novices, and it is not improbable that Richard of Wallingford lived at Gloucester College during his first Oxford period as well as his second.[28] From the St Albans *Gesta* covering the rule of Abbot de la Mare, in the second half of the century, we learn that there were students of more than one nation. The Benedictine rule continued to hold for all. Long absences from the home monastery put young men into the way of temptation, and efforts were made at Gloucester College to continue the monastic routines of the order. The prior was ordered not to permit the monks either to study or converse with secular clergy. They were to be punished for absences from divine service; to observe all required disputations in term time; to preach often, in Latin and English to fit them for preaching in their monasteries; to undertake all university exercises and take degrees only under one of their own religious order. It was as a special concession that they were allowed to hear lectures outside their own college, or to dispute in the university schools.

After Richard of Wallingford's death, the institution grew in standing. Following papal edicts of 1336 and 1338, the monasteries were ordered to send one student to a university for every twenty monks in their houses. In the case of St Albans, this meant five students, although their number at Oxford was often higher. By fits and starts the English Benedictines at last began to take university education seriously, and for the next two centuries the foundation prospered. At least thirty-eight monasteries associated themselves with it, and St Albans eventually made its house the finest of all. Around 1420 Abbot John Whetehamstede—a former prior of students and a scholar with wide European connections—gave money towards a new chapel and pictorial glass windows, and provided a vestiary and fine library with books. He was described then as the college's second founder, and it was the arms of St Albans over the main entrance that greeted visitors. During the following century the monastic colleges flourished as never before; and then quite suddenly there came that cold wind from London that quickly put an end to them all.

Rival Institutions

The Oxford of the early fourteenth century was much smaller and more closely defined than now, although the plan of its central streets has changed little. It was a walled city that for four centuries had provided a useful meeting place for royal councils. In addition to the castle there was the residence later known as Beaumont Palace, built by King Henry I just outside the town's North Gate. Henry's grandsons Richard and John, future kings of England, were both born there, but by the end of the thirteenth century the palace was no longer in use as a royal residence. It underwent some changes of ownership and was partly dismantled before a substantial portion of it was granted to the white friars, the Carmelite neighbours of Gloucester College, in 1318. We have already seen how this worked to the advantage of the Benedictines.

The way from the North Gate to Gloucester College was still mostly a walk across fields. The situation was hardly different when Hollar engraved his map in 1643 (Figs 7 and 8). Immediately beyond the college there were extensive water-meadows, so that it was suitably insulated from the temptations and general lawlessness of a city already crowded with townsmen, clerks, and—if we are to believe a constant complaint—thieves and vagabonds posing as clerks. After the dissolution of the monasteries, the buildings and grounds were leased for a time to royal servants before being turned into a house for the bishop of Oxford. On the first bishop's death this was vacated, and degenerated quickly, until it was eventually acquired by St John's College as a hall for its students. Known as Gloucester Hall, it was a sad token of the buoyant institution of Richard's day, when it bid fair to rival Oxford's earlier monastic institutions of learning.

With the dubious exception of the Durham foundation, for northern black monks, and Exeter Hall, founded with some of the qualities of a college in 1312, there were still only three institutions meriting the name of college in Oxford, if we except the friaries: University College and Merton College were within the town walls, and Balliol College was just outside. University College was poorly endowed and supported very few fellows. Those at Balliol were obliged to leave after taking the master's degree, which explains why some of the best of them migrated to Merton. The bulk of the university was in its student population, then approaching two thousand, mostly absorbed by the numerous halls. The name sounds grander than the reality: some were houses, few of them very large, but others were no more than apartments. There were then about 120 of them, spread across most of the walled town, with a few more to the north and south outside the walls. The young Benedictine scholar would have felt rich by comparison with most of the student members of halls. What would have given him pause for thought, however, as he walked round the town, were not the halls but the four principal convents of mendicant friars—his nearest neighbours the Carmelites, the Augustinian hermits also on the north side, and the Franciscans and Dominicans to the south.

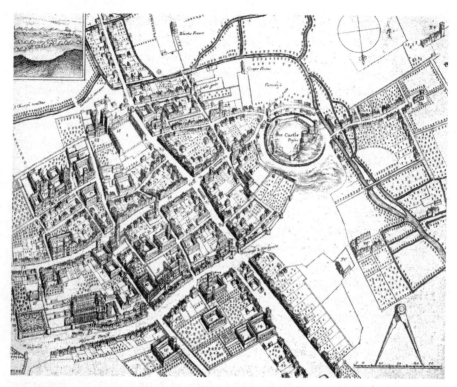

7. Part of a view of Oxford engraved by Wenceslaus Hollar in 1643. By then there were many more collegiate and university buildings than in the fourteenth century, but the plan of the town was little changed. Some of the key points relating to the earlier period are shown in the next figure.

The mendicant (begging) friars had taken Oxford by storm many decades earlier. They had changed the face of learning right across Europe during their short existence. The Franciscan Roger Bacon refers to the mendicant orders as 'the student orders', and there is much truth in the implied distinction between them and the regular orders of monks. From the mid-thirteenth century, the Franciscans aimed to have a doctor of theology in every priory. By the rule of the Dominican order, every priory was a school, and its friars were to study throughout life. The Italian Thomas Aquinas, the greatest theologian of the age, was a Dominican who studied in Paris under another great theologian and natural philosopher, Albertus Magnus. Few disputed that the Oxford teaching and writing of the mendicant orders was of a high quality. The chief fault to be found with them was not in this, but in their thrusting behaviour.

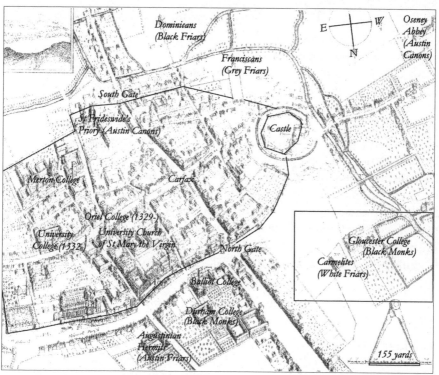

8. Some of the principal points of reference in the Oxford known to Richard of Wallingford (but here related to the previous figure). The rectangle at lower right, with Gloucester College, was shown in more detail in Fig. 4. Other lines show the wall of the Norman castle and the defensive wall then still completely surrounding the town. Note that the houses of monks and friars were outside the town walls, with the exception of St Frideswide's Priory, which was partly inside. There were thirteen parish churches, not here named. The streams are only minor tributaries to the Thames (in Oxford called the Isis only since the sixteenth century), which falls just off the upper and right-hand edges of the plan. The two college dates here added are for occupation of their respective sites, not foundation.

The Franciscans, the grey friars, had a strong presence in Oxford. Poverty and humility had been the weapons by which St Francis had wanted his friars to regenerate Christendom, but true poverty did not last long. Within a couple of generations of their founder's conversion in 1209, they too had begun to acquire royal privileges, to raise taxes on behalf of the pope, and to erect sumptuous buildings. Complaints on this score raised by Matthew Paris, the St Albans historian, sound like a case of Benedictine sour grapes until one discovers that—as recent excavations have shown—the Oxford church of the grey friars was three hundred feet long, no mean achievement for a mendicant order.

Complaints about the avarice, pride and arrogance of friars in general echoed from pulpit to tavern. At the one extreme of seriousness there were ongoing controversies between the secular clergy, who had care of the populace at large, the monastic orders, and the friars. Their disputes were very often about their respective rights in hearing confessions and giving general absolution—the forgiveness from sin for which a payment was often exchanged as penance. At the other extreme there were ribald comments on such matters as the Augustinians' love of fancy shoes, and their habit of hitching up their gowns to show off their well-shaped legs. Late in the fourteenth century, Chaucer's friar on the pilgrimage to Canterbury, Brother Huberd, illustrates the stereotype perfectly. He is a man who profits by his begging and wears clothing worthy of a master or a pope:

> ... he was nat lyk a cloysterer
> With a thredbare cope, as is a povre scoler,
> But he was lyk a maister or a pope.

The friar was 'a wantowne and a merye', strong but gentle, devious but courteous and sweetly spoken. 'In alle the ordres foure'—the four mendicant orders—there was none that knew so much 'of daliaunce and fair langage'. We are not told to which of the four orders Huberd belonged, but his silver tongue was not in question. It was such that he could squeeze a farthing from a shoeless widow. Capable of charming 'yonge wommen' and paying for the marriages of those he had doubtless earlier seduced, he carried around with him little presents to give to 'faire wyves'. The range of his female acquaintance extended from 'worthy wommen of the toun' to every barmaid, every 'tappestere'.

Too close an intimacy with women was a common charge against male clergy of all kinds, and the black monks, the 'cloisterers', were not immune from it. The stereotype, however, was not quite that of the friar: Chaucer's Monk was typically fonder of eating and hunting. There was, however, a more principled bone of contention between the different classes of religious. This too was pointedly noted by Chaucer, in his account of Huberd, who 'hadde power of confessioun', or at least claimed as much, 'For of his ordre he was licenciat'. Who could, and who could not, hear confessions and give absolution? There was to be an occasion, later in Richard of Wallingford's life, when he would count himself fortunate to know the answer to this question.[29]

As a newly-arrived student, perhaps even 'with threadbare cope', Richard would soon have been made conscious of fast-moving social currents in the stream of learning, and of their relevance to his advancement. One lesson he learned at Oxford was how to be detached from his fellow men when things were not going his way. The universities had modelled themselves to some degree on the craft guilds, corporate associations for protecting themselves against interlopers and guaranteeing profit. The history of Gloucester College, like that of the university and its subdivisions to this day, is one of a struggle for corporate existence, protection, and privilege. Charters of privileges granted by kings and popes were often invoked when the mendicant friars began to establish schools of theological teaching, for the friars were much resented by the established doctors in theological faculties everywhere. The friars somehow managed to infiltrate them to the highest level. They were finally brought to order, but only when they were well entrenched in the system. In Oxford, the Dominicans had been first to arrive, in 1221. Soon there was three-cornered contention between them, the Franciscans, and the university. It was less serious than it had been in Paris, although we should not forget the involvement of Franciscan and Dominican rivalry in the Tempier affair of 1277 and its Oxford aftermath. Trouble with the friars flared up again from time to time in Oxford in the later fourteenth century, and in many instances it centred on a trial of intellectual strength.

On entering Oxford, Richard would soon have discovered the reputation of the mendicants for learning. Robert Grosseteste, former lector to the Oxford Franciscans, was still revered as a great luminary, not only in theology but in the natural sciences. While Grosseteste was not himself a Franciscan, Roger Bacon and John Pecham later in the century were, and there were other Franciscans who were making comparable reputations in the natural and mathematical sciences in Richard of Wallingford's day. Across the road from Gloucester College, the erudite Carmelite scholar John Baconthorpe was studying, a close contemporary of his who would later achieve renown in Paris as 'the resolute doctor', or even 'prince of the Averroists'.[30] There were many other mendicant scholars of note. Independent and quarrelsome though the mendicants often were, their scholarship was the product of a vital movement that must have filled the young Benedictine with envy.

Nine Long Years and More

Richard was in Oxford to learn, and to that extent he had to follow a rigorous university scheme of learning, a curriculum combined with rules of advancement. Then, as now, it was what held the university together as an institution. The university had at first no endowments and no buildings of its own of any importance. The masters taught in borrowed churches, or in rooms hired from monasteries or other landlords. Two halls next door to Merton College, for example, one of them going under the name of St Alban Hall, were actually owned and rented out by the nuns of Littlemore. The main business of the university was done in the church of St Mary the Virgin, which in the late thirteenth century augmented its Saxon and Norman buildings with a fine tower. During the period when Richard was first studying at Oxford, a magnificent spire was added to the tower, and it has dominated the High Street ever since. While the university had no official rights over the church of St Mary, as time went on it channelled ever greater sums of money into its fabric. During Richard's second period at Oxford he would have attended meetings in the new Congregation House, which was built on the north-east side of the church around 1320, with support from Thomas Cobham, bishop of Worcester. This two-storey building, which still survives, had a room for a library on the upper floor.[31] When the faculties congregated, they did so in the church itself, each faculty being allocated its own area. Students for the bachelor's degree disputed and debated in the porch of the church and in a room over the porch, known as the parvise. Most important university examinations and sermons took place within the building. It is hardly surprising that the parish of St Mary became the hub of academic Oxford. Parchment makers, scribes, and bookbinders settled in the parish in large numbers. The church even spawned a new college: between 1324 and 1326 its rector, Adam de Brome, persuaded Edward II to found Oriel College on a pattern similar to that laid down by Walter de Merton. Yet again, advanced learning was in de Brome's thoughts, rather than the needs of young students.

Like the great majority of Oxford clerks, Richard began his studies in the faculty of arts. Relatively few ever went beyond that stage to study in the higher faculties of law, medicine and theology, but for those who wished to do so it was virtually necessary to qualify first as master in the arts faculty. The friars tried to by-pass that requirement, but even they bent the knee, and by Richard's time the Oxford faculty of arts had gained almost complete control of university government, and of the overall pattern of university study.

The word 'arts' did not have its modern resonance in the middle ages. It referred to the seven liberal arts, three or four of which we should now describe as sciences. Grammar, rhetoric, and logic ('dialectic'), that together made up the 'trivium', have already been mentioned as subjects on which a student might have made a good start before enrolling in the university. The 'quadrivium' covered arithmetic, geometry, astronomy, and music—but music studied in a formal and mathematical way. Students working towards either the bachelor's or master's degree were expected to take part in highly formalised disputations with other students under the guidance of a more senior person, usually a master, who summarised and weighed their arguments. There was also a system of 'repetitions' daily and weekly, to ensure that the doctrine expounded in ordinary lectures had been well understood.

The first four years of what was generally a seven-year course took the student as far as the bachelor's degree. He would typically 'determine' in his fifth year, proving himself through suitable exercises, swearing that he had followed prescribed lectures for four years and had satisfied certain other conditions. (The word 'determination' referred to a public summing up by a master of the student's disputation.) He could be expected to 'incept' as master, enter into the office of master, after his seventh year. Regular attendance at lectures was demanded, since few students could be expected to be able to buy books, and memory played a much more important part in education then than now. 'Ordinary' lectures usually involved a master expounding a text and considering 'questions' arising from it. Other types of lecture could be given outside specific texts; and then there were free general discussions, held perhaps twice a year, at which any question could be put to the master, although the procedure for debating them was no less formal than before. Such affairs must have constituted a chastening ordeal for most of those who took part, even though many were doing so chiefly to advertise their own brilliance. To increase the numbers of students and teachers in attendance as far as possible, all other forms of instruction in the university were suspended. The victim had to defend his position in response to questions raised by anyone at all (*a quolibet*) on any relevant topic at all (*de quolibet*). Surviving written reports of what was said on such occasions, known as 'quodlibetal questions', provide important historical evidence of the cut and thrust of the scholastic life. The fact that university scholars were so often prepared to try out their day-to-day techniques in unlikely territory—logic in the theology of the Eucharist, for instance—is surely not unconnected with the training that they had in asking, and responding to, quodlibetal questions.

Once he had determined in arts, perhaps in 1314, Richard of Wallingford left Oxford for a time. When he returned, in or around 1317, he presumably took on the lecturing duties that were a necessary part of the course for the master's degree. He might even have begun this phase before leaving. All seeking that degree still had to attend appropriate lectures by those regent masters who had already qualified, and in whose hands the government of the faculty was placed. When a man finally incepted as master, he had to swear that he had followed lectures on an established series of subjects. Most of the subjects related to classical or well-established texts, but some of them were related to books specially written within the university.

At length, after the typical candidate's seven or more years of study—Richard's period of absence would have lengthened this—and as the most important of the various conditions for granting him the licence to incept as master, nine regent masters (not including the one who presented him) had to declare that his knowledge was equal to the title. Five other regent masters then had to testify to his religious orthodoxy. This was the occasion for an inception feast, which was a festive but also a religious affair, with its own special ritual.

Even now the clerk was not home and dry. He was at the beginning of full membership of the faculty of arts, but he had duties to perform. He must dispute within the university for forty days, and lecture for two more years, a period known as 'necessary regency'. Then at last he was given permission to leave the university or begin study in a higher faculty, or to continue lecturing in arts as a regent master, which allowed him to earn fees and at the same time to do more advanced work. In Richard's case it was then that he threw himself into a study of the most advanced mathematics and astronomy available to him.

'Nine years or more.' From the time of his arrival in Oxford, in Richard of Wallingford's case, twelve years might have elapsed. For most clerks, this was all too much. Very few scholars had a college stipend. In order to survive, many had to beg or perform menial duties for wealthier students. Some were simply lacking in intellect or stamina, and fell by the wayside. Others—such as the poet Richard Rolle—were disillusioned with the entire character of university teaching, and especially with what were seen as its hair-splitting vacuities. For all its faults, however, the medieval university satisfied the greatest needs of church and state—even with the help of clerks who had never managed to finish the course. It supplied parish priests, parish clerks, and candidates for a whole range of administrative posts where basic literacy was called for. At the highest levels it provided an elite to serve not only church and state but learning itself. Like many of the leading academic lights of his time, Richard of Wallingford would eventually serve in all three capacities.

5

An Astronomer Among Theologians

RICHARD OF WALLINGFORD broke off his study before he had incepted and before entering into the duties of a licensed Oxford master. Instead he left for St Albans, in a well-calculated move. He needed support for further study. The abbey was rich, and it is quite likely that the prior of Wallingford would not or could not support him further without a greater religious commitment on his part. In 1314, at the age of about twenty-two, he took the vows required of a Benedictine monk.

Cause for Regret

St Benedict, the founder of his order, who was not an ordained priest, did not intend his monks to be priests, even though he made allowance for the possibility when drawing up his rule. Over the following centuries, however, the proportion of monks who entered the priesthood rose steadily. To become a priest, a man had to pass through various stages, although some of them might be combined—acolyte, subdeacon, deacon and finally priest. In early fourteenth-century England, only about a quarter of all black monks who began the process completed it. Richard was ordained deacon on 18 December 1316 and priest on 28 May 1317.[32] In the autumn term of 1317 he was able to return to Oxford, supported financially by the convent and by Abbot Hugh of Eversdone. As a priest, his standing on his return to Oxford would certainly have been raised, but more important to him was the fact that he was able to return at all.

Once there, the chronicler tells us, 'he studied philosophy and theology continuously for nine years, until by the general decree of the masters he was adjudged worthy to be licensed to lecture on the *Sentences*'. The reference to philosophy tells us that he fulfilled the obligations that allowed him to proceed to the degree of master. The statement that he was licensed to lecture on the *Sentences*, the standard work of Peter Lombard, meant simply that he went on to become a bachelor of theology.

He later admitted that his first love was not theology, but the thought that it was not filled him with a profound feeling of guilt. His guilt might go far towards explaining his perseverance, but the hope of personal advancement no doubt also played a part. The formal study of theology at an advanced level was not particularly common, even in universities like those of Paris, Oxford and Cambridge, where the subject was highly prized. (There were other universities that did not even have theology faculties when they were first founded.) The Oxford course for the degree of bachelor of theology was so long, on top of the master's course in arts, that few pursued it—and not a few of those who did so died before finishing it. The course was structured around the *Sentences*. In collecting together key passages from the Bible, the early Church Fathers, and some early masters, Peter Lombard's aim had been to reconcile their inconsistencies, and so systematically to draw useful general principles from them. It was no longer the rather naive basic text that counted, however, so much as the commentary that the theologian himself wrote on it. To read some of those commentaries is to be left with the feeling one might have had hearing Liszt being forced to qualify as a composer by practising his virtuoso technique on 'Three Blind Mice'.

Since medieval commentaries on the *Sentences* offer us an insight into some of the best philosophy of the age, not excluding natural philosophy, it is unfortunate that nothing by Richard of Wallingford on this subject is known to survive. He was certainly required by university statute to produce his commentary, even if he was more deeply concerned with things that would not easily fit into this literary form. The St Albans chronicler Thomas Walsingham writes down this personal report—clearly by a predecessor, since Thomas never met Richard of Wallingford himself:

> We have often heard him complain, not without a sigh, on two separate counts. First, that he left the cloister for study far too quickly, and at too early an age, and even before his training was complete. Second, that he disregarded other philosophical studies and paid great attention (in fact more than was fitting) to mathematical pursuits, in which he was particularly learned, namely to speculation on the propositions of arithmetic and geometry, astronomy and music. His writings and work on instruments are evidence of this. His attention thus distracted, he studied theology and the rest of philosophy all the less—a fact that caused him grief.

For all his obsession with these wayward subjects, he was still given the high honour of the degree in theology, as the St Albans writer goes on to note with much pride.

Oxford Theologians Abroad

We have no means of knowing the names of scholars who guided Richard of Wallingford's theological studies, or of those he heard lecture, or even of those who befriended him, but he was at Oxford during a remarkable period in the university's history. In the previous century, most of the leading English scholars in theology and philosophy had studied and taught in Paris for at least a part of their careers. From the second decade of the fourteenth century this became less common, a clear sign of change in the perceived status of Oxford. Several of the Oxford philosopher-theologians to whom Richard might have listened, or with whom he might have debated, were known by repute outside England, not always entirely for reasons of their scholarship. From the beginning of the period we have Henry of Harclay, a notable theologian who had studied at Paris and was elected chancellor of Oxford in 1312. He died at the papal court in Avignon in 1317, while representing the university in its violent dispute with the Oxford Franciscans over the requirements for graduation in theology, but his reputation may be judged from the very frequent reference others made to his writings—for example, to his commentary on the *Sentences*. Harclay had been in Paris at the same time as John Duns Scotus, a Franciscan who was the most influential theologian of the day. Scotus had himself earlier studied and taught in Oxford. He was no longer there in Richard of Wallingford's time, but his name echoed through the Oxford schools—for example, in the lectures of the Franciscan John of Reading, when violently opposing the teaching of yet a third notable Franciscan, William of Ockham. Philosopher, theologian and political writer, Ockham is now widely acknowledged as one of the most original intellects of the middle ages, but his career was anything but conventional. Having qualified as bachelor of theology, lecturing on the *Sentences* in 1317-19, he moved to the London friary in 1320 whilst waiting to incept as master. It is not certain that he returned—he taught in London for a time—but he certainly aroused much hostility in Oxford in the early 1320s and later. The excellent Franciscan scholar Walter Chatton kept up a fruitful debate with Ockham, and was greatly influenced by him. When Ockham was summoned to the papal court in 1324, to answer charges of heresy, he had the misfortune there to suffer an unfavourable report by one of his old Oxford antagonists, the former chancellor, John Lutterell. By 1333 Chatton too was adviser to the pope. Little by little, the arm of the university was extending its reach and its power.

There were Oxford philosopher-theologians who achieved scholarly fame without any strong personal continental connection. The Domini-

9. Merton College looking south, as engraved by David Loggan and published in his *Oxonia Illustrata* (Oxford, 1675). The lower text wrongly includes Richard of Wallingford among famous Mertonians. He would, however, have known the impressive chapel, the library (but without the dormer windows) in what became known as Mob Quad, beyond it, and the college hall (centre ground, left). The foreground buildings had been remodelled.

can Robert Holcot was one, of about the same age as Richard of Wallingford. He would be reviled long afterwards in Paris for his thesis that God is the cause of sin, without being its author. Thomas Bradwardine was another, and a third was Richard Kilvington, slightly younger than they. Kilvington's turn of linguistic philosophy, and his analysis of sophisms in particular, did much to advance a style of thinking in the physical sciences in which Oxford was especially strong, and the same could be said with still greater justice of Bradwardine. There were also notable Benedictine theologians at Oxford, although they are now less well remembered. Robert Graystanes was perhaps the best of them at this period, a Durham monk at Durham College when Richard of Wallingford was at Oxford. After Richard became abbot, Graystanes suffered a crushing reversal of fortune. Elected and consecrated bishop of Durham in 1333, he was forced to step down in favour of another former Oxford man, the powerful political figure Richard de Bury, whose royal nomination for the position was hastily sanctioned by the pope.

Richard de Bury had been known to Richard of Wallingford for at least five years before this episode, and perhaps much longer. He had never actually incepted at Oxford, but that does not seem to have done him any harm—indeed, the same was true of Ockham. Richard de Bury held numerous church appointments and went from strength to strength through royal service, including diplomatic service abroad. By 1334 he had become chancellor of England. Today he is best remembered for a book he wrote—probably with the help of Robert Holcot—under the title *Philobiblon*, 'In Praise of Books'. In his own time he was respected as a rich and influential patron of scholars. Bradwardine left Merton College in 1335 to enter his service. Richard FitzRalph of Balliol College, a gifted theologian with a reputation which led Pope Benedict XII to request his advice, was another who spent time with Richard de Bury soon after his election to Durham. In the 1340s Kilvington also entered Richard de Bury's service; and there were several others of comparable distinction who had done so in the intervening period.

Since Merton College (Fig. 9) was the most important centre for scientific study at this time, it is natural to scan the horizon for senior members of the college whom Richard might have met. One such man was John Maudith, a versatile scholar who shared his interests. Maudith was yet another who later entered the de Bury household. Walter Burley was a more renowned Merton College theologian, logician, and natural philosopher, seventeen years or so older than Richard. He had studied and taught at Oxford, but Paris had exerted a stronger attraction, and he had left Oxford by the time Richard arrived. It was in Paris that Burley wrote most of what turned out to be one of the most important commentaries of the century on Aristotle's *Physics*, so that it cannot be presented as an entirely typical specimen of Oxford thinking of the period. Walter Burley had social ambitions in his native country. He was an acerbic critic of some of Ockham's doctrines, which did him no harm. He became the English king's envoy to the papal court at Avignon, at about the time Richard of Wallingford and William of Ockham were both there. The nature of Burley's commission suggests that he had acquired a reputation as a theologian: he was there to ask the pope to institute an inquiry into the sanctity of Thomas of Lancaster, with a view to his canonisation.[33]

It was almost certainly in Paris that Walter Burley met Richard de Bury, who was there in 1325-26 in the young Prince Edward's service. Using service with de Bury's as stepping-stone, within ten years he had moved to the new king's household. These comings and goings may seem of no great consequence to a history of thought, but they illustrate an important point. Like so many other scholars of the first rank, here were men who did not believe that university life was an end in itself, much less that the world came to an end at the boundaries of scholarship. A display of genius was one thing, a powerful friend was another. When, a victim of the Black Death, Bradwardine died in 1349, it was as archbishop of Canterbury. Walter Burley ended his days in Italy, with occasional visits to Avignon, obtaining and exchanging non-residential benefices in England, on his own account and for friends and kinsmen.

These were some of the men Richard of Wallingford must surely have met, commanding intellects linked together by powerful social chains. During his time at Oxford Richard showed no sign of rising to a position of power, as they were all doing in their different ways—even if it was only power within the university. He probably shared, nevertheless, some of their values and ambitions, despite the moral constraints of the Benedictine rule. From an intellectual point of view, there was one great difference between him and them. Not one of the luminaries mentioned here, even among those who so influenced the course of European scientific thinking, was deeply motivated by the mathematical and astronomical sciences, at least in a conventional sense. For the time being, the number of Oxford scholars working in the higher reaches of the latter subjects was very small indeed. A decade or two later, the Oxford situation had changed considerably; and Richard of Wallingford takes much of the credit for that fact.

'Mathematical Pursuits'

The nature of the activities Richard of Wallingford pursued during the nine years between his return to Oxford in 1317 and his final departure in 1326, at the age of about thirty-four, must surely be explained by his personal skills and tastes, however they were formed. His first taxing exercise during that period was to compose a set of 'canons', rules explaining how to use a certain collection of trigonometrical tables (with an astronomical purpose) that had been compiled in 1310 by John Maudith. We know that Maudith produced star catalogues that rested in part on his own observations, so it comes as no surprise to find that he took a strong interest in instruments with which observations could be made, such as the astrolabe and astrolabe quadrant.[34] He was something of a physician and a theologian, and in the first capacity astronomy would have stood him in good stead. He was lecturing at Oxford during Richard's first period there, and also during the first two years of his second period, and the fact that their interests were so close makes it likely that the Merton scholar helped to form the younger man's love of mathematical astronomy, and perhaps even set him the task of writing the canons for his own tables. In fact Richard seems to have written two versions, one lengthy, the other abbreviated, soon after his return to Oxford in 1317.

Richard's appetite was whetted by this exercise, and his next piece of writing made an extremely important contribution to his chosen field of study. Its Latin title was *Quadripartitum, sive quattuor tractatus de corda versa et recta* ('A Work in Four Parts, or Four Treatises on Right and Versed Chords'). It should be noted that its shortened title *Quadripartitum* was one commonly used for any four-part work, notably for Ptolemy's work on astrology. Richard of Wallingford's *Quadripartitum* was very different: it amounted to the first substantial treatise of the Latin middle ages that can be reasonably described as trigonometry. Like most medieval works of mathematics, it was mathematics with a purpose, its initial purpose being to ease the task of astronomical calculation. Long afterwards, when Richard was abbot, he wrote a work with the title *De sectore* ('On the sector figure'), a revised version of his *Quadripartitum*, having in the meantime learned of some important writings by a Spanish mathematician, Jābir ibn Aflaḥ. In other words, his first great enthusiasm stayed with him throughout his life.

Both the *Quadripartitum* and the revised version of it contained what we should now describe as trigonometrical equations or identities, and other analytical expressions, all of them extremely complex, but all obtained without the aid of a convenient notation. A single equation needed to be written out in consecutive prose which might occupy a whole paragraph, often a very long one. In many cases, a diagram would have been of no great help to the author or the reader. The whole exercise meant keeping very much more in the memory than is necessary now, for a page filled with words does not lend itself to easy manipulation. Reading them today, we can scarcely refrain from translating them mentally into our own notations as we proceed: in short, we cannot even enter the minds of those for whom they were originally intended. Theirs was a culture which made use not only of a great deal of memory, but of mental processes of the kind where on occasion the author is even unable to say exactly his answer was obtained.

The universities relied chiefly on the spoken word for transmitting knowledge of whatever kind, and for the highly formal debate that the student had to practise. Examination at all levels was also oral. The fourteenth century saw a steady rise in the demand for books, but since books long remained outside the reach of poorer students, the universities kept to their oral traditions, even after the invention of printing. What was true of the university world was even truer of the monasteries. The level of monastic copying of books, and indeed of reading, is easily exaggerated. We occasionally hear of the monks being castigated for not propagating knowledge of the Bible, but it is as well to remember what that entailed: the parchment alone in a fine Bible, even allowing for the shorthand script of the day, represented a flock of perhaps three hundred sheep. Some of the classic scientific texts of antiquity, such as the *Almagest* of Ptolemy, were more than half as long as the complete Bible. When books were loaned outside a monastery, it was quite usual to take a large sum of money, the value of the book, as surety. Few monasteries would have owned advanced scientific treatises, and a young monk would rarely have been allowed to take them away. Being able to borrow lengthy books was a rare privilege which in the university at this period was more or less restricted to the wealthier houses. The best endowed of them being Merton College, it would be pleasant to think that John Maudith helped Richard of Wallingford in some way with texts, as he evidently did with ideas. It is extremely unlikely the Oxford Benedictine house could have helped in the same way.

Astrology and the Calendar

Richard wanted to master mathematics so that he could make use of it in astronomy, but what drew him to astronomy? Like almost every other student in the university, he had been obliged to study the elements of the subject as part of the course in arts, and there he had clearly shone, but there was more to the blandishments of astronomy than intellectual pleasure alone. Astronomy was considered to be an important tool of all servants of the church, to the extent that it was needed to make proper religious use of the calendar. It also opened the way to astrology, a highly esteemed science—for science it was, by the lights of the time. From beginning to end, the standard introduction to the astronomy of the quadrivium, the text by John Sacrobosco, contains no hint of astrology, no hint of how useful the astronomy of the sphere might be to the astrologer. That it would be useful was something that went without saying. Richard of Wallingford's own astrological concerns were unambiguous, and were revealed at an early stage when he wrote a treatise *Exafrenon pronosticationum temporis*, usually known by its abbreviated title *Exafrenon*. Its full title translated into English would have been 'A Work in Six Parts on Forecasting the Weather'. (The last word of the title has to be understood in the sense of the French *temps*, rather than the English 'time'.). At first sight it is a manual explaining the chief tenets of astrology and their significance, but woven into the text are references to floods and droughts, lightning and storm, and trends in the weather, general and specific—all of them supposedly decided by the arrangement of the heavenly bodies. The tides are also introduced—notice that the word in English also means 'time'. Throughout, there are hints of celestial influence on human fortune, which therefore connects yet again with the theme of 'times'. The work was a concise and well-ordered compendium drawn from earlier writers, much of it ultimately from Ptolemy's classic treatise on astrology, *Tetrabiblos*—in Latin another *Quadripartitum*. Richard drew on Arabic writers, such as the ninth-century Albumasar (Abū Ma'shar), and for literary effect made slight use of a work called *Centiloquium*, wrongly thought to be Ptolemy's. Some of *Exafrenon* came verbatim from a text by Robert Grosseteste. No treatise by the great bishop Grosseteste was to be taken lightly in Oxford. *Exafrenon* improved on its sources to the extent that it had a logical thread of sorts running through it, and made clear use of the necessary mathematical materials, which most ordinary writers were content to leave to the imagination.

This was an age in which almost all scholars, even the most orthodox of theologians, accepted the idea that the stars, Sun, Moon, and planets, influenced the material and spiritual life of mankind. The pope of the day, John XXII, seems to have accepted not only astrology but magic—something at which Richard quite certainly drew the line. There were standard scholastic debates about some of the implications of astrological doctrine for the freedom of the human will, but those who took part in them usually left the basic tenet of celestial influence untouched. Wherever the university clerk looked, he was constantly reminded of astrological belief, through symbolism in sculpture, wall decoration, church windows, and of course in various types of illustrated manuscript, such as church calendars. What none of that imagery could convey was the elaborate network of doctrine lying beneath it, and for this the *Exafrenon* offered what many found a worthy introduction—as we can judge from the fact that it was translated into English in the later middle ages on at least three separate occasions.

It was probably during his second period at Oxford that Richard of Wallingford composed a work on the mathematical principles underlying the calendar. His work on the subject (then known as 'computus') is now lost. It would have touched on matters with which every priest was supposed to be familiar, for at its core were rules for computing the date of Easter and the feasts of the church that have no fixed date (as do saints' days) but that depend in some way on the date of Easter. Richard would later compose at least one calendar, for a queen, requiring him to apply the principles of his *Computus*. Like most calendars of the period, his contained a great deal of astrology as well, but of a personal sort that *Exafrenon* generally managed to avoid—not that weather and climate are entirely without human implications.

At the very end of *Exafrenon* a story is told that comes with a few slight changes from Aristotle's *Politics*. It concerns an otherworldly astrologer, the philosopher Thales, who was so angered by the way 'lovers of worldly riches' scorned his subject for its uselessness that he made them eat humble pie. Having predicted—by astrological calculation—that there would be an abundance of olives after a prolonged shortage, Thales hired all the olive presses that were to be had, in winter, so that at the time of harvest all were forced to buy oil from him. He then went to the Capitol and exclaimed: 'Behold, fools, and see how rich philosophers may be if they wish. And he gave away all his gains and returned to his old poverty'. The ending gives a suitably Benedictine twist to an old and somewhat distorted story.

New Instruments: The Rectangulus and the Albion

In 1326, towards the end of his second Oxford period, Richard of Wallingford composed two works, both of them describing astronomical instruments, both highly original. The treatises had very different fates. The first, on what he called his 'rectangulus', was later re-edited by Simon Tunsted, so it cannot be said to have had no following. On the contrary, Richard, a Benedictine, would have been delighted to think that, twenty years or so after his death, both of his chief writings on instruments would be receiving the homage of the provincial minister of the Franciscans. The treatise on the rectangulus describes the construction and use of an instrument by which one might perform, more rapidly and more easily, the sort of calculations normally done using the methods explained in his own *Quadripartitum*—and of course elsewhere. The rectangulus, was a mechanically ingenious system of

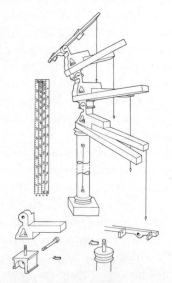

10. The rectangulus. For a fuller account of this, see Part Four.

graduated rods, pivoted to allow movements in three dimensions. There were plumblines to indicate angles, with the help not of circular scales but of non-uniform graduations on the rods. It was a masterpiece of three-dimensional thinking, but using it, or even understanding what it could do, was for most ordinary mortals—perhaps even for most of those who had already studied astronomy—simply too difficult. We have no means of deciding how many were ever made, but it seems that not a single example of a medieval rectangulus has survived to the present day.

Richard of Wallingford's most important achievement in astronomy was to design an entirely different kind of instrument, but again one that was intended to simplify many standard types of computation. He called it his 'albion'. Like his treatise on the rectangulus, that on the albion is dated

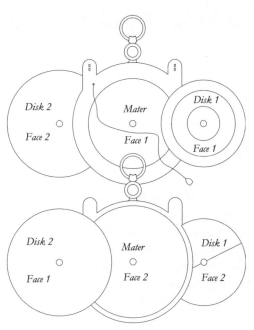

11. The albion, discussed more fully in Part Four. The disks of the albion and the body of the instrument (the 'mater') were divided into eleven principal zones, in which no fewer than sixty-seven different instruments or component scales were engraved. Note the fixed sighting vanes, which were used together with a plumb-bob when the albion was needed for the observation of angular altitude.

1326. While in appearance the albion was utterly different from the rectangulus, the two have qualities in common. Both, if they are to be used with understanding, make great demands on the user's ability to visualise what is going on; and yet in both cases—as with a modern personal computer—people of lesser ability can in principle use the rule book and achieve results without having any real understanding of what they are doing. The great difference between Richard's two instruments is that whereas the principles of rectangulus could probably be taught to students through a lecture and demonstration, the albion could not. It embodied certain mathematical ideas that are far from obvious, and that would have required not only a fair knowledge of medieval planetary astronomy but an understanding of various intricate graphical techniques. The albion was historically important, because it made astronomers look at certain mathematical questions in new ways, for many generations to come, to the benefit of western mathematical science beyond the mere realm of computation.

Richard's 'Treatise on the Albion' (*Tractatus albionis*) describes both the construction and use of the instrument. Its chief use was as a mechanical device for calculating as quickly and painlessly as possible the positions of the planets in the heavens, for any moment of time. That of course was information needed by every practising astrologer. The problem itself, stripped of its astrological aspect, was one which had already engrossed mathematically minded astronomers for two thousand years. As viewed at the time, the problem was a very grand one, concerned with nothing less than the rationale of the material universe. The planets move against the background of the fixed stars in highly complex ways. The search not only for broad explanations of their movements but for satisfactory parameters to fit to the explanatory models had been pursued with extraordinary success in early Greece and Mesopotamia. In the Hellenistic world it had culminated in the work of Claudius Ptolemy, in the second century of our era. His great astronomical work *Almagest*, in thirteen books, remained canonical until the new system of Copernicus in the sixteenth century—and even Copernicus modelled his text on Ptolemy's in numerous ways. Throughout the middle ages, the broad picture remained that set out by Ptolemy, although in eastern and western Islam, and later in the Latin world, numerous improvements were made to the Ptolemaic models and new methods of calculation were devised. Richard of Wallingford was heir to a European tradition which had benefited greatly from the astronomy of Muslim Spain. Like all who wished to calculate accurate planetary positions, he used astronomical tables. Not only were the tables lengthy—and thus expensive—but their use was time-consuming. The albion was aimed at simplifying and shortening the various types of calculation.

Consider the case of a person who wished to calculate a horoscope and place the Sun, Moon, and all the planets on it. Many scholars who would dearly have liked to do this would have found the task beyond them, despite the smattering of astronomy they had been forced to imbibe from their compulsory courses in arts. Others might have taken days to work out the answers. A practised expert with a talent for computational arithmetic—something that few then had—would have taken perhaps a couple of hours. A substitute for those accurate but slow and painful techniques was desperately needed. The generic term for the instrument meant to provide it was *equatorium*—a word also used in English. The design of an equatorium presents several theoretical and mechanical problems, for which various alternative solutions were found. Richard of Wallingford's albion was in its day by far the most ingenious medieval instrument in its class, added to which the treatise he devoted to it gave a masterly survey of theoretical astronomy in general, far in advance of the usual university texts, and more compact than most. Whether or not a student had any plans to make the instrument as instructed, had he worked conscientiously through the entire treatise he would have become extremely well qualified in astronomy. He would also have gleaned new mathematical insights which he would have found it hard to obtain anywhere else, insights into mathematical functionality, for instance. Later generations of astronomers benefited greatly from those insights, and the influence of the treatise can still be detected in the seventeenth century.

Recalling how difficult it would have been for the ordinary scholar to have built a rectangulus, it must have come as a welcome surprise to Richard of Wallingford's readers to discover that he had devoted several pages of his treatise on the albion to guiding the would-be instrument-maker through the necessary stages in making one. Of course most of what he wrote concerned drawing the scales and graduating them, but even this he described in great detail, paying attention to such small matters as how to place marks that could be rubbed out later. The two discs were to be held together by an ingenious pivot, with a sliding catch that would allow for its rapid removal. Such a gadget on a modern household device would not be given a second glance, but the care with which he described it, and explained how to form the polished brass discs that it holds together, reminds us that its author was no stranger to working with metal.

It is unthinkable that Richard of Wallingford would have refrained from making his own albion. As it happens, not only is he shown inscribing a circular instrument in a portrait of him in the *Gesta abbatum* (Fig. 1), but a marginal note in the best manuscript copy of his treatise on the albion refers to the abbot's own instrument. Naturally enough, his treatise was much copied and edited by later scholars. The Franciscan Simon Tunsted produced one version in the mid-fourteenth century, and another was prepared by John of Gmunden—a famous Viennese master who helped to make his university an influential astronomical centre in the fifteenth century. The Merton scholar William Rede, a late contemporary of Richard's, owned one and gave it in due course to his college. A hastily and rather carelessly produced version of the treatise was prepared by one of the greatest of Renaissance scholar-astronomers, Regiomontanus; and other influential southern German astronomers made great use of certain of its properties.[35]

Why the name 'albion'? The word, the St Albans chronicler informs us, meant simply 'all by one' ('al bi on'), and this etymology was dutifully repeated for centuries thereafter. Was there not a touch of immodesty in the name, perhaps even twice over? It is hard to avoid the feeling that it was aimed at impressing not only the world of learning but the monks of his abbey. 'Albion' was a name that had been given to the island of Britain from Roman times: Pliny and Ptolemy had used it. The name was also close to 'Alban', his abbey's patron saint and England's protomartyr. If Richard of Wallingford ever used the name to gain the sympathy of his fraternity, there can be little doubt as to the occasion. It would have been in the autumn of 1327, shortly after the treatise was completed, when he returned to his abbey to ask for money to allow him to graduate as bachelor in theology at Oxford. The situation he discovered in St Albans was not at all what he was expecting—unless certain monks were right when they said that he was capable of prophesying by astrological means.

The Astrolabe

Richard of Wallingford's two instruments will be considered in more detail in Part Four, but this is a suitable point at which to introduce another instrument, namely the common astrolabe, since it is intimately linked with the design of the St Albans clock. The albion incorporated the elements of one, as well as an astrolabe of a very different type—one that was 'universal' in the sense that it could be used for calculations appropriate to any place on the Earth's surface. The ordinary astrolabe lacked this advantage, but was easier to comprehend. Neither instrument owed anything to Richard of Wallingford, although the universal instrument was a source of inspiration for some of his minor writings. That it was hard to understand no doubt explains why surviving examples are rare—and perhaps also why Richard was so fond of it. The astrolabe of ordinary type, however, was familiar to very many scholars of the middle ages. Every serious student of astronomy would have wished to possess one, and those who could not afford an example in metal could at least make a passable version with parchment, or parchment-covered wood.

Crucial to an understanding of the ordinary astrolabe is an awareness of the fact that it carries two images, a moving representation of the stars and a stationary set of lines against which the movement of the stars may be judged. One might imagine oneself at the centre of a glass dome through

12. An exploded view of an ordinary astrolabe. The parts are held together by the pin at the bottom, through which passes the horse-headed wedge at the top. The bottom rule (alidade) is a sighting rule for taking the altitudes of the stars or the Sun or Moon. The rule at the top is for laying off angles when calculating with the main instrument. The essential parts for calculation are (1) a fixed plate of local coordinate lines, drawn for the appropriate geographical latitude; and (2) a rotating star map (the rete, the fretwork disc at the top). The tips of pointers on the rete represent major stars. There are here two plates shown, each usually double-sided, for four geographical latitudes in all, and the interior of the body of the instrument (the mater, at the bottom) makes for a fifth. When used for observation, the assembled instrument is suspended from the thumb by the ring on the mater so that it hangs vertically. The altitude of the object observed is read off a peripheral scale on the mater. The back of the mater (not here visible) usually carries a calendar scale, allowing the user to place the Sun on the rete for the date in question. The Sun will be somewhere on the ecliptic ring, which is conspicuous on the rete.

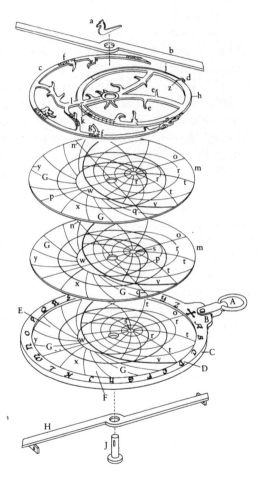

which the slowly rotating sky is observed, the framework of the dome providing fixed reference lines. To an observer, it is of no consequence whether the stars are at various (very great) distances, or on the glass of the dome. The idea of such a star sphere is perfectly reasonable, when the distances are large. (It is less reasonable when we are concerned with a nearby object like the Moon.)

Rather than using such an arrangement in three dimensions, the stars may be mapped on to a flat surface, projected from some suitable centre so that the north pole (around which the sky turns) is at the centre of the map. (The centre of projection was usually the south pole of the sky, and the plane was the plane of the equator.) The local reference lines, the frame of our glass dome, may likewise be projected on to the same plane. The most significant of the fixed reference lines is the horizon line, below which nothing is visible. Other lines could be added, for instance, at five-degree intervals above the horizon, finishing at the zenith overhead. Another fixed line of interest is the meridian, on which (in the northern hemisphere) the north pole of the sky lies, as well as the southernmost point of the horizon. As for the star map, it may have various circles superimposed on it. The most important of them is the ecliptic, the path through the stars followed by the Sun in the course of a year.

If a pin is inserted through the two maps at the pole of each, the uppermost being transparent, it becomes possible to picture the daily rotation of the stars, of their risings and settings at the horizon, and their culminations on the meridian line. Such devices are still sold, one of the two maps being printed on card, the other on transparent plastic. In the middle ages it was usual to put the stars on a rotating fretwork of brass known as the rete, which was pinned at the pole so as to allow it to rotate around the pole of a solid brass plate, with the local coordinate lines engraved on it (Figs 12 and 13). On the rete, the rotating star map, there were usually twenty or thirty familiar bright stars, represented by the tips of small pointers, and named.

The astrolabe could be used for calculation and for observation. When suspended by a ring from the thumb, observations of the altitude of the Sun by day, or of one of the bright stars or planets or the Moon by night, could be made with the help of a sighting rule (the alidade). Assuming that the object under observation is identifiable with a point on the rete (such as one of the named star pointers), from the measured altitude of the object (its angular separation from the horizon) the rete may be set correctly for the time in question. Various conclusions may then be drawn from the position of the rete. If, for instance, one can say where the Sun is on the rete (this changes from day to day, but is something an astronomical calendar will tell us), then by relating the position of the Sun to the meridian we may deduce the hour and minute of the day. This is just one of dozens of ways in which an astrolabe could be of assistance. Entire treatises were needed by those who wished to master them all: one such treatise, in English rather than in Latin, was written at the end of the fourteenth century by the poet Geoffrey Chaucer for his young son. By no means everyone who set eyes on the St Albans clock would have known how to read the time off its dial, but most would have known that it incorporated a grandiose astrolabe. Even those who did not understand it would have been dimly aware that they were looking at an abstract representation of the heavens, such as they had seen on small manual instruments—on that owned by the abbot, for example.

13. A fourteenth-century astrolabe, shown assembled. Compare the previous figure. The plate for the locality is visible behind the rete. *(Oxford, Museum of the History of Science)*

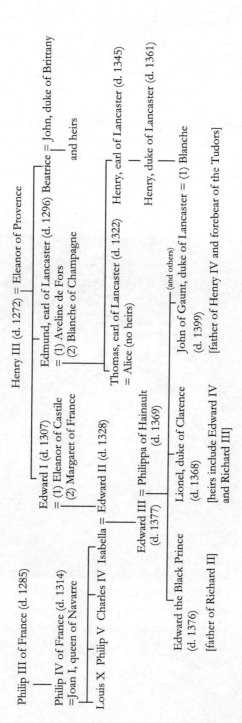

14. The English succession (house of Plantagenet) and its connection with the French royal house through Isabella, queen of Edward II.

6

The State of the Kingdom

WHEN RICHARD OF WALLINGFORD returned from Oxford to St Albans in the autumn of 1327 he was leaving behind him a town that was sharing in the turmoil of the country at large and heading for another where the situation was even worse. Social agitation was widespread, but it had an extra dimension in monastic towns like St Albans, where public discontent was directed against the monastery around which the town was organised, and more particularly against the feudal privileges of the abbot. Discontent had been fermenting for decades in many monastic towns, but it suddenly became dangerous when political events weakened the constraints that had been keeping it under control. One of the more serious outbreaks of violence was at Abingdon, where in May 1327 the townsmen, abetted by scholars from Oxford, stormed the gates of the abbey. For the townsmen this was less a question of high political principle than a protest against specific tolls and taxes on fairs and markets, and the abbey's control of them. The townsmen obtained a charter for the town from the abbot, under duress, although this was revoked once the revolt was quelled by the sheriffs of Oxfordshire and Berkshire. As Richard would certainly have known, the much more numerous St Albans townsmen had forced similar concessions from his own abbot, Hugh of Eversdone, only a few months previously. He was probably not aware, when he set out for his monastery, that a commune had been set up in the town, challenging the abbot's jurisdiction, and that unsuccessful attempts had been made to storm the abbey gates, for the defence of which large numbers of armed men had been needed. What he certainly did know was that this local turbulence was as nothing compared with the civil war which had recently split country and university, and in which the king and queen were on opposing sides.

The Fall of Edward II

For five or six years past, the long-standing attempts by the barons to limit royal power had been dividing and redividing loyalties, at all levels of society. Isabella of France, the queen of Edward II, had become increasingly alienated by his behaviour towards her and towards the faction of nobles she favoured. The final breakdown in their marriage followed the battle of Boroughbridge in 1322, at which Edward defeated a rebellion led by his cousin Thomas, earl of Lancaster. The earl was brought before the king at Pontefract, and there, in his own castle, was sentenced to death as a traitor and rebel. His execution followed quickly, and Edward brought back into office his old friends and allies, the exiled Despensers, father and son (both named Hugh). The Despensers were by any standards corrupt and rapacious, and highly unpopular. To the opposition, Lancaster's execution was an even more potent symbol of the king's personal and political inadequacy and misjudgement, and it lost him much support among the waverers. After Thomas of Lancaster was beheaded, the monks of Pontefract priory removed the body and buried it before their high altar, where before long it became a place of miraculous cures, and so of pilgrimage. After the king's own death, attempts were even made to have Thomas canonised, as though he had been the embodiment of all the virtues which the king had lacked.[36]

The queen, without seeking the part, came to be regarded as the nucleus of the opposition party. More than her pride was hurt when, in 1324, the king's supporters forced her to hand over some of her English estates, on the grounds that they represented a threat in the case of a French invasion. In this case the two men chiefly responsible were the king's chancellor, Robert of Baldock, and his treasurer, Walter of Stapledon, bishop of Exeter and co-founder of what was to become Exeter College in Oxford. The meddling of Oxford men had much to answer for in what followed. One of the queen's most active supporters was Adam of Orleton, bishop of Hereford. The king's instincts were sound when he tried—in vain—to persuade the pope to exile Orleton from England. The involvement of senior churchmen added a potent extra dimension to a struggle that was beginning to divide civil society at every level.

Orleton was an Oxford master and doctor who in 1314 had arbitrated in the university's dispute with the Dominicans over the requirements for graduation in theology. He was a scholar, but not a political lightweight. He was later entrusted with several diplomatic missions to France and the papal court. When he eventually moved to the storm centre of English political strife it was with the intention of mediating between the king and the discontented barons. If he belonged to any party at the outset, it was the church hierarchy, but in the end he sided with the rebellious barons at Boroughbridge. As a friend and protégé of one of the king's most dangerous enemies, Roger Mortimer, he was subsequently charged by parliament with treason. There was no English precedent for a bishop to be tried by a lay tribunal, and he refused to plead. His fellow bishops secured his safety, but not Mortimer's: it was through Isabella's influence that Mortimer managed to escape from imprisonment in the Tower of London and flee to France. In 1325 she herself sailed for France. Her chief purpose was to negotiate over English possessions there, and in this she was successful: lands in Gascony and elsewhere were to be retained by Edward on condition that he pay homage to the French king, Isabella's brother, Charles IV. The English king ceded the lands to his young son Edward, who crossed to France to do homage for them.

At the court of France, Isabella joined an influential group of exiles, including Mortimer. He had fought loyally for Edward II in Scotland and Ireland, but quarrels with the Despensers in 1321 had led to his downfall. Soon the queen became his mistress; then, with her son by her side, she openly defied her husband and refused to return as long as the younger Despenser remained at his court. On Despenser's advice, Edward responded by demanding that her brother, Charles IV, return her to England. To avoid this, she now moved secretly to the Low Countries. There she arranged for the betrothal of her son to Philippa, daughter of William, count of Hainault, Holland and Zeeland. Isabella and Mortimer, with a small army and a fleet of ten ships that she had persuaded the count of Hainault to provide, then invaded England, landing in Suffolk on 24 September 1326. The commanders of the English navy so hated the Despensers that they refused to oppose her. Violence erupted in London in support of the queen, the 'She-Wolf of France', and her growing army. Orleton and many English nobles now openly joined forces with them.

Isabella's triumphant progress took her to Bury St Edmunds (where she 'borrowed' money that had been deposited at the abbey by her husband), Cambridge, Barnwell, and Baldock (where she plundered the property of her enemy the chancellor, Robert of Baldock). Hearing that her husband had fled London, she journeyed westwards, first to Dunstable and then to Oxford. Her son, Edward, not yet fourteen, was as yet only a pawn in the game. Humble Benedictine though Richard of Wallingford was, he was a scholar of some standing—we recall that he had just completed his works on the rectangulus and albion—and it is quite possible that he was a member of the university congregation when the young prince heard an increasingly bellicose Bishop Orleton castigate his absent father and plead the cause of the queen his mother. The text of Orleton's treasonable Oxford sermon was 'I will put enmity between thee and the woman and between thy seed and her seed'. This was certainly no polite university exercise in rhetoric.

From Oxford, Isabella went on to Wallingford. On that same day, 15 October 1326, open rebellion broke out in London. Her opponent, Bishop Stapleton, was caught by the London mob and beheaded with a butcher's knife. It is said that, when the queen later received the gift of his head, she remarked that it was an excellent piece of justice. In the meantime she pressed on again westwards, first to Gloucester, her support growing along the way. The Despensers were soon separately captured and gruesomely executed—without facing trial. Numerous supporters of their faction eventually suffered a similar fate, including many senior university men. Bishop Orleton saved the life of his opponent Robert of Baldock, 'the brain of the Despensers', but not for long: the London mob imprisoned him and he died of ill-usage in May 1327.

The king himself was captured in South Wales and imprisoned in Kenilworth Castle. In January 1327 he was deposed by a parliament whose resolve had scarcely needed stiffening by another of Bishop Orleton's sermons, this time on the text 'A foolish king shall ruin his people'. The younger Edward rose by degrees from keeper of the realm to king. On 25 January 1327, at the age of barely fourteen, he was proclaimed king. Isabella and Mortimer effectively ruled in his name, at first with great popular support, although for what they fought against, rather than for themselves. Before the year was out, the deposed king was murdered—at Berkeley Castle in Gloucestershire.

Edward III and the Downfall of Isabella

Edward III came to the throne at a crucial stage in Richard of Wallingford's career. Insofar as he was later to have dealings with the crown as abbot, Edward III was the only king he would know. Edward's marriage to Philippa of Hainault—he was fifteen and she fourteen—was solemnised at York in January 1328, after which time he slowly began to shed the influence of his mother and her paramour. In 1330 Isabella and Mortimer added the execution of Edward's uncle, the earl of Kent, to the murder of his father. The man who was nominally the king's guardian—Henry earl of Lancaster, the queen's uncle—was also imprisoned by the ruling couple. A treaty concluded at Northampton, recognising Scotland's independence, was unacceptable to Mortimer's fellow barons, and helped to ensure his final demise. This came in 1330 at the hands of the young king, advised by Sir William Montagu, Richard de Bury, and other close friends. Edward struck at the time of a meeting of the great council of state held at

Nottingham Castle. With Montagu and other young noblemen, he entered by an underground passage at night, and had his mother and Mortimer seized. Mortimer was conveyed to the Tower, condemned by his peers in parliament, and hanged and drawn at Tyburn as a traitor.

Edward's *coup* made him ruler in fact and not only in name. His inheritance was a frightening set of political problems, although it was not the task of the St Albans chronicler to consider them, since they touched on the lives of the abbots only indirectly. Edward first turned his attention to Scotland, where he had already fought two costly campaigns. He was seeking for glory in war and revenge for Bannockburn—where his father's army had been so disgraced by that of Robert the Bruce in 1314. After a series of false starts, in 1333 the young king led an army which crushed that of the Scots at Halidon Hill, and so, for a time, he was a hero in the eyes of his subjects. Thinking that Scotland was now his, he left it to his Scottish ally Edward Balliol, but a treaty of the following year proved to be short lived. The year of Richard of Wallingford's death, 1336, saw Edward forced to lead yet another campaign into Scotland, and it was not the last.

From the beginning of his reign, Edward had been haunted by the spectre of a French ambition to seize his duchy of Aquitaine, parts of which they had seized in 1329. A French treaty with the Scots, agreed in 1327, had already complicated his relations with Balliol, and a sermon by the archbishop of Rouen in July 1335 made public the French king's intention to send an army in support of the Scottish faction opposing Balliol, that led by David Bruce. Edward's own claim to the French throne, through his mother Isabella, provided him with excuse enough for repudiating old treaties, liege homage, and other promises—there were French and Scottish precedents in plenty, as he would have said. His claim to the French throne meant that he regarded any territory he could win in battle from Philip of France as his by right. Richard of Wallingford, however, died two years before the logical consequences of that philosophy were evident, with the onset of what we now know as the Hundred Years War.

Closer to the life of the monastery were the financial and commercial consequences of the king's actions, in particular, his desperate need for taxes to fund his various campaigns. The church was all too often expected to make such grants outside parliament, which led to much discontent, as did the ever tighter controls on the export of wool. More sinister was Edward's political use of an embargo on the export of wool. (His hope was that the Flemish weavers who relied on the wool would bring pressure to bear on their rulers to take up arms against the French king.) Edward III's reign saw more important changes, but they owed relatively little to him.

One was the slow evolution of the parliamentary system, and parliament's division into two houses. Another was the growing popular hostility to the church's control of wealth, and its administrative powers. If Richard had lived longer he would have been deeply concerned by these trends. As it was, most of his relations with the crown were of a formal kind.

If the king's motives could rarely be described as principled or altruistic, his mother's were even less so. Before Mortimer's execution, Isabella took far more of the revenues of the crown than were allotted to her son. After his death she was forced to give up most of her estates, although even then she was compensated with an enormous allowance—£3000 per annum, more than three times the income of the St Albans monastery. She settled for the rest of her life at Castle Rising in Norfolk, being visited annually by her son and allowed to move about as she wished. Her last years were spent in a nunnery of Poor Clares; and, on her death, in 1358, she was buried in a Franciscan church in London. When, at a later date, the great middle window in the church was blown out in a storm, her son paid for its replacement, 'for the repose of the soul of the illustrious Queen Isabella'. The mendicants knew how to organise such things.

At some time in the summer of 1327, while Isabella's husband Edward II was in prison—'that great friend of this monastery', the chronicler called him—and while she was still at the top of fortune's wheel, she had visited the abbey of St Albans. There she had witnessed for herself the unrest among the villeins who owed feudal service to the monastery. To get her support, they brought their wives and mistresses and—so they claimed—children who had resulted from rape by the monks, and waylaid the queen. She understood English so badly that she had to ask why the women were protesting in this way with bared bosoms. One of the barons accompanying her replied that the women were acknowledging themselves to be adulteresses and whores. Isabella ordered her carriage to be on its way. If the queen was not amused, the monks clearly were. The story as told in the *Gesta abbatum* was of course not meant to illustrate the growing disillusion of the labouring classes, or the stirrings of anti-clerical feeling, or indeed any other social, constitutional, or cultural thesis. It was simply presented as an amusing skirmish in the war between right and wrong—the monks' right and wrong, not ours. The French-speaking Isabella was merely a foil. Some years afterwards, Richard of Wallingford brought a lawsuit against her. There is no hint of triumphalism in the chronicler's report of that, but in view of Isabella's extraordinarily acquisitive nature, it was perhaps judged superfluous.

Part Two

An Abbot's Rule

The abbot should do all things in the fear of God and observance of the rule, knowing that he will certainly have to render an account of all his judgements to God, the most just judge.

Rule of St Benedict, chapter 3

The English form of government, at the period we treat of, was but a rude and imperfect cast of the admired and matchless constitution which we enjoy at this time.

Peter Newcome, *The History of the Abbey of St Albans* (1793)

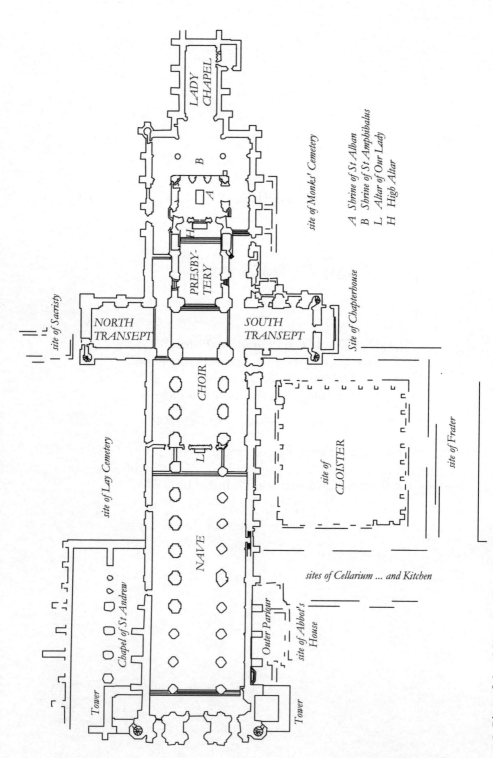

15. Plan of the abbey church of St Alban, with the sites of some of the principal monastic buildings.

A Shrine of St Alban
B Shrine of St Amphibalus
L Altar of Our Lady
H High Altar

7

A New Abbot

NO SOONER had Richard of Wallingford arrived in St Albans than Abbot Hugh died—on 7 September 1327, the eve of the feast of the Nativity of the Blessed Virgin Mary. 'There are some', wrote the chronicler of Richard, 'who say that he predicted by the constellations of the stars that Abbot Hugh would die and that he would himself become abbot.' So much for the higher reaches of astronomy, the world of albion and rectangulus. After three days, Hugh's body was interred with due solemnity and a delegation was sent to the king in the customary way to ask for formal permission to elect a successor. This having been obtained, letters were sent out to all the priors of the cells of the abbey to attend an election on 29 October, the day after the feast of St Simon and St Jude. The election was by a system known as 'compromise', and was held in the abbot's chapel above the parlour. There were nine 'compromissors' or arbitrators: the priors of Tynemouth, Hurley, Wymondham, and Wallingford (this last post was no longer held by Richard's adoptive father), an archdeacon, a scrutineer, a confessor, a hosteler, and a scholar. The scholar was Richard of Wallingford.

The chronicle has much circumstantial detail throwing light on Richard's character, and it plainly comes from someone close to the event. On the day of the election Richard preached to the assembled monks, taking the text 'Choose from amongst you the worthiest man'. The scriptural passage, when quoted at greater length from the first book of Samuel (17:8-9), is more illuminating. It is taken from the challenge issued by Goliath to the Israelites, the challenge that was accepted by David:

> Choose a man from among you and let him come down to me ... If he is able to fight with me and kill me, then will we be your servants; but if I prevail against him, and kill him, then you shall be our servants, and serve us.

The previous day, another of the compromissors, Richard of Tring, the confessor, had addressed the convent using the text: 'I have chosen you'. More fully quoted (from John 15:16), this continues 'You have not chosen me but I have chosen you', a far more pacific message than Richard of Wallingford's, on whom the choice of the electors eventually fell.

He is said to have stated at a mass held on the day of the election—in honour of Saints Alban and Amphibalus and the saints whose relics were kept at the altar of St Amphibalus—that he felt it in his heart that the election would fall on the most fitting person. Did he remind all concerned of his scientific genius, and in particular of his masterly instrument the albion, so close in name to Alban's? Do the references to his having predicted that the old abbot would die indicate that he openly advertised his astrological expertise? Did he promise the abbey an astronomical horologe? Or did he lay emphasis on his advanced qualification as an Oxford theologian? We can only guess.

'The newly elected lord abbot', we are told, 'Richard of Wallingford, having been led into the church with trembling and respect, as is proper, was made to stand before the great altar, and the election was announced to the people by brother Richard of Paxton, archdeacon. It was beyond the expectations of everyone, and especially of the laity.'

Richard claimed to be perplexed when informed of the convent's choice, and it is conceivable that here he was following custom, although some of the monks rather waspishly suggested that his hesitation to accept the office of abbot was feigned. Though he said long afterwards that he regretted that he had so readily consented, he was reported as having said at the time, and later, that in accepting office he was 'afraid of affronting the Holy Spirit'.

Goliath

In the light of his chosen text, Richard of Wallingford undoubtedly saw himself in the role of David. He can hardly have meant to present himself as Goliath. Who then was the Goliath of his sermon? Those who had witnessed the civil violence of previous months would have thought twice before stepping into Abbot Hugh of Eversdone's shoes. Did Richard see the villeins of the town as a collective Goliath—those who only a few weeks earlier had laid siege to the abbey and had petitioned Isabella in vain? The humiliation caused by the earlier concessions Abbot Hugh had made to them seems to have precipitated his death. Or was Richard referring obliquely to the debts Hugh had left as his legacy? Hugh's predecessor had left a similar legacy, and the abbey debt had been accumulating steadily for decades. Wars had reduced the value of many of the abbey's possessions, especially those in the north of the country. There had been famines. The early years of the century had been especially unkind, with the weather growing colder in northern latitudes generally.[37] In 1315, incessant rains had ruined the harvest across the whole of Europe, and on one occasion in that year Abbot Hugh, a close friend of Edward II—perhaps this is why the chronicler, Thomas Walsingham, was so hard on him—had to face the shame of not being able to feed the visiting king and his retinue. When Richard preached to the assembled convent, was he tempted to quote from his treatise *Exafrenon*, on the astrological causes of such things? Did he hint that with the help of his 'all by one' and his astrological knowledge he might predict such events, bringing profit and honour to the house of Alban? Quite apart from natural disasters, however, the monks must have realised that the abbey farms had not been well managed, and of course they were conscious of the disastrous collapses of large sections of the abbey buildings in 1323. Despite the desperate measures taken, the buildings had still not been repaired as they listened to Richard of Wallingford's address. The giant Goliath might have provided a metaphor for any or all of these things.

In the month of March before Hugh's death, the young king—or, rather, those acting on his behalf—had appointed commissioners to inquire into the dissipation of the revenues of the abbey of St Albans. Their report had been severely critical. The new abbot, therefore, lived frugally during the first months following his election, eating simply and not offering gifts to all and sundry, as new abbots were often tempted to do. He must have been anxious not to be blamed for the abbey's debts, should the commissioners return soon. Those debts amounted to some 5000 marks, not to mention a large number of corrodies or pensions. (A mark was 13s. 4d., that is, two-thirds of one pound sterling.) Hugh had tried to raise money from the St Albans cells, and in doing so had guaranteed the hostility of all the priors. It seems that only three of the priors were now prepared to make the customary present to Richard his successor; and none of the obedientiaries—the chief holders of office under the abbot—did so. One prior gave ten marks, another two pounds; a third gave a cup worth sixteen shillings. The new prior of Wallingford, William Heyron, gave nothing. The new abbot, however, who owed so much to that priory, must have been quietly jubilant when he accepted his new responsibilities, for all that his eyes were moistened with tears. He was at the top of Fortune's wheel.

Avignon

The most pressing business for a new abbot was to obtain confirmation of his appointment, first from the king, and then from the pope. As in the case of the succession to the see of Durham, where the king forced his will on the monks, such things did not always go according to plan. Richard of Wallingford set out on his formal visit to the young king, then at Nottingham Castle. As a symbol of his frugality, he took with him only a small retinue. Having obtained confirmation on 29 October, he began at once to prepare for his long journey south to the papal court. Again he kept the number in the party to a minimum, determined to make the journey at the lowest possible cost. There were in the party three men described as chaplains: Nicholas of Flamstede, who was prior of the cell at Hertford, Richard of Paxton, the archdeacon, and Richard of Tring, confessor. An archdeacon would have been regarded as chief of the abbot's administrators and attendants, while the confessor was his spiritual adviser. All had been electors, and perhaps Richard believed that all were favourably disposed to him. They took a single squire (*armiger*, a man bearing arms) as a bodyguard, a clerk, and a salter (*salinarius*); none of these is given a personal name in the chronicle. Seven horses in all. The party was completed with two other clerks not belonging to the abbey, to make up a safe number. This was not the great train that would normally have been expected of the abbot of such an important and traditionally wealthy monastery, but these were not normal times, and Richard was determined to cut his coat according to his cloth.

Their long and difficult outward and return journey lasted about twenty weeks in all, more than twelve of those weeks spent on the move. They set out on 23 November 1327, reached the papal court in Avignon on 4 January 1328, and arrived back at the abbot's manor of Crokesley (Croxley, near Rickmansworth) some time around 12 April 1328. From beginning to end, Abbot Richard had the unstinting support of Nicholas of Flamstede, a faithful friend who also kept meticulous accounts. From them we know that the expenses of the journey were precisely £953 10s. 11d. This was considered a prudent total, and yet it was of the order of half the abbey's revenues for a whole year, and increased the abbey's debts by nearly 30 per cent. Prior Nicholas, about ten years Richard's senior, acted as his counsellor throughout, and raised his spirits when necessary. It was indeed often necessary, for—to use an expression coined by a twelfth-century predecessor of the abbot—this was a journey into the land of 'the insatiate sons of the horse leech, ever athirst for money'.

Why Avignon?

That was a perennial complaint. In the mid-thirteenth century the pope had persuaded Henry III of England to pay large and unsustainable sums of money to Rome, so large that an angry group of lay barons eventually forced the king to agree to a meeting of parliament at Oxford in 1258 to discuss the constitutional question. The papacy's need for money, however, had not diminished. When Richard of Wallingford was in Avignon its financial circumstances were even more parlous than in Henry's time. Division and turmoil in the church hierarchy during the three previous decades had begun with an attack on Pope Boniface VIII by agents of Philip IV ('the Fair'), king of France. The French king had levied taxes on clerical income without papal approval, and in retaliation the pope issued a bull forbidding the clergy to pay taxes to any lay ruler. Boniface then excelled himself with another bull in 1302, asserting in a notorious closing sentence that it is necessary to salvation that all human beings be subject to the Roman pontiff. Taxation as the price of Paradise was poor economics, and Philip retaliated by bringing a rich assortment of charges against the man who seemed to wish to set himself up as supreme ruler—they included heresy, sodomy, and sorcery. He called for a church council to judge the pope. Then, to forestall a promised bull of excommunication, he arranged for anti-papal Italian troops to seize the pope himself, in his summer retreat at Anagni, not far from Rome. The citizens of the town were incensed, and managed to free him. Not before 1378 would another pope be at all complacent about residing in Rome. (Gregory XI then made Rome the papal capital once more. His action made matters even worse, dividing the church and giving rise to the so-called Great Schism.)[38]

After the Anagni incident of 1302, Philip turned the whole affair into a national crusade. At Boniface's death in 1303, the cardinals—more or less equally balanced between French and non-French—had failed to elect one of themselves as his successor. Their choice, after an inconsequential papacy lasting only nine months, eventually fell on the archbishop of Bordeaux, who took the name Clement V. That new pope, who agreed to Philip's wish that he be crowned in the king's presence in Lyons, chose to establish his court in Avignon. In due course more and more French cardinals were created, and during the whole of Richard of Wallingford's time as abbot of St Albans the Catholic Church was in many respects an arm of the French state—although the pope was even-handed towards England when he was not under pressure from France. Strictly speaking, Avignon was governed by the counts of Toulouse and Provence, and was a fief of the king of Naples and Sicily, but in culture and political sympathy it was to all intents and purposes French. Six popes in succession ruled the western Christian Church from Avignon, all six of them French. It was with the most forceful of them, Pope John XXII, that the little party from St Albans had to deal.

Pope John XXII (1316-34)

John was the second of the six Avignon popes. At heart, and by university training, he was a lawyer. By the time of his election in 1316, at the age of about seventy, he had acquired much experience of the ways of the world and of the church—he had been both chancellor to the king of Naples and bishop of Avignon. He had been born Jacques Duèse, to wealthy parents in the southern French town of Cahors, then an important financial centre. There was never any doubt as to where his political loyalties lay. Of twenty-eight cardinals created by him, all but eight hailed from southern France. That only three were members of his family must have seemed an act of great self-control, bearing in mind his predecessor's record of five. Both showered material and monetary gifts on their families and old friends.

John had more than his share of enemies. Some were enemies by accident. One of his first acts concerning England, for example, was to send two cardinals to mediate between English political factions and to consecrate the bishop of Durham. Elements in the disgruntled north had them ambushed under the impression that they were King Edward's men. This was not an affair of any European consequence, but it illustrates the pope's political pretensions, which were almost his undoing. He soon achieved notoriety for taking sides in a bitter and long-standing conflict between the two main factions in the Franciscan order. Pope John favoured the Conventuals, who wanted a more stable community life, which they thought would be better fitted to study and preaching. The Spirituals, on the other side, were those who wished to follow strictly St Francis' rule of poverty. With the help of scriptural evidence, John argued that Christ and the Apostles had owned property: it was a comforting thought for the well-housed vicar of Christ on earth. In a series of papal edicts, he came down ever more strongly in favour of Conventualist principles. He also actively persecuted those Spirituals who raised their voices against his decisions. He handed some of the dissidents over to the Inquisition, with the result that in 1318 four were burned at the stake. They would have taken small comfort from the pleas of future historians that their cases were theologically complex and that their fate was not unusual.

Two scholars of note now by chance came together to oppose Pope John in Avignon itself. First to come to the papal court was William of Ockham, the leading English philosopher of the age, a Franciscan whom we met briefly in connection with Oxford theology. He had been summoned in 1324 to answer charges of unsoundness in some of his theological and philosophical views, on which the self-seeking John Lutterell, an embittered man who had been dismissed as chancellor of Oxford University, had reported adversely to the pope. (Lutterell's critique followed the lines of Dominican theology, another instance of the real and sometimes bloody effects of rivalry between the religious orders.) Ockham had overlapped in time at Oxford with Richard of Wallingford during his period of theological study, and it is very likely that they were at that stage at least aware of one another's existence.

The second notable opponent of the pope's teaching was the Franciscan minister general, Michael of Cesena. He was called to Avignon to answer for his opposition to John XXII's condemnation of the doctrine of absolute evangelical poverty. The two Franciscans grew increasingly alarmed at John's actions, and his wish to revoke earlier papal pronouncements on the subject of poverty whenever they conflicted with his own views. Michael escaped from detention in May 1328. Together with Ockham and other members of their fraternity, he fled to the safety of the court of Ludwig IV of Bavaria, Pope John's bitterest political enemy. Ockham spent the rest of his life engaged in political polemics against the Avignon papacy.

Whether or not Richard of Wallingford encountered Ockham in Avignon it is impossible to say. They would not have been thrown together by their religious affiliation, and it is unlikely that the new abbot would have felt sure enough of himself to enter into any controversy that touched on the pope's autocratic behaviour. The question of evangelical poverty played only a small part in the much wider conflict between pope and emperor that must have been the principal topic of conversation in Avignon whilst the St Albans party was there.

Like everyone in the neighbourhood of the papal court, they would all have known the circumstances of the wider dispute. They would have known of how the designated Holy Roman Emperor, Ludwig IV of Bavaria, had been in dispute with Frederick of Austria and had refused to accept John XXII as a competent arbitrator between them. They would also have known of how John had taken the Austrian side and refused to crown Ludwig emperor, with the result that Ludwig explicitly denied papal authority over imperial affairs. He also defended the Franciscan Spirituals, so rubbing yet more salt into John's wounds. Two other distinguished philosophers at the imperial court, Marsilius of Padua and John of Jandun, gave supporting arguments to prove that the authority of a church council was superior to that of a pope.

John XXII retaliated by excommunicating the emperor. Ludwig then took the quarrel to Italy itself, and was bold enough to oppose the pope there. By siding with opponents of the pope, he found most of what he had wanted seemingly falling into his lap. He was offered the iron crown of Lombardy in Milan on 31 May 1327. He then marched on Rome, where on 11 January 1328 the imperial crown was presented to him by representatives of the people of the city. News of this crisis reached Avignon during Richard of Wallingford's stay there. He might have pondered the relative merits of a small English baronial rebellion and this dispute on a European scale. He might have reflected on the precarious nature of authority, and his own powers and responsibilities. He must certainly have wondered what the future held in store for him and his party in Avignon.

News of later developments would have reached him only after his return to England. In a decree of 18 April 1328, Ludwig announced that he was deposing the pope. The nearest he could get to actually doing so was when a straw effigy of 'Jacques of Cahors' in pontifical robes was solemnly burned. For the remainder of Richard's rule as abbot of St Albans he had to live with a situation in which the pope to whom he had sworn allegiance was under attack by a man who could command not only great military power but an impressive army of intellectual clerics. The emperor inevitably tried to set up his own pope. A Franciscan from the Spiritual wing, Peter of Corbara (Pietro Rainalducci), was elected antipope in John XXII's place, and assumed the title Nicholas V. In April 1329, however, abandoned by his patron, Nicholas decided to submit to John. He was pardoned but imprisoned for life—comfortably so, and with a large pension—in the papal palace at Avignon.

The emperor kept up his attack on John until the latter's death in 1334. The church, even so, had one abiding advantage over the empire: it depended less on the person of the pope than did the empire on its emperor. The empire was relatively formless and unstructured, by comparison with a church that for centuries had been given its traditions in forms settled by scholars, not only theologians but experts in canon law. The pope could claim to be heir to St Peter. The emperor was a man above other men, but a mere man none the less. The church could spare a few of its more brilliant firebrands and still remain intellectually rich. The abbot of St Albans was in good company after all.

The Road

The journey to Avignon to obtain papal confirmation of the new appointment was of symbolic value, and the church required it, but it was a journey that offered experiences of a sort that few ordinary monks could ever expect to have. Perhaps the reluctance of the priors to give the new abbot the customary gifts had something to do with their disappointment at not being included in the party. The journey had of course been made often enough by others who could offer advice, including monks from their own abbey. They would have been warned to clothe themselves in the expectation of a cold winter. Language would have been no great problem: they would have had Latin for clerical use; and we might expect one or more of the monks to have had a knowledge of Norman French, which was still spoken by monks from noble families. (It was the preferred language of Richard's predecessor in office, and it was used by his successor too.) They would have been told which roads to follow, and which rivers were to be bridged and which forded, and where; and they would have received further guidance from the houses of black monks and other places at which they lodged along the way. The travellers would have learned from their hosts where the most wholesome drinking water was to be had, and where the best relics and architectural marvels were to be seen. Great abbeys often had guest-houses where travellers were housed and fed, even hospices for sick travellers, but on occasion the party would have needed to lodge at inns, where fodder for the horses would often have cost more than their own fare.

One of the hazards of travel was that of taking a wrong turning, although much of the journey would have been in the company of other travellers. This made for greater safety in places where there were large numbers of thieves and vagabonds, ranging from destitute peasants to knights and clergy who had turned to crime. The easiest part of the journey would have been along English roads to the port. As senior clergy, with the king's confirmation of Richard's election, they would not have needed further written permission to leave the country. There were many ports they might have used, including London itself, Gravesend on the Thames estuary, and Sandwich and Winchelsea on the south coast. They most probably crossed from Dover to Calais. With good fortune, the crossing would have lasted much less than a full day, although storms at sea could lengthen the crossing to a day or more. Even if they crossed to Calais, they might have done as most pilgrims did, and have ridden through the ancient port of Boulogne to pray in the abbey of Notre-Dame to a miracle-performing Virgin.[39] If that was their route, perhaps they proceeded on the road south to Amiens, where pilgrims commonly called to worship before the head of John the Baptist. It is likely that they would have visited the abbey of Saint-Denis outside Paris before entering the great city itself. No matter how pressing their mission, and how anxious the new abbot was to avoid unnecessary cost, they would have visited the chief architectural sights of the place. They would have seen Notre-Dame, Sainte-Chapelle,

Saint-Germain-des-Prés, and any one of a dozen other churches, not to mention the royal palaces—although the latter they would have seen only from the outside. Richard of Wallingford was familiar with the writings of many Parisian university masters, and no doubt knew a handful of English scholars who were then teaching in Paris. Whilst his curiosity must have been whetted by the opportunity to look inside their halls and chapels, he is unlikely to have found the time to do much more than that.

A route followed for centuries by many English pilgrims would have taken the party from Paris through Rocamadour in Guyenne—English territory—but that would have meant travelling far out of their way. Instead, they might have joined the upper Loire at Orleans or Nevers, have gone upstream from there by barge, and then have ridden overland to pick up the Rhône at Lyons. Whether they did this, or reached Lyons entirely by road, they would surely not have missed the cathedral of Saint-Jean in Lyons, or the many fine churches near at hand. How much use they made of river transport we cannot say. It was generally much more expensive than travel by horse, but the temptation to travel down the Rhône would have been great. A mixed company of horsemen and pilgrims on foot could cover thirty miles a day for a week or two. Allowing five days in seven for travel over the total period of six weeks, the distance covered daily by the St Albans party would have been roughly twenty-five miles. They could be forgiven for time spent dallying on so beguiling a journey. There is more than a suspicion that the return journey lasted a week or two longer, but of course the Rhône runs against the northwards-bound traveller, and that may help to explain the slower return.

The Throne of Costly Grace

Their long outward journey completed, the St Albans monks were greeted by an imposing sight of Avignon—less so than it would become over the following decades, but certainly impressive. Surrounded by ramparts, the city was dominated by a precipitous rock, the Rocher des Doms, on which stood the newly-built palace of the bishops and the twelfth-century cathedral of Notre-Dame des Doms. The tower of the cathedral was then being raised higher. To its side were the beginnings of what in the later fourteenth century would become a vast fortification, enclosing the papal buildings. John XXII was creating a new palace simply by enlarging the old palace of the bishops. There were already other fine buildings to be seen. Looking out from the ramparts on the rock they would have seen the River Rhône below, crossed by the famous bridge that is still celebrated in the familiar song. Four of its arches survive to this day, as does the chapel of St Nicholas on one of its surviving piers. It was not as massive as London Bridge, but more elegant. According to local tradition, it was built in eleven years by volunteers from among the common people, in response to a dream by St Bénézet, who convinced them of his mission by moving enormous stones effortlessly. No Wallingford man could fail to be moved by the sight of a long bridge.

The new and sumptuous buildings, with their spectacular painted interiors, were matched by the pomp of the papal court. Having arrived in Avignon, the St Albans party would have wasted no time in searching out the appropriate office. They must have been bemused by the hordes of clergy and high church officials, but their situation was a common one, and the affairs of the church were well regulated. They had certain formalities to complete. They were not to go to those places where regular income from all quarters of western Christendom was being collected with unprecedented efficiency; nor to those covert quarters where fees could be paid for pardons, absolution, and indulgences at the lower end of the market, or for simony (the sale of offices) at the other extreme. On the day following their arrival in Avignon, Richard was given an audience with the pope himself.

After the customary triple genuflection and kissing of Pope John's right foot, Richard presented him with what was plainly a well-rehearsed request. He quoted Hebrews 4:16, 'Let us therefore come boldly unto the throne of grace, that we may obtain mercy, and find grace to help in time of need'. The 'throne of grace' in this passage, as he explained to Christ's Vicar on Earth, was his most holy name. And lest the pope miss the allusion to the Hebrew meaning of the name 'John', Richard went on to enlighten him:

> For it is interpreted 'merciful Lord', or 'Grace of God', added to which it bears witness to the most agreeable reports of your fame, spread abroad by the lips of those who have put it to the test throughout the whole world.

He then asked John to confirm his election in the customary way. Having obtained leave to visit the cardinals, he withdrew. No doubt others were waiting in the corridor with similar requests.

He had done what he could to mollify the supreme pontiff, and he was perhaps proud of his oratory, but he must have known that the game had only just begun, and that money was the key to its completion. Pope John had already done much to put the finances of the papal treasury on a firm footing, but in ways that most churchmen found callously indifferent to their needs at home. The English church hierarchy was resentful of the heavy burden of taxation, but those who had experienced the distant splendours which it was being used to subsidise must have felt doubly indignant. They knew that the church needed finance, but hated the sensation of helplessness: since the pope was ultimately in control of church appointments, he had almost total control over the fees paid by those appointed. The old complaints against the system multiplied as new fees were invented. John issued a scale of charges for no fewer than 145 types of document that the papal chancery were empowered to issue. It was a lucrative monopoly that was still virtually intact when the English state finally broke with the Roman Church in the time of Henry VIII. Catholic historians often judge John's reign favourably by the simple fact that he left to his successor an inheritance of 24 million ducats. Richard of Wallingford would have said that they were missing the point.

The stolid English monks were, unsurprisingly, no match for the mercurial papal lawyers. Mistakes were soon found in the wording of the St Albans electoral decree, and Abbot Richard was politely requested to renounce his election, 'voluntarily and of his own free will'. This tautology could not hide the fact that he had no real choice in the matter. He was not an abbot but an ordinary monk once more. After consulting with the cardinals, however, on the state of the monastery and the person elected, the pope agreed to reappoint him abbot.[40] The new abbot was happy to wait upon the pope a second time, to thank him for his kindness. One may assume that there were fees paid to officials in the papal chancery at every stage in this charade. Some days later the new abbot received the benediction of the cardinal bishop of Oporto, after which, 'with a clear conscience', the time was ripe for him to visit 'the cardinals his friends' to give them fitting presents.

So much for the opening moves in this stately minuet. It remained for the abbot to present himself to the chamberlain of the college of cardinals and the chamberlain of the Holy Father to settle the question of the tax due from the monastery on the occasion of the vacancy. The grandeur of the treasury buildings might have served as a warning of the powers of treasury officials. From this, as from other parts of the account of the Avignon adventure, it is clear that the information comes from one of the party, perhaps the unnamed St Albans clerk. Whoever he was, his account of exchanges with papal officials—who were the very prototype of the eternal taxman—is tinged with faint irony. They opened the meeting with reference to a tax owed by his monastery, whereupon Richard interjected that he knew of none to which they were liable except one ounce of gold every year by way of exemption (*nomine exemptionis*). The officers ordered him to be silent. They found marked in their register an arrear of 720 marks, said to be the equivalent of 3600 florins, due for a visitation by virtue of a vacancy. Richard was struck dumb. When he recovered his speech, they warned him against protesting, and compelled him to swear on the Bible that he would pay these sums at certain stated periods. For good measure, they added a threat of excommunication if he failed; or if, in case of failure, he did not return to the papal court within five months.

All of this was put into writing by the papal notary. The officials, expert in more than mere arithmetic, apparently left the fledgling abbot feeling elated at having to pay such a small sum, for he was told by those who had been present when his predecessor came to the papal court that the tax had then been much greater. What was paid took no account, of course, of the numerous lesser payments that he was required to make, and which were more carefully inventoried in the *Gesta abbatum* for the earlier visit of Abbot John IV to Rome in 1304. There were then presents of goblets, plate, a ring, or (in most cases) simply money to the examiners, the referees, the 'agents of the lord pope's pyx', various named procurators, notaries and other agents, those who wrote out the documents and those who corrected their work, the writers of supplications, the registrars, those who drafted the bulls, and then the fees for the bulls themselves, and for registration. The pope's new scale of 145 charges might even have endeared its author to Richard, with his training in the use of precise astronomical tables, for at least he knew more or less where he stood. At all events, he was happy to think that his successor would benefit by the low precedent being set. And to seal his peace of mind he was advised not to object to the sum being charged, since the officers of the court were at liberty to put absolutely any figure they chose on the tax.

Five or six days later, Richard of Wallingford was at last allowed to swear a standard oath of obedience to the pope. How the St Albans party spent their days, as they waited on one official after another, we shall never know. They were perhaps guests of the black monks in Avignon, who might have warned them of the dangers of the city. They would have learned that this papal territory had a reputation as a place of asylum for heretics, for mercenaries who lived by blackmailing the church, and for plain criminals; and that it was 'plague-ridden when there was no wind, and wind-plagued when there was'. It is altogether unlikely that they would have met the young Italian poet Petrarch (Francesco Petrarca), father of Renaissance learning in Europe. He lived for much of his life in Avignon, however, and was already making the place famous in literature. He had recently taken minor orders there, and lived in the town as the protégé of the bishop of

Lombez and in the service of the influential cardinal Giovanni Colonna. The most famous event in Petrarch's life was when he saw Laura—the subject of his love lyrics—for the first time in the church of St Clara on 6 April 1327. The St Albans party were not to know this, but at least they would have known the church.

How long they kicked their heels in Avignon we cannot say, but before leaving the city to return to England Richard of Wallingford tried his own hand at the courtly game and submitted certain petitions to the Holy Father. We are not told what they were, but the English party was no match for the astute lawyer pope. Richard's requests were entirely ignored, the excuse being that the court was in a state of confusion on account of the death of the king of France, Queen Isabella's brother Charles IV. Charles was about two years younger than Richard of Wallingford. He died at the age of thirty-four on 31 January 1328, two days before the pope finally confirmed Richard's status as abbot. Papal messengers would have brought the news from Paris in about six days. It would undoubtedly have been disturbing, but not enough to interfere with the collection of papal taxes.

Fortune's Wheel

When Richard of Wallingford returned to England he did not announce his arrival to those at the monastery; instead he headed for the abbot's manor at Crokesley. He must have felt that the world was at his feet. Seven months before, he had been one scholar monk among many; now he was Richard II, twenty-eighth abbot of the proudest monastery in England, a place where he could put his talents to good use and achieve great things. He had been filled with new ideas. He had seen strange customs, strange sights, fine bridges, new machines, great water-mills, fine manuscripts, and many of the wonders to which men and women could be moved by faith. Most impressive of them all were the architectural splendours of the age. Having journeyed a hundred miles to a great city, one did not leave it without worshipping in its finest churches. Richard had travelled across Europe to within thirty miles of the Mediterranean Sea, and his experience had swelled tenfold. He had met the Vicar of Christ on Earth and his cardinals. He had every reason for jubilation.

But then, on the night of his return to Crokesley, he felt a burning sensation in his left eye, followed by great pain. In anticipation of what was to follow, the chronicler tells us at once that in the course of the next two years his eye would gradually worsen, until a veil had obscured it entirely. For the time being, however, he could hope for a cure.

The next day, the new abbot wrote to instruct the priors of his outlying cells, and the obedientiaries, those who ran the domestic and external affairs of the convent, to meet him at his manor of Langley. There he chose those who were to be members of his household. With a fitting complement of attendants, he could wait in state on the king at Northampton to pay homage to him. (A parliament had been held there in February 1328.) And still Richard did not give the prior and monks of his abbey, or even his chaplains, advance warning of his plans. Instead he rode back from Northampton and into the town of St Albans with his train, taking those in the monastery entirely by surprise. There is no reason for thinking that this was other than for dramatic effect. The monks rose to the occasion, donning suitable dress quickly and ringing out a peal of bells to welcome him. For his part he preached a sermon on the text 'The man is returned and sits in his place'. He presented his papal documents and, after the kiss of peace, again took everyone by surprise by announcing that there would be dinner in the refectory—the cellarer had been let into the secret. The abbot, determined to begin on a friendly note, ate with the monks in the refectory, while the seneschal and the cellarer dined in the abbot's chamber with the grand people who had travelled with him that day. The day ended with the abbot receiving, as was his due, the keys of the monastery.

The chronicler's account of all this reads like the first act of a tragedy. It is the light before the dark. We have been warned very briefly of appalling things to come, and yet for the time being all is as it should be. Once the keys have been handed over, however, the cold realities of monastic rule slowly begin to present themselves.

8

Reprove, Persuade, Rebuke

WHEN MONASTICISM was first established, the cloister was the only proper place for the monk who would live a holy life, but as time went by, and the state and universities called on the cloister for its expertise, the monks were more often seen in public, and the various monastic orders became more worldly. The recognition of expertise of any kind—whether it is in responding to a theological argument, judging a case of law, or building a church—gives rise to elitist distinctions, even when there is a democratic system for electing the elite to high office in the first place. To elect is to judge, and Christian abbots could be judged by Christian principles, but the election of an abbot happened only once in a while, and between elections the abbot was liable to be challenged from many different quarters.

In the early years of the fourteenth century, those without rank or privilege who provided the base of society inside and outside the walls of the monastery were becoming increasingly conscious of their collective power. They found that while they could not always reverse the actions of those in high office, they could at least often temper them. Above them loomed the great lay barons and their ecclesiastical counterparts, bishops and abbots, who were often as rich or richer than neighbouring barons. The chief difference between those two classes of overlord was that the churchmen were not aiming at dynastic advantage. They were, however, just as determined to advance the cause of their ecclesiastical communities, and since many of them—especially the bishops—had risen through service to the crown, they tended to take the side of the upper echelons of society rather than the lower. Like the great lay landowners, they held at least part of their inheritances in chief directly of the crown, by feudal tenure. They in turn granted out the land to tenants. The lord was called 'mesne-lord', and all of his holding was known collectively as his 'manor'. It included manor farms, waste lands, appointments to minor church livings, and rights of jurisdiction over tenants. The abbot of St Albans had this manorial status. Whether or not the lord was a mesne-lord in this sense, his tenants were bound to render services to him in payment for the tenancy. In some cases they knew in advance what the services were—they might be required to plough or manure the lord's land for so many days in the year, for instance—but in some cases the tenants were virtually serfs, not knowing what the morrow would bring, and what would be asked of them. In both cases, however, they were tenants at the will or whim of the lord, and could be told to quit without right of redress. The law as such would not change materially on the vexed question of fixity of tenure for another century and a half, although conditions improved as a result of the Black Death in mid-century, for it made labour scarce. Even the labour of the monks was put at a premium as a result of this calamity. Abbot Richard did not live to see it, but his abbey lost its abbot and forty-seven monks, including the sub-prior and prior, within the space of a few days in 1349.[41] New monks, when they eventually came to fill the gap, could be depended upon to follow the old in matters of monastic obedience. The villeins, however, were in a different human category. There had always been a natural desire to cast off the uncertainties of villeinage.[42] This desire was joined with new hope as one of the results of the upheavals of 1327, that is, at the very time when Richard of Wallingford rose to lordly status. The feeling throughout the land that social injustices might soon be remedied was contagious, and it would deeply affect the entire course of his life as abbot.

Discordant Notes

Monastic towns had grown up quite naturally around the walls of most of the great abbeys. We have already seen something of the ways in which the villeins and burgesses of such towns as Abingdon and St Albans had expressed their wish to abolish the abbots' seignorial privileges and rights, and to remedy the charters which the abbots had dictated for the towns. Inside the monasteries too there was often dissatisfaction. The monastic rules had been designed from the outset to be strict but fair. Not all in the monastery had humility enough to accept even a strict and fair regime, but there had always been such dissidents, and the community could keep them in check. The monastic rule of St Benedict gave the abbot scope to play the part of the autocrat. Richard of Wallingford quickly learned how to do so, in the course of settling questions of privilege, and he did not always pay as much attention to equity as the rule recommended. When the rules were perceived to have been unfairly applied, the monks asserted their individuality in ways that directly mirrored protest in the world outside.

Outside the monastery there was also a rising sense of collective power in the upper reaches of society. Parliament was beginning to grow in importance, and almost imperceptibly the knights from the counties and the burgesses from the towns were establishing themselves as essential to it. The shires had been represented by their knights since Edward II's accession, but the king's want of money to make war led to a system in which the wealthier citizens were also allowed a voice, as a palliative to increasingly heavy taxation. As it happens, it was at St Albans in 1295 that the very first parliament was held to which an English king summoned two burgesses from every city, borough and important town. During Richard of Wallingford's abbacy, St Albans was the only borough that sent two burgesses, although it seems unlikely that he liked the idea, judging by an undated petition by the burgesses from around the time of his death, in which they complain that the sheriff will not summon them, his excuse being that he owed homage to the abbot.[43]

As for the abbot of St Albans himself, to the outsider he must have appeared blessed by exceptional good fortune. He held a high position in parliament. While many churchmen were hardly better off than the peasants around them, the abbot of a rich house had revenues enough to keep great state—far greater than most rich knights. On the other hand, he could have been expected to share the standard regrets of the higher clergy, one of which was that the church was expected to subscribe to the costs of government and foreign wars. As abbot of St Albans he owed the service of six knights, whom he could nominate himself if he wished, although he usually left the choice to those twenty-two of his tenants who between them had to pay the knights' fees.[44] Then there was the debt owed to the highest court of his church: quite as a matter of course—and doubly so, in view of his experiences in Avignon—Richard would have bemoaned his obligations to the pope, who by church law as well as by tradition was supposed to have a claim on the property of all churchmen. Abbot Richard's debts to king and pope were curiously interlaced. Direct papal taxation had begun in 1199, to pay for the Fourth Crusade. Thereafter it had become a regular requirement, tolerated by English kings only as long as they could take a share of the proceeds. By Richard's day the king was often taking more than nine-tenths of the proceeds, three times the sum that the clergy voted directly to him. (For some years after Richard's death payments to the pope ceased altogether.) Most English churchmen also resented the pope's power of appointment, which was thought to interfere with native liberties, and before long Richard of Wallingford would have good reason to agree on this point too.

He soon found himself caught up in several fast-running streams of discontent. In part he was unfortunate in his local circumstances, but in part his troubles stemmed from his love of confrontation. His first trial of strength concerned a knight of the shire and came only a few weeks after his return from Avignon, on the feast day of the Passion of St Alban, 22 June. The abbot had sent out letters of invitation to friends and neighbours to attend an informal meal, described as 'not the kind where names are called out as guests entered'. Despite all the informality it was held in the great hall of the monastery, into which came Sir John Aignel, who promptly stood at the head of the table and claimed to be cup-bearer (butler) to the abbot, a privilege that he said had been granted to his ancestors. It was evidenced, he said, by the fact that he held land in Redbourn called 'the Dispenser's Land'.[45] He claimed the abbot's goblet as the traditional fee.

Abbot Richard immediately contradicted the knight. He had never heard of the tradition; he did not wish to be served by Sir John, who was only making the claim in order to get the cup; and, in any case, the feast was in honour of the martyr Alban, not of his own return from Avignon. Asked to desist in his petition, Sir John was silent for a time, but later told his neighbours at the table that he had touched the goblet, and therefore had a right to it 'in seisin'. This was a legal expression, here implying that the goblet was a token of possession of his land. The whole affair had clearly been carefully planned in advance. The facts now seem quite trivial: part of the 'Dispenser's Land' was held by this man in return for rent-service and part by the convent. The convent kept a horse on it, for use by the abbot whenever he visited Tynemouth in the north, but this was only one of eight such horses, placed on the land of six separate tenants. Documents held in a chest in the abbot's chapel were duly consulted, and they showed that an ancestor of John Aignel had surrendered some such rights as he was now claiming. A storm in a goblet, perhaps, but it pointed to something more: Richard of Wallingford would need a lawyer's acumen if he was to hold his own as a guardian of property and privilege.

A Visitation

Richard soon began to prove that he had not only a lawyer's mind but a puritanical spirit to go with it. He set to work at once to make a 'general visitation' of the monastery and its cells, and of what was under their jurisdiction, and to make recommendations for the reform of all shortcomings. He had the assistance of two notaries, Richard, prior of the Tynemouth cell and a loyal supporter of the new abbot through thick and thin in the years ahead, and brother John of Sulsull, monk. They completed their examination in five days, but the visitation proper took well over two years, thanks to the fact that some of those involved absented themselves discreetly when they were to be visited, and others dissembled. There were things to be put right. No one could doubt Abbot Richard's intentions when they received from him his new edition of the statutes of the Benedictine order. The version in use was that confirmed on separate occasions by the papal legates Otto and Ottobon in the previous century. The new version had about sixty chapters, which were to be read out in the cloister for all the monks to digest twice every year, and Richard insisted that he would not make his visitation until they were properly observed.

He had of course been a St Albans monk himself, so he knew the convent of old, and he knew where to look to discover its defects. The visitation uncovered many ills; and 'on account of human wickedness' they were not all put to rights. Many of the monks were accused of sins of the flesh, and some cleared themselves of the charge, although—as the chronicler remarks—'how they did so God only knows'. Some who confessed, or pretended to do so, were treated compassionately by the abbot, and little by little excused. Some who were charged with ownership of property—contrary to the Benedictine rule—were punished, but with a degree of mercy. Those who had confessed to having entered the monastery by simony, the purchase of their place, were given two months in which to enter another religious order, hopefully by a more correct route.

'And so there came about a revolt.' Even some of those who had agreed in private with Richard that he was right joined in the protest. The protesters were mostly young monks who had been accused by others who were standing four square with the abbot. The ringleader was none other than Richard of Tring, chaplain and confessor. The confrontation must have been felt as a betrayal on both sides, for they had shared the hazards of the long journey to Avignon and back. It is hard to avoid the suspicion that Richard of Tring was a man disappointed at his own failure to be elected abbot; and, as a confessor, he was a potentially dangerous person. The protesters made unspecified accusations against their abbot, but—we are told—he greeted their crime of disobedience with charity, and for the time being protected Richard of Tring and at least some of the others.

That Richard of Wallingford had learned something from the pope's treatment of his own case is perhaps to be seen in his final decision. Those who did not conform to his rules must publicly renounce their order; but they could, if they wished, be considered for readmission.

The Abbot's Dues

Having first faced down a layman of substance, and having then purged the ordinary monks of the convent, the new abbot was not inclined to rest. He went on next to censure a group of the most powerful men within the walls of the monastery, the obedientiaries, and in doing so he came close to orchestrating his own downfall.

He had already discovered that he had exchanged the life of the scholar, not for that of a man of prayer and contemplation, but for that of an administrator. In due course he would learn the gentle art of making time for his old passions, but not yet. There was a chasm separating him from his confrères of years past, one created not simply by rank but by the fact that he had control of the abbey's wealth—although not total control. To rule an establishment the size of St Albans required a division of labour, a division of the abbey into administrative departments, each with its own officer in charge, its own income, its own duties and accountability. These officers were the 'obedientiaries'. The sacristan and sub-sacristan looked after the sacred objects, vestments and bells; the cellarer and sub-cellarer looked after housing; the chamberlain after clothes and bedding; and so on down the list. Obedience was expected of them—as their name implies—and cooperation too. Some departments yielded an income from semi-commercial activities—by selling surplus grain, for instance—but each was allocated an income from manors, estates, rents or other dues, and each was expected to provide the abbot with the wherewithal to pay taxes, 'tithes', that were due from the abbey to the king. These tithes (tenth parts) are not to be confused with the tithes paid by the laity for the upkeep of church institutions generally. The latter represented money moving in the opposite direction, a part of the income the obedientiaries might receive.

These monks spent little time in the cloister. In order to administer the affairs of the convent, internal or external, they needed servants and financial powers. To all intents and purposes they had incomes of their own, although of course nominally they owned nothing. They could raise loans, and they controlled their own households within the monastery. By the nature of the system, these men often became powerful in their own right, not only by virtue of the wealth they controlled but because they had certain powers of jurisdiction. The abbey cellarer, for instance, was tenant of the manor of Park, and from the surviving 'court book' of Park, with its records of the fines and taxes he imposed, we can see how a proud man in his position might easily have imagined himself a minor lordling. Such monks could be dangerous if they were self-willed. Richard soon discovered that some of the obedientiaries, after the death of Abbot Hugh, had ceased payment of the tithe due to the abbot. There were five offenders, all of them key figures in the day-to-day running of the abbey: Hugh of Langley, sacristan; John of Woderove, chamberlain; John of Tywynge, refectorer; Richard of Hetersete, almoner; and William of Winslow, kitchener. Having failed to heed a warning issued by the abbot, the five were punished in ways that men in their positions must have found it very hard to accept, no matter how meek their rule required them to be. Their failings were announced publicly, they were degraded, stripped of their offices, and removed from their stalls in the choir, in the chapter-house, and in the refectory (also called the frater). Added to that, they were condemned to perpetual silence and declared unfit to hold any office or benefice until they had obtained dispensation through humble penitence. Abbot Richard then excommunicated the five in writing, separating them from the communion of the brethren, until such time as they deserved forgiveness. Finally, for good measure, he decreed that they should receive corporal punishment on Thursdays and Saturdays of every week.

'The result of it all was great sadness in the convent', wrote the chronicler, in a fine specimen of loyal understatement. After the abbot's outburst in the full chapter house, he retired to the parlour with the prior of the abbey and some of the older brethren. Having given a vehement rebuttal of what he took to be the insult offered to him, he was at last persuaded by the senior men to amend his sentence. This he did—whether promptly or not we cannot say—ordering the five to do secret penance instead, and to sin no more. But the damage was done. The main storm seemed to pass away quickly, but clouds of resentment lingered, in some cases for many years.

The Leper

The convent was not only saddened, its members began to ask questions about their abbot's state of health and fitness to rule them. 'From that day', writes the chronicler, 'there began a conjuration, indeed, it would be more correct to describe it as a conspiracy, among certain of the false brethren, against the status of the abbot'. The word 'conjuration', with its legal overtones, might even have been used to suggest that a group of dissidents banded together by oath, and not only with a tacit understanding that they would correct an injustice.

The reference to the abbot's health was an allusion to his bodily condition, not to his state of mind, and takes up the thread in the chronicler's narrative which began with the pain in Richard's eye on his return from Avignon. As one of the charges laid by the protesters makes clear, he was now recognised as being under attack by leprosy. The very word still carries terrifying overtones of the unclean and stigmatised outcast, whose life would have been even less bearable had Christ himself not set an example of associating with lepers. When Christ's enemies were plotting to kill him, he dined in the house of 'Simon the leper'—or 'Symond leperous', as he was called in the English of the time (Matthew 26:6 and Mark 14:3). On the basis of translated religious texts and the reports of returning crusaders, many Christian ecclesiastics had become aware of the relatively enlightened treatment accorded to lepers in Islam. For such reasons as these, the church considered it a duty to build leper-houses—or 'lazar-houses' as they were known, after the leper Lazarus in the New Testament (Luke 16:20). They were only partly for the comfort of those afflicted: they also provided a means of isolating lepers and restricting the spread of the contagion. There were some thousands of these hospitals across Europe by the late middle ages, a large number of them having been founded in England between the late eleventh century—when there was a notable epidemic—and the early thirteenth century. A large town might have several such places, some of them treating other diseases too. St Benedict's rule laid great emphasis on care for the sick in general, and it is not surprising that two twelfth-century abbots of St Albans, Abbot Geoffrey (1119-46) and Abbot Warin (1183-95), founded leper houses. The first of these was the hospital of St Julian, on the London road, not far from the town. Run by a master and four chaplains, St Julian's was intended to house six leprous men, and from its later written constitutions, we know much about the life of its inmates in the fourteenth century. The second hospital was at St Mary de Pré (St Mary des Prez). Designed for leprous nuns, it was placed at a fitting distance from its precursor.[46]

In the middle ages, various diseases that are now treated as distinct went under the one name of 'leprosy'. The main cause of what now goes under that name—the bacillus *mycobacterium leprae*—has been known for little more than a century, and even today a definite diagnosis is often difficult. True leprosy was widely present in Europe between the tenth and fifteenth centuries, after which—for unknown reasons—it declined rapidly in the north, except in Scandinavia. Since the incubation period is between two and five years, Richard evidently picked up the disease long before his Avignon journey. The chief symptoms are a loss of sensation at nerve endings, defects of vision, and the gradual destruction of blood vessels, skin tissue and bone. With all of this there are the attendant sores and ulcers, and the loss of limbs and other parts of the face and body that so alarm those in the presence of the leper. There are other diseases with similar symptoms—for example syphilis (whether this was present in Europe at the time is a subject of disagreement) and tuberculosis. Since it was not possible then to distinguish between the various alternatives, perhaps the best definition of medieval lepers is that they were those who were called such, and treated as such.

To be treated as a leper was a penalty almost as cruel as the disease. There was some comfort in being put under the church's protection—being given distinctive garments and utensils, for example, each blessed before the leper received it, rather as in the ritual of ordaining the clergy. That lepers were forced to announce their presence with a bell would have been found much less agreeable, for it underscored their isolation, but that rule was not universal. Many degrees worse was the experience of being declared legally dead, or of being forced to undergo a ritual of separation from one's home and one's past, a ritual that in many cases was modelled on services for the dead. Cases are known in which the leper, before being cast out, was made to stand in an open grave while the funeral service was read.

There is not the remotest suggestion in the *Gesta abbatum* that any such stigmatisation was ever mooted in Richard of Wallingford's case. What was said in private conversation within the ranks of the disaffected monks might of course have made reference to any one of a dozen commonly held beliefs—such as that leprosy was a punishment from God for personal sin, especially the sins of pride, envy, wrath, and even simony. Customs varied much with time and place. Attitudes towards lepers tended to harden over the centuries, and yet there were always lepers who continued to live in their original surroundings. This was more common the wealthier the leper. There are two notable cases of kings who continued their rule as lepers. Baldwin IV, who died in 1185, was king of Jerusalem, where, as it happens, the Knights Templar had already made a compromise with those of their number who developed the disease: they were made to leave the order for the isolation of a separate Order of St Lazarus, while nevertheless being permitted to fight with the army in battle. Much nearer to Abbot Richard in time and place was the great Robert the Bruce, the king of Scotland who went far towards securing his country's independence . He lived for fifteen years after his great victory at Bannockburn, continuing to campaign against, and negotiate with, the English during those years, despite the steadily worsening leprosy from which he died in 1329. Richard of Wallingford showed similar courage in confronting the world as he fought against fate. John Leland alleges that he had to withdraw from the monastery to a house that he had built in the town, but there is no other evidence for this, and it seems likely that Leland was simply misremembering what he had read about the abbot's relations with the hospital of St Julian.

A Good Shepherd?

What the chronicle does tell us is of the charges laid against him—not all of them a consequence of his affliction—and of the proposed remedies. Some of the conspirators thought that he should be committed to the care of a custodian, the implication being that his powers would be in the hands of a regent, but that it would not be formally necessary to elect another abbot. Several made a proposal that 'the house and barony' should be transferred to the care of the king. When the tumult was at its height, some even conspired to deprive the abbot of his pittance from the convent kitchen, the extra allowance of food or wine distributed to all monks on feast days and other special occasions. After all, one of the five who had been famously chastised by Richard was William of Winslow, the kitchener responsible for pittances.

There is reason to think that this last proposal was not quickly forgotten by the abbot; much less was his response forgotten by the monks, for the chronicler harks back to the subject on a later occasion, to illustrate Richard's character. He tells of an incident that he knew only by hearsay. The abbot brought the older monks together to discover his rights as regards pittances from the convent kitchen. Were they his by grace or by right? What was their value? He brought the older monks together in the parlour next to the prior's chamber to settle the question. There he was told by the sub-prior, brother William of Nedham, that pittances of the sort under discussion were granted on a personal basis to Abbot John of Hertford (1235-60) without the intention that they would be continued to his successors. He added that his successors had claimed them as of right, even though they had been merely given as a grace. When the sub-prior had finished speaking, each and every one of the monks present agreed. Richard, however, was not satisfied, and went at once to the infirmary to consult the older monks there, namely brother Thomas of Boningdone and other bed-ridden monks of great age. Alas, all answered as the sub-prior had done. With this, the abbot flew into a rage and withdrew in a state of great anger, saying that he would sooner break off one of his fingers than agree to this. It was later said that it would have been better for the convent if they had yielded to his request, or at least responded in a more fitting manner; but that he had not meant to keep his promise.

What was the ideal response to such attacks as he was forced to suffer in the early months of his rule? Contrition on his part would have been regarded as a sign of weakness. The rule of St Benedict offered guidance, and making allowances for the changed social circumstances of the age, Richard of Wallingford seems to have followed its advice up to a point. The second chapter of the rule told him that at the Last Judgement the shepherd will have to answer for his sheep; but that if the abbot-shepherd has done his utmost to amend the vicious ways of an unruly flock, it will be the sheep who are judged adversely. He must follow the apostle's teaching: Reprove, Persuade, Rebuke. Furthermore:

> Those of gentle disposition and good understanding should be punished, for the first and second time, by verbal admonition; but bold, hard, proud, and disobedient characters should be checked at the very beginning of their ill-doing by the rod and corporal punishment.

The St Albans chronicler favoured a somewhat different line, judging the abbot as one might judge courage on a battlefield. He evidently admired Richard's eventually cool response. The abbot, as we are told, seemed to be neither disturbed nor moved, and showed himself to be quite indifferent to the schemes of his opponents, although, to be sure, there were aspects of the case that distressed him. Those who were closest to him often heard him say that he was most disturbed by the thought that a change of prelate would mean extra expense for the monastery. This rings true to Richard's general obsession with the abbey's debts, but his words were doubtless also a rhetorical device.

Whatever the truth of the matter, he was soon forced to admit that shrugging his shoulders was not enough, The rebel actions had been contrary to his former declarations, so he called a meeting in a chapel known as the Chapel of the Picture. Two St Albans priors were present, as well as the faithful prior of Tynemouth, certain monks, and a few secular clergy. That he had brought outsiders into the monastery shows that he considered the matter serious. He proceeded to formally excommunicate each and every person who, in private or in public, had tried to remove him and had advocated transferring the temporalities of the monastery to a custodian appointed by the king.[47] 'And he pronounced them conspirators!' Conspiracy is a common theme of the Old Testament, and it is not difficult to imagine the rhetoric with which he embellished this particular accusation.

We shall find that, in the end, Richard of Wallingford's declining health did force him to share his powers, and that in February 1333 he finally appointed Nicholas of Flamstede as coadjutor, an assistant of the sort appointed by many an old and infirm abbot or bishop before and after him.[49] We might say that on this point battle honours were fairly equally shared.

Rights

After his visitation of the monastery, Richard carried out visitations of all the cells, either in person or through his commissioners, monks, and deputies skilled in law. He held a solemn assembly of clergy, a synod, of the kind that established local rules to supplement the general rules of the chapters. He travelled around the abbey estates, manor farms, and churches under his jurisdiction, and received the homage and fealty of his tenants. He collected the tithes due, which allowed him to carry out repairs on some of his churches, and to pay off a few of the many debts left by Abbot Hugh. For others he negotiated terms for future repayment. Hugh of Eversdone had not only left a legacy of debt, he had lost property and lands belonging to the monastery. Richard created a network of influential friendships by which old rights might be restored—the chronicle is not very specific, but as we shall discover shortly, he cultivated that pivotal royal servant Richard de Bury.

In matters of discipline, the proximity of the town of St Albans was a permanent source of trouble. Adopting Caesar's principle of 'Divide and conquer!', the abbot removed to the most distant of the abbey's cells those monks who were closely linked—by family or less honourable liaisons—with people in the town. A century earlier, Abbot William of Trumpyngtone had bought houses in London and Great Yarmouth as retreats for his monks. Perhaps consciously following his example, Abbot Richard used the cell at nearby Redbourn as a means of removing some of the pressures from the lives of monks who had not gone astray. Redbourn, like St Albans, is on the river Ver as well as on Watling Street, then an important thoroughfare. It was a special place to all members of the convent, for the priory was built on the site of the supposed discovery of the bones of St Amphibalus, the priest who converted St Alban. There were two gilded shrines there with the relics of St Amphibalus and his companions, and monks were appointed in turn to watch over them night and day. Three monks at a time were sent out 'in a decent carriage' from St Albans, and they stayed for a month before being replaced by the next trio.

Whatever the intention, judging by the improvements decreed by a later abbot, life at Redbourn in Abbot Richard's time seems to have been unusually spartan. The food was cold, and could not be resold in the town, and the grant of fuel was so small that life there in winter was very hard to bear. If Richard's rules for the place are anything to go by, it was not easy to prevent the monks from paying visits to neighbours when taking the air, or even staying away for the whole night. Perhaps it was only their way of keeping warm. They were warned against other things—in particular, damaging their neighbours' hedges by hunting, or perhaps we should say poaching. Only the obedientiaries were officially allowed to keep greyhounds, but there are many instances of monks being reprimanded for neglecting their religious duties in favour of coursing. It may be recalled that Chaucer's well-fed Monk on the Canterbury pilgrimage had a weakness for the hunt, and the strictures of the poet John Gower on monks' love of horses, hounds, and hare coursing can be parallelled many times over in fourteenth-century writings. That members of the order must not hunt was a rule spelt out explicitly by a general chapter of black monks in 1328, and was put side by side with rules for celebrating divine service, for sharing food with the poor, and for avoiding bloodletting in Lent. The neighbour who complained most about damage to his hedges was seemingly none other than Sir John Aignel, the would-be cup-bearer.

The abbot's legalistic frame of mind may seem surprising to those who think of him as a story-book academic, an otherworldly scholar seeking after truth. If there ever was such a creature, it was certainly rare in his day. He had reached his position by a hard road, and now found himself on an even harder one. Was he merely a mandarin, a product of his scientific, scholastic, formalistic, mathematical outlook? He certainly had a rigid sense of his rights, which is to say a narrow sense. In some cases he clearly believed—and the chronicler often agreed—that he had superior knowledge that entitled him to override the decisions of others. One such occasion was when, after the death of their prioress, the nuns of the dependent priory at Sopwell voted sixteen to three for Alice of Hakeneye as her successor. Richard promptly forced on them the appointment of Alice of Pekesdene, with whose merits he was familiar. A more memorable occasion, as far as the monks of his convent were concerned, was when he routed certain Dominican and Franciscan friars who claimed the right to preach and hear confessions within his jurisdiction. They were in fact insisting on their rights under a papal bull issued by Pope Boniface VIII in the jubilee year 1300.[50] Richard's tactics were to ask one of them to tell him which cases were reserved to be heard by a bishop or abbot having the necessary jurisdiction. The friar was completely taken aback by this unexpected question, which made him seem an idiot. Before he could collect his senses, the question was followed by another, asking him how he could give absolution when he was ignorant of the times when he had the power to do so and the times when not. Fifteen different cases are solemnly listed in the papal text. The friar and his group were apparently confused and impressed by this Benedictine show of learning. They beat a hasty retreat and afterwards showed more caution. The chronicler leaves us in no doubt that the black monks of St Albans were not disappointed in their choice of abbot, on that occasion, at least.

9

The Visitor Visited

HAVING BEGUN with a policy of retrenchment and reform, the irate abbot pressed on with the correction of abuses. Appalled at the laxity of the monks' discipline, he carried out a threat to banish the worst of the troublemakers to distant cells. He then resorted to the chief disciplinary weapon of the church, and 'excommunicated' brethren. This could mean many things, even denial of social contact with fellow Christians, although as described in the rule of St Benedict it could mean nothing more than exclusion from limited aspects of the common life—the common table and the oratory, for instance. It usually implied a denial of the right to receive the sacraments, or, in the case of a priest, to administer them. The pope's wholesale use of it was giving rise to so much friction with the English crown that royal officials often refused to implement it at that higher level. Locally, it was being used increasingly, and quite inappropriately, for civil offences, as when an Oxford chancellor in Richard's day excommunicated townsmen for leaving rubbish in the streets. In the fourteenth century there was also the option of imprisonment. Strictly speaking, to imprison the excommunicated person needed the rights of a bishop, but Richard of Wallingford obtained that right by royal charter. He did so with the help of Richard de Bury, who was at this time keeper of the privy seal.

An Abbot in Parliament

There was nothing out of the ordinary in the idea of a great abbot moving in the same circles as Richard de Bury, who in the strictest sense was outranked by him—until 1333, when the latter was made bishop of Durham. The office of abbot ensured that he was invited to those national assemblies that we now regard as embryo parliaments. Edward III held forty-eight parliaments over his reign of fifty years. The records of those held in Richard of Wallingford's lifetime are fragmentary, but their general pattern is known. The members of highest dignity were the lords spiritual: bishops, abbots and priors. The archbishop of Canterbury sat on the king's right hand, then the archbishop of York and other bishops, all in a planned order of precedence. Below the bishops came the abbots and priors, first the prior of St John of Jerusalem, who symbolised the very centre of Christendom, and then the abbot of St Albans, the premier English monastery. (This was occasionally a bone of contention with the abbots of Westminster and Canterbury, but the principle was generally accepted.) There was a distinction drawn between those who were summoned individually by writ, and those who were summoned indirectly. It was considered an honour to be in the first class, as was Richard of Wallingford, who 'held of the king through a barony'. There were between twenty and thirty black monk abbots and priors summoned by writ, and fewer than ten from all other orders together. As far as the king was concerned, it was not the honour of individual abbots that mattered but their power to raise taxes for him.

On the king's left hand sat the lords temporal, the dukes, earls, barons and others. Their ranking and privileges, even their rights to being summoned, depended to some extent on territorial strife, fortune, and royal favour, and was not clearly defined until long afterwards. After the lords temporal came a variety of officials and administrators, chief justices and other judges, legal advisers, barons of the exchequer and the king's principal ministers—his chancellor, treasurer, keeper of the privy seal, chamberlain and steward of the household. Theirs was the real responsibility for the running of affairs. They were expected to keep a low profile, and not to arouse resentment among the great nobles. Richard de Bury, in parallel with his many church appointments, was still rising steadily through the ranks in the royal service—king's clerk, marshal in the household, keeper and treasurer of the wardrobe. Between 1329 and 1333 he was keeper of the privy seal, and by 1334 chancellor of England. Also brought to parliament were the clerical proctors (often as supporting officials for the lords spiritual), knights (representing the shires), and burgesses (representing the boroughs). These commoners were summoned indirectly.

Even by modern standards, the proceedings of these gatherings were casual affairs, many members arriving late or not at all. There was as yet no elected Speaker, and the reason for the summoning of parliament was left to be explained by an appropriate noble or official, but was often left unclear. Parliaments gave the king a chance to discuss affairs of state, wars, even crusades, with those upon whom he ultimately depended. Affairs of the church and relations with the papacy were often broached. Taxation, monetary policies and commercial questions were discussed, and by Richard of Wallingford's time the broad pattern of direct parliamentary taxation was fixed. Parliament had a role in framing legislation, and yet again this often touched on relations with the papal court. It might also be called upon to interpret former legislation. Parliament was the highest court of English justice, a court for the trial of the nobles by their peers, and occasionally a court of appeal where the lower courts could not decide cases. This function meant that experts were needed, and they were to be found in council, chancery and exchequer, but in general the amateurism of parliament was in marked contrast to the professionalism of its counterpart in France. It is hard to believe that Richard of Wallingford, with his love of the rule book, rejoiced in the easy-going ways of the assembly.

While by his office Richard was one of the principal members of parliament, little is known of the part he played in its meetings, and he can hardly have attended more than five. The abbot of St Albans had his own manor house in London, but parliaments were rarely held in London. The first summonses he could have answered were for those held at Lincoln and York in 1328. He might have been at Winchester early in 1330 or at Nottingham in November of the same year—when the king finally threw off the yoke of his mother and Mortimer. In March 1332 there was a London meeting. It is almost certain, bearing in mind his rapidly deteriorating health, that he was not present in person at any later parliament. From December 1332 to Richard's death there were five parliaments in York. He had indirect dealings with the first of those, but did not attend.

There is some mystery about Richard's movements in the early years of his abbacy, since in February 1329 he begged for royal permission to live away from the abbey for three years, to save the expense of entertaining guests and maintaining the great state of an abbot. Permission was granted 'in consideration of the depressed and indebted condition of his abbey, to dwell for three years in such places, either within the realm or beyond seas, as shall appear to him most expedient for the purpose of avoiding the burden of too great expenses; provided that he depute in his stead prudent and discreet persons to rule the abbey'.[51] There is no evidence of his absence from the neighbourhood: he could not possibly have conducted his various litigious actions from a great distance. It seems not unlikely that his request was a stratagem. Perhaps he quickly changed his mind, and convinced himself that the abbey could afford his presence, or that there was no one fit to take his place.

Early in his rule as abbot, if not before, Richard of Wallingford made friends with two of the men closest to the king. One was William Montagu (or Montacute), who in 1327 campaigned against the Scots at the young king's side, and who in 1330 was leader of the band of armed men who seized Mortimer in Nottingham Castle. In 1337 Montagu was made the first earl of Salisbury. The other friend, hand in glove with Montagu, was

Richard de Bury. He was more at home with a book than a sword, and yet Edward III was much beholden to him. He had been tutor to the future king when they were together in France with Queen Isabella. As treasurer of Guyenne he had provided her with funds, and had been forced to flee for a time to sanctuary in Paris for his pains, pursued by the agents of Edward II. De Bury held so many different posts that it would be surprising if he remembered them himself, but before he was made bishop of Durham he spent some time as chaplain of the papal chapel in Avignon. There he met Petrarch, who wrote kind things about his ardent wit. One wonders whether that pleased him more than the twenty chaplains and thirty-six knights he had in attendance in Avignon—a far cry from Richard of Wallingford's humble St Albans party of a few years before. At all events, de Bury and Montagu were powerful friends for an abbot under attack to have on his side.

Balancing the Books

The harsh punishment meted out by Abbot Richard to the worst offenders among his monks led to a sense of outrage in the convent which is easy enough to understand, but it was also fuelled by a gift he made personally to Richard de Bury. The gift was of four books, belonging not to the abbot but to the convent. In view of the obsessive character of the bibliophile who received them, they would have made an effective bribe, but if that was their purpose, what were they meant to secure? It is said that they were meant to guarantee the interests of the convent at the court. We have already mentioned the fact that Richard de Bury obtained a royal charter for the abbot, granting him the right of imprisoning excommunicated persons, a right normally reserved to bishops. This was hardly a privilege the convent would have considered in their own interests, but perhaps there were others.

The author of *Philobiblon* had collected books from an early age. He let it be known that no present was more acceptable to him than books. He borrowed them to have them copied; he had agents buying them from libraries and booksellers in France, Germany and England; and he begged them from friends. His numerous appointments in the church merely enhanced his passion and his opportunities. He planned a library at Oxford connected with Durham College and hoped to endow it with most of his vast library, but in this his executors were thwarted by the heavy debts he left behind him, and they were forced to sell his valuable collection of books. It filled five large carts. This, however, was all after Richard of Wallingford's death.

The passions of the rich collector of books were not those of the monks who copied them laboriously, and whose cloistered lives depended on them for solace, or for a sidelong glance at the allurements of the outside world. The library at St Albans was especially rich in classical writers.[52] The anger of the monks when they discovered that the abbot had given away four of their valuable books can be easily imagined. They contained works by the Roman playwright Terence, the Roman poet Virgil, and the Roman authority on rhetoric Quintillian. There was also a book by the translator of the Bible Jerome (writing now against an old associate Rufinus), in which he tells of a dream in which his Christian belief had to struggle with his love of pagan literature. These works had been given away, and all—it was supposed—to promote the interests of the convent at the court of the king. 'An absolutely abominable gift', wrote the chronicler, who could have quoted from the monastic rule of St Benedict about caring for the monastery's property.

There was an anticlimax, however. The writer continues with an account of how the members of the chapter of that time were almost as bad, since they agreed to the sale of thirty-two books to the same man, Richard de Bury. And there is a suitably unrestrained report on that affair too: the sale yielded £50 (fifty pounds of silver), and of this the abbot kept half. The other half he divided between the refectorer (£15) and the kitchener (£10) as subsidies, so converting spiritual sustenance into satisfied stomachs.

And then finally, the dénouement. When Richard de Bury was made bishop of Durham in 1333, 'guided by his conscience', he returned many of the books to the monastery, with his name inscribed in them. And after he died, with his estate heavily in debt, his executors sold other St Albans books back to Michael of Mentmore, the abbot of the time. In return, prayers were said in the convent for bishop Richard de Bury's soul. There is no mention of Richard of Wallingford at this point in the narrative.

Enemies and Friends in Adversity

Richard de Bury's assistance in obtaining the royal charter on excommunication perhaps dates from the second half of 1332. In that year, Richard of Wallingford's leprosy began to worsen markedly, so that not only people outside the convent but also those within the house saw it as increasingly burdensome. News was spread far and wide of the seriousness of the condition of the man—astrologer, wise councillor, a man of great heart, and yet attacked by this affliction before his time. The word used at this juncture was 'elephantiasis', which refers to the elephant-hide appearance of the skin in leprous patients. It suggests that he had the disease in a form known as cutaneous leprosy, where the disease often spreads to the face, causing thickening and corrugation of the skin. As the disease worsened, he virtually lost his voice.

Some offered him their sympathy, but others, harbouring ambitions of their own, were much pleased. Not—the chronicler hastened to add—that they were from the monastery of St Albans. They were black monks, but outsiders. One such person was especially disruptive, a certain Richard of Ildesle (or 'Hildesley') from the Benedictine monastery of Abingdon. The name of the perfidious monk suggests that he came from East or West Ilsley, hamlets five miles south west of Wallingford and ten due south of Abingdon. It is not unlikely that his and Richard of Wallingford's paths had somehow crossed at an earlier date. At all events, the monk addressed letters to the papal court drawing attention to Richard's serious condition, claiming that he was incapable of performing his duties adequately. The result was that, in a bull dated 3 November 1332, John XXIII instructed the bishop of Lincoln to carry out an inquisition into the abbot's infirmity. It had been reported, according to the letter, that the abbot 'could not keep company with those of sound health without giving offence'.[53]

The bishop—within whose diocese St Albans fell—appointed three commissioners to investigate. One of them, the papal nuncio Icherio of Contoreto, did not attend. The others were John of Offord, a canon of Lincoln, and Robert of Bromley, an Oxford professor of civil law. They visited the abbey on 15 January 1333. The grievances mentioned at this stage included a claim that the abbot could not meet the brethren in the chapter or choir; that through lack of care the furnishings of the monastery were destroyed; that immovable goods (lands, houses, and such) were let out to the laity; and that the fabric of the monastic buildings was in ruins. They were asked to decide on the truth of the claim that if the pope failed to take action the abbey would soon be irredeemably destroyed. Richard of Ildesle had a gift for overstatement, if nothing else.

If the inquisitors thought that their task would be easy they were mistaken. The abbot was a master in the art of confrontation. He was by turns indifferent to their interrogation and resolute, and he impressed his monks not a little with his tactics. He said he put his faith not in man but in his patron, the Blessed Alban, and he quoted the prophet Jeremiah (17:5): 'Thus saith the Lord, Cursed be the man that trusteth in man, and maketh flesh his arm, and whose heart departeth from the Lord'. He put his trust in one man, to be sure. Having heard of the approach made by Richard of Ildesle to the papal court, he sent his prior to the king at York, who was then holding a parliament, to ask whether the king was prepared to tolerate the idea of an outsider monk encroaching on his monastic preserves. He noted that the king's ancestors had been patrons, protectors, advocates, and even founders, of the St Albans monastery. In front of parliament the prior made the point that Abbot Richard had been severely hampered by the burden of debt left to him by his predecessors and by the sums of money that were to be paid to the pope. He was sure that by the labours and intelligence of the abbot, if he were left in peace, the monastery would recover.

At this, 'certain ambitious men' once again reiterated the point made in the papal bull that the abbot's infirmities would impose irreversible damage on the monastery. The king's chaplains thus petitioned the king to appoint an assistant to the abbot.

The effect of laying this petition before parliament was not at all what was planned. It directed the anger of the king's council and other nobles towards those who had invited the pope to interfere. Was papal interference not burdensome enough already? The offenders had brought in the pope 'with the desire to rule'. But worse, the letter to the pope had been sealed with the king's seal, by none other than Richard de Bury, keeper of the privy seal, who was duly reprimanded. He replied, nervously, that he had acted under coercion. We are not given any names, but the remark puts a different complexion on the letter from the envious Abingdon monk. Was he egged on by his abbot, William Cumnor, perhaps, or even by someone close to Richard of Wallingford?

We know only the outcome, which was that Nicholas of Flamstede, the prior, was sent back to the St Albans convent on 14 February 1333 with a recommendation that they elect him coadjutor. This was duly done, and letters were sent at once to the pope by the abbot and convent on 2 March, requesting that this appointment be recognised. The king sent a similar letter. The text of the first letter survives, and from it we glean the information that Nicholas of Flamstede was about ten years the abbot's senior. The letter answers the points made in the bull sent by the pope to Lincoln, putting a brave face on the abbot's infirmity, and underscoring the powers he had left to him.

The letter also hints at the ambitions of those who first set the wheels in motion, while the chronicler has more to say on this point. It seems that Richard of Ildesle was in the town, perhaps to witness the abbot's downfall. The brethren threatened that if he were to come into the abbey to find out the opinion held of him there, they would butcher him as a horrible example to all with similar ideas in the future. We are further told that there were in the monastery many 'of great stature and strength' who would probably have done this. The monk of Abingdon's conscience told him that this was not a sufficient reason for martyrdom, and he fled the town of St Albans once and for all. Whether or not such homicidal threats had anything to do with the fact, the abbot and convent were henceforth left unmolested.

This must have given one of them in particular food for thought. Nicholas of Flamstede, Richard's coadjutor after the spring of 1333, had been one of the causes of the convent's disgruntlement at the abbot's earlier behaviour towards them. When the previous prior, Robert of Norton, had died, Richard of Wallingford had recommended Nicholas—then prior of the cell at Hertford—as his successor. The monks were shocked, and greeted the proposal 'with a great silence', whereupon the embarrassed candidate began to excuse himself, 'more from humility than from truth, as everyone knew'. The abbot then asked whether anyone there present knew good reason why Nicholas should not be raised to the office of prior; and several acknowledged that they did not. The day was won, and within a few minutes, and a pretty speech making judicious reference to chapter 65 of the rule of St Benedict—which Richard caused to be read out in full—the appointment was made. The chapter in question begins with the words 'It frequently happens that the appointment of a prior gives rise to serious scandals in monasteries', and goes on to recommend that the abbot have the main responsibility for the appointment.

The abbot and his new prior, we are told, were of one heart and one mind in all they did. They had been companions on the journey to Avignon, when Richard had relied heavily on the advice of the older man. Their relationship changed, however, after the prior's appointment as coadjutor. Richard was now blamed for making too many demands on the convent, in particular for the return of surplus food and clothing. He considered the prior to have shown himself ungrateful for his elevation, and to have become less supportive afterwards. Whether or not the prior sinned in this matter is something the chronicler leaves to the reader's judgement: he only reports what he heard from the older monks. Perhaps the writer is not quite so even-handed as he wishes us to believe. He likes a touch of scandal, and takes the opportunity to report the story about the abbot's pittances, as told in our last chapter. He uses the oldest excuse in the biographer's repertoire: he believes that it throws light on Abbot Richard's mind—by which he means his legalistic turn of mind in regard to his rights. He insists that he was not responsible for this piece of tittle-tattle. He was, after all, only reporting what he had heard from the older monks.

10

The Litigious Abbot

THE MEDIEVAL TOWN of St Albans had grown up around the abbey, on the east, west, and north sides. The chief road runs north-east, roughly in line with the abbey tower and starting at the market place, which was in that position at least as early as the tenth century. Divided up into spaces for stalls of the various trades, the market place had an Eleanor Cross at the abbey end.[54] The market place was the centre of the townspeople's town, but the right to hold a market—granted in the first instance by the king—added greatly to the town's commercial value, and so to the prosperity of the lord abbot. By Richard of Wallingford's time, finer and more permanent shops and houses were being erected in and around it, and there was a growing sense of self-esteem in the community, having little to do with the grand monastery in its midst.

The abbot of St Albans controlled many aspects of the lives of those outside the convent walls. There were a few local customary laws, but generally speaking the common law of England dictated his position, as it had done since the Norman Conquest. We have already seen that land was held directly or indirectly of the crown, and that below the king there was a hierarchy of tenure. All were tenants, not owners. Those below the king at the top of the pyramid owed services in return, but they could themselves have tenants owing them service—and owing service to the king as lord paramount. The great landholders, of which the abbot was one, were mesne lords, and the entire holding, with waste lands, smaller manors, rights of jurisdiction over tenants, presentation to church livings, and so forth, was his manor. Outside the monastery walls there were three main classes of person beholden to the abbot, lord of the manor in this sense of the word: the free tenants, the villeins by tenure, and the villeins (or bondsmen) by blood.

Relations with the first class were reasonably good. The free tenants were a kind of minor squirarchy, with a large degree of independence, but performing services for the abbot. Such a person might owe military service, for example, for forty days in the year, while others might owe a fixed rent, paid in money. Such services were arranged either at the abbot's court, traditionally held under the great ash tree in the abbey precincts, or at the court of the district (the soke) in which they lived.

Most villeins by tenure were less privileged, but here there were two distinct classes, and when the chronicles speak only of 'villeins' it is rarely possible to distinguish. All were bound to render services, such as ploughing the lord's land, but the more favoured class knew what those duties would be. They held 'certain tenure'—for example, they might be committed to ploughing for so many days in the year, and giving service at haymaking and harvest. Villeinage by 'servile and uncertain tenure' was a very different state of affairs. The tenant—villein by blood, in the sense that he was born into his condition—was now bound to do what his lord demanded. In earlier days he had been effectively a serf, who often had no idea what the morrow would bring. This lower order slowly began to disappear, however, after a number of legal reforms under King Edward I. A number of court decisions gave them a more secure status, although they were still expected to perform duties 'according to the custom of the manor'. This gave rise to much contention about what customs actually had been in the past, as Richard of Wallingford was soon to discover, but at least there was henceforth a more open attitude to handing on land to others (by the system later known as 'copyhold'). There was in any case an economic difference which cut across the legal classes of villeinage. The more prosperous villeins, with 25 or 30 acres ('virgaters'), were expected to pay more service, and render more in kind, than the less prosperous cottagers, with five acres or less. Most virgaters, besides many cottagers, were nevertheless villeins by blood, traditionally bound to the manor, unable to leave without the lord's licence, and liable to various fines and taxes. The profits of the manorial court were the lord's, and mills were usually his monopoly.[55]

The villeins were not the only class troubled by the law as it related to the holding of land and ties to the manor. Edward's statutes were meant to deal with the thorny question of land-holding by religious persons, who were 'dead in law' (hence the word *mortmain*, 'dead hand'). The main legal problems here were those relating to the services that they owed to any 'lord of the fee' who had sold them land. The educated religious were generally clever enough to ride roughshod over the law. The unlettered villeins had no such advantage. They harboured many resentments of their own, one of the most deeply felt, as we shall see, being that which concerned the law on the milling of flour. A single entry from the court book of the manor of Park speaks volumes for the hurt it could cause: 'Margaret Newman fined sixpence because she used another mill than the lord's mill'. Another resentment concerned labour service. There were men bold enough to refuse it on several occasions during the thirteenth century, within the St Albans jurisdiction. They did so again in the period 1318-27, when their refusals grew steadily more threatening, culminating in the riots around the time of Abbot Hugh's death. The irksome taxes and restrictions on their lives were another source of rancour, as were the measures taken to prevent them from poaching. Again, the court book of Park provides numerous examples: 'Fulk de Spitalstrate set snares to catch hares in the warren', 'John Smith, John Howe, Henry Prat, and John the beadle caught fish with their nets in the lord's private waters', and so forth. To country dwellers, poaching and taking wood was not theft but something sanctioned by natural law, and it is as well to bear in mind their frequent experiences of famine. Held fast as they were in the lord's legal nets, their only compensation was that in times of great hardship they were his responsibility, and that, by and large, abbots and bishops took that responsibility seriously.[56]

The experience of most townsfolk was naturally different from that of the typical virgater or cottager. Despite the abbot's unquestioned legal and fiscal powers over them, the villeins in the town had a long memory, and they knew that those powers had not always been very forcibly asserted there. From small beginnings at the time of the Domesday survey, the town of St Albans had grown steadily in prosperity, until in the time of Henry III it was affluent enough to be ranked with the larger boroughs—if we are to judge by demands made on the free tenants for service to the crown. St Albans merchants produced dyed woollen cloth for sale at home and in France; and under the rule of King John they were prepared to pay a tax to continue this lucrative trade. Relations with the convent began to decline, however, with the abbacy of Roger of Norton (1260-90), after which time they became steadily more strained, with faults on all sides. This decline coincided with the beginning of a long period of immigration from the countryside, which was making the town still more affluent, but which spurred on a wish for independence. At the same time, the convent was becoming more worldly, and more conscious of its own wealth and privilege. Conflict was inevitable.

The townsmen were as yet without the clear leadership or organisation needed to deal on anything approaching equal terms with their feudal master. They were not above trying to set up an independent commune of some sort, however, possibly copying a model tried earlier in London. They appointed twelve burgesses, men of substance (*majores*), as their representatives, and they instituted a common chest for their own internal levy of money for community purposes, a form of local taxation. Trials of strength began to follow at regular intervals. The townspeople were battered and bruised repeatedly by the lord abbots, and not least by Abbot Richard. Looking forward to later history, it is easy to understand why so many men from St Albans joined Wat Tyler's rebellion of 1381. It was at St Albans that an example was made of the Kentish preacher John Ball, who in that year was tried and executed in the presence of the sixteen-year-old King Richard II. Three St Albans men followed the preacher to the gallows on that occasion. For centuries, each step of this kind turned out to be just one more on a seemingly endless and hopeless road. Bearing in mind his plebeian roots, was Richard of Wallingford not betraying his own estate? As a religious, he would have answered that he had put himself outside the world, and outside the classes into it was divided, except by reason of the duties that tradition pressed down on him. He could have mentioned his duty to the king, and to Christendom, but there was more to his conduct than duty alone can explain. The law of the land offered a personal challenge to him: it was like a mathematical problem, something to be enjoyed for its own sake.

Justice Within Whose Law?

The smaller manors owned by such abbeys as St Albans held regular courts—the 'hallmote' courts of a lord of the manor, held in the lord's hall two or three times a year. They were presided over by one of the obedientiaries, or even by the abbot himself. Embarking upon the administration of justice in his domain, Richard of Wallingford soon discovered that handling the mettlesome inhabitants of the surrounding town was going to be difficult. There was a struggle for power in most English towns in which abbeys had commercial and financial privileges, and naturally enough, the abbot's officers there became resented symbols. In St Albans, for example, the abbot's cellarer was controller of the market. The abbot appointed the keeper of the town prison. Richard later added other appointments to these, tightening his grip on the townspeople at every stage. As we shall see later, he eventually reorganised the town, keeping the peace with a constables and two pledges in each of four wards—almost as though he had planned it on an astronomical diagram.

The situation in St Albans had not remained constant over earlier centuries. At first the borough had been responsible for a court of its own, although it had been a court presided over by the abbot's reeve, and in any case dealt only with such minor matters as the assize of bread and ale. For the more serious questions of crime and plea and plaint, the townspeople had to go to the abbot's court—'the hundred court of the abbot's liberty'—and the fact that it was held in a romantic setting, under the spreading branches of the great ash tree in the abbey courtyard, is not likely to have cut much ice in the township. In 1253 Henry III gave the town a measure of independent justice, but the machinations of successive abbots helped to nullify this. The abbot's reeve, for example, was replaced by a townsman, a bailiff; but since the latter was appointed by the abbot, and was usually prepared to take the abbot's side, this was little consolation. In the early fourteenth century even the borough court seems to have been somehow quashed by Abbot John of Maryns (de Marinis). As we shall see, it was later to be reinstated in another subservient form by Richard of Wallingford, although it was destined to disappear again after the Peasants' Revolt of 1381. At the heart of dissension was the question of those liberties given to the burgesses by Henry III. Immediately before Richard's appointment they had petitioned the king in parliament and, during a long legal discussion in St Paul's Cathedral, had obtained an important agreement. It included the right to a separate court. They were to be entitled to attend the hundred court only when summoned by writ. Richard of Wallingford seems to have considered it a point of honour to reverse this arrangement.

The wealth of most abbeys depended to a great extent on the monopolies they held. Through all of the struggles between Richard and the townsmen, the monopoly on grinding corn and fulling cloth remained a source of friction. To the townsmen, bypassing it was a question of deep moral principle, but of the law too. Had not Henry III granted them those 'liberties'? Abbot Richard took a lawyer's view of the line dividing justice from morality; and when he acted to secure justice, it was justice first and foremost for his own community. He seems to have been happy to ignore Saint Benedict's rule, as it concerned the monastic use of property. He could have pointed out that circumstances alter cases.

The Mills of St Albans

There were windmills within the St Albans jurisdiction in Richard of Wallingford's day, but they seem to have played no part in one of the bitterest of contests between the abbot and townsmen concerning the use of mills—he with his elaborate and expensive water-powered mills, and they with their humble, manually operated mills. In this case, the abbot was truly Goliath. In the Domesday Book, there are more than 5600 mills listed for the area south of the Trent alone, their scale roughly indicated by their annual worth. In the vill of St Albans, for instance, there were three

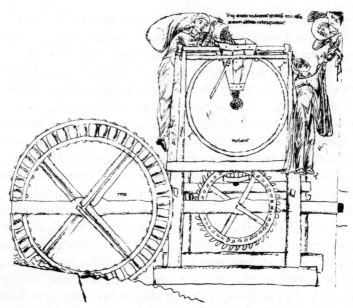

16. A corn-mill driven by an undershot wheel (on the left). This drawing was copied in the nineteenth century from a manuscript, since destroyed, of the twelfth-century *Hortus deliciarum* of Herrad of Hohenbourg. Despite the drawing's poor perspective, we can see how the main shaft drives a gear parallel to the water wheel, and how that gear engages with a smaller one in a horizontal plane, coupled in some way to the movable millstone. The woman in the centre tips a sack of grain into the hopper. On this there is a lever to control the fall of the grain into the hole in the upper millstone, and so into the space between the stones. How the milled grain (flour or meal) is collected is not shown.

mills yielding 40 shillings, while the abbot's two mills in Norton yielded only 16 shillings. There were four main sorts of water-mill in use in the middle ages. One crude kind, now known as the 'Norse mill' or 'Greek mill', was used only in hilly regions.[57] The other three types were larger and far more important, all of them with a vertical wheel on a horizontal axle which drove the machinery. In one, the 'undershot wheel' (see Fig. 16), the water flowing under it turned by exerting a pressure on the vanes at the bottom. In the second, the 'overshot wheel' (Fig. 17), the water was channelled over the top of the wheel and fell on to its vanes, or better still into troughs (buckets), on the far side. In this more efficient case, the falling weight of water provided most of the power. A breast-wheel used a mixture of the two principles: the water flowed into buckets on the nearside of the

17. An overshot water wheel, as used in a mountainous setting in connection with mining. Detail of a woodcut from Georgius Agricola, *De re metallica* (1556).

wheel about half way up, with overspill flowing down and under it. In any of these three cases, in addition to the main wheel, simple gearing would have been needed to transmit power from the axle shaft to the millstones (or to some other driven device). Such gearwork would almost always have been of oak or other wood. Even today there are mills, water and wind driven, working efficiently with purely wooden gears.

There was little that was new in this technology. The Romans were using water-mills, perhaps of all three main sorts, on a large scale in the later empire, with wheels of the order of five or ten feet across. It is likely that the techniques they used were never entirely lost. Overshot and breast wheels were probably still relatively uncommon in the fourteenth century, but it would be a mistake to think that they were useless with the sort of placid rivers found in southern England generally—and the river Ver in St Albans is placid enough. One way of using them very effectively was to dam the stream to create a mill pond. A head of water of as little as six or eight feet has been used quite effectively, even in recent centuries, in this way. A sluice gate is opened to release water when it is needed, and closed at other times—in particular at night, the chief time for replenishing the head of water. St Albans had dams, but we can only speculate as to the mill type.

Most early mills were used for grinding grain. In this case, no matter what the source of power, the lower stone was usually the fixed stone (the 'bed stone'), and the upper stone rotated on top of it. The grain to be ground was introduced through a hopper into a central hole in the top stone, al-

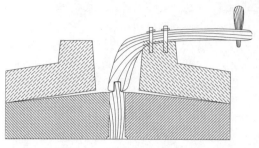

18. A simple hand-mill, or quern.

lowing it to pass between the two stones, in which were grooves to help it to pass outwards. There were then various simple arrangements for transferring the meal or flour into sacks. The arrangement of the millstones in windmills was much the same. By the fourteenth century, water-mills were being used for many other purposes, such as fulling cloth (which involved beating it with trip hammers, to clean and thicken it), sawing wood, crushing seed for oil, and crushing ore for smelting.[58] They were also used in a few smithies of advanced design—chiefly in operating the bellows, or hammering the hot iron. It would not have been out of character had Richard of Wallingford introduced such ideas into a St Albans smithy. Water power was certainly used at St Albans in the preparation of malt, although the details are not known to us.

A household of small means could in principle have ground all of its own grain in a simple hand-mill. Cutting and trimming millstones is not easy, but there were tradesmen skilled in the art, and if they could not be afforded, time and industry could supply them. An ambitious home miller might even have built a wooden framework, allowing him to harness a horse, ass, or ox to the task of turning the upper grindstone. Either way, the work would have been slow, and few households would have managed to process much more than ten pounds of grain in an hour. Some of the abbot's mills would have processed grain at twenty or thirty times that rate—and water does not tire. A smaller type of hand-mill, a quern, was little more than a couple of small millstones with a handle fitted to the top stone (Fig. 18). With stones of the order of eighteen inches in diameter, such a mill might have ground four or five pounds in an hour. Most of the references to hand-mills in the *Gesta abbatum* were almost certainly to querns. Slow and inefficient they no doubt were, but a more serious objection to them was that no matter how small they were, they were being operated outside the law, the abbot's law. There were great profits to be made from the abbey mills, but as Richard would have noted, large capital sums had been invested in them. Park was one of six manors whose mills were renovated or replaced by him; and £100, a colossal sum, was spent on a horse-mill there in 1331. The villeins' hand-mills were labour-intensive, but they were small, cheap, and numerous; and they undermined the abbey's economy.

Hand-Mills and Liberties

A lawsuit brought in the year 1275 between Roger of Norton, the twenty-fourth abbot, and the townspeople of St Albans, may be used to illustrate a pattern of protest which was set for the next half century. A certain Michael Bryd was charged with having built a hand-mill in violation of the privileges of the abbot and his convent, whose mill he should have been using. The monastery then owned at least three large water-mills in the town, and others elsewhere, and insisted on its feudal right to enforce their use. Payment for their use was made with a fraction of either the grain being ground or the flour produced. (The word 'multure' was used to describe that toll, or the right to exact it.) Michael Bryd, however, claimed that he and his forebears had the right to take their grain wherever they wished, and that they usually used the mills at Toreham and Wheathampstead. One of the St Albans abbey mills was for fulling cloth, and a case was heard in the same year against one Henry de la Porte, who was charged with not having used the fulling mill of the abbot and convent, to full a length of russet cloth. He too claimed established custom. Both were found guilty of the offence, and of misreporting past tradition, and each was fined one mark.

A year later there was a very similar conflict with other named parties, who admitted the offence, and professed themselves contrite for using their own hand-mills. They offered the abbot ten tuns of wine in compensation, and he generously took only five, besides not pursuing the matter at the hundred court. On this occasion we get an insight into master-servant relations: the abbot made his millers and millers' boys take oaths of fealty, presumably hoping to have them on his side to ensure that customers would not slip through the net in future.

A generation passed. Then, around 1299, the water-mill of the chamber-

19. A more ambitious form of hand-mill than the simple quern. After a fourteenth-century Romanian manuscript.

lain, at Redbourn, was burned down from unknown causes. We learn that the mill was surrounded by trees, which prevented the fire from spreading. Mills are prone to catch fire, especially when a forgetful miller allows the stones to grind without grain between them, and on this occasion there was no suggestion of arson. What happened after another fifteen years, however, in the time of the easy-going Abbot Hugh of Eversdone, was certainly no accident.

The townsmen flatly refused to grind their wheat at the abbot's mills. Robert of Lymbury was arraigned, as Michael and Henry had been before him, but—carrying sword, hatchet and cutlasses—he had resisted arrest by the abbot's bailiff when the latter tried to remove the millstones from his house. The abbot on this occasion was awarded damages of 100 shillings, an enormous sum, and the man was gaoled—a sentence later exchanged for a fine of 20 shillings. Similar actions were then taken against three other townsmen who owed homage to the abbot. From this time on, the same form of protest against the abbot's privileges recurred with increasing frequency. The townsmen beat up one of the monks and damaged a house owned by the abbot in the town. The townsmen of Watford fished in the abbot's fishponds. They had, however, all underrated Abbot Hugh. He brought a series of successful legal actions against them, before his death took him to a higher court.

He died, as we have seen, not long after these episodes erupted into rebellion, under cover of the general lawlessness at the time of Queen Isabella's insurrection. The demands made by the townsmen might have seemed painfully familiar to Abbot Hugh, but, in the new challenge to the feudal relationship between abbot and town, fundamental moral principles began to be voiced. The townsmen wanted elected representatives to parliament. They asked for the right to supply twelve townsmen as jury in trials before itinerant justices; that twelve of their number should have the assize of bread and ale; that they should have the common lands, woods, waters and fishponds named in Domesday; and that they should have hand-mills and be allowed to repair them when necessary. All of these things they claimed as due to them by ancient custom.

Abbot Hugh temporised. He gave his agreement verbally, but not in writing, and this fact caused great anger. The monastery was besieged by the townsmen of St Albans for forty days, and suffered at least two major assaults. During that period no food was allowed in, and attempts were made to set fire to the buildings. Two of the abbot's men were forced to pay large sums of money to be allowed into the town. The king's bailiff was captured and imprisoned, and would have been beheaded by the townsmen had he not paid a ransom of sixty acres of land. The idea caught on, for four of the abbot's men were later separately forced to pay money to avoid execution. The townsmen withdrew after their first assault, under a royal threat. On the second assault they seem to have come off worse, with one of their number being captured by the monks. Sending now to London, the townsmen petitioned the king, and he, citing Magna Carta, forbade the abbot to molest them further in the enjoyment of their traditional liberties.

The question then arose: what were those liberties? Domesday was consulted—it mentioned three mills in the town, but was otherwise unhelpful—and other documents. In due course the king ordered that their demands be met. The horrified monks left the chapter and at first refused to seal the agreement. Eventually, having registered their protest before notaries, they agreed. The struggle seemed to be going their way.

One of the first actions of the newly liberated burgesses was to demand that Barnet Wood be open to them. Granted their wish, they immediately tore down hedges and branches of the trees there, and raided the abbot's rabbit warren and fishponds. Those things were easily done in the heat of the moment. What required much more deliberation was their final gesture: they set up eighty hand-mills. This was part of Richard of Wallingford's exasperating inheritance from Abbot Hugh.

Morality and Bloodshed

The hostility of the town did not of course cease with Hugh's death. It remained a source of great concern to Richard, who was not of a mind to be on the losing side in any battle. One of the townsmen's illicit acts at the time of the rising against Hugh had been to make a seal of office, a symbol of their wish to be independent of the monastery. Like the town charter, the new abbot wanted it back. He used various stratagems, not all of which went as he had wished, but one of his successes fell into his lap without his having to take the initiative.

On St Margaret's Day, 20 July, in an unspecified year, the townsmen refused to implement the system going back to Saxon times known as 'frankpledge'. Its purpose was to guarantee a measure of collective responsibility among the populace at large. At this period it was applied to all except members of the church and the highest reaches of society, and freeholders. Under this system, each individual was deemed to be a member of a group (usually of ten or twelve households) that could put up a surety for the good behaviour of any of its members. Serfs and the landless in general were thus kept more or less under control without an elaborate system of external policing. A member of the group offering surety for another was a 'pledge', a group of ten was called a 'tithing', and ten tithings formed a 'hundred'. If a member of a hundred was accused of a crime, and the others could not produce him for trial at the shire court, all were responsible for a fine.

The court held in St Albans every St Margaret's Day, known by the name 'view of frankpledge', was an occasion for the group to represent its members, if called upon to do so. The townsmen now not only refused to act as jurors on the court, they refused to hold the court at all. The abbot therefore, with his own council, instituted a new system. The town was divided into four wards, each with a constable of the peace and two chief pledges, to oversee order in the place. The abbot took over some of the functions of the older court. Persons detained in the St Albans gaol, for example, were no longer to be released on bail ('mainprised') except on the say of the abbot.[59] The townsmen had clearly miscalculated, if they thought they were dealing with another Abbot Hugh.

Despite these successes, five or six years of his abbacy passed before Richard of Wallingford was finally satisfied with his performance against the townsmen. Taking back the privileges they had wrested from Abbot Hugh under duress was not, he believed, an abstract moral question. It was not merely about the loss of a native liberty that came cost-free. There was no such liberty. What the townsmen had snatched was at the expense of the liberties of the monks—for privilege was in their eyes their liberty, despite St Benedict's injunction 'to avoid worldly conduct'. So how could the balance be redressed?

The abbot's reasoning reflected his view of his pastoral responsibilities. The townsmen's refusal to obey the abbot's dictates had led to a serious decline in morality, so he would attack them on that front. It was his crusade, and it rested on a right that he believed no one could challenge. He boasted that he felt compelled to show the townsmen his horns. This was a pun, more appropriate to an Oxford lecture room: the Latin phrase also meant that he was 'mitred', which of course as an abbot he was. He declared that he would 'wield his sword against the fortress of impiety, establish his own stronghold, and prepare for battle with public sinners'. In fact, in a tactic reminiscent of his electoral sermon, he began by launching an attack on one man who could have been seen as a Goliath, a person of great influence in the town, a certain John Taverner. In May 1330 the abbot indicted Taverner on a charge of adultery. In October of the same year he continued with his mission, excommunicating another four leading citizens for their moral laxity, and obtaining their arrest for contempt of his authority.[60] It was the Taverner case, however, which gave most hope to those of the townsmen who longed for the abbot's downfall.

A summons was served on John Taverner, alias Marchal, on 13 May 1330. It was served by Walter of Amundesham, who was the abbot's marshal, and two of the abbot's clerks. Physically assaulted by Taverner, the marshal was obliged to respond—the chronicler's account is sympathetic to him, of course—by striking 'a defensive blow' at the wrongdoer, who later died. The angry mob in the market-place retaliated by attacking the marshal with sticks, stones, swords, forks, arrows, and—as though that were not enough—with 'various other weapons'. Needless to say, he too died.

To the abbot's dismay, the coroner in the case rounded on him and his clerks for causing the deaths of both men. Things were not going according to plan, and worse was to come. The abbot, his two clerks, and his archdeacon were indicted for the deaths, and orders were given that they all be taken into custody by the sheriff of Hertfordshire. (The sheriff was the king's executive officer in the shire.)[61] It seems that the abbot alone escaped this humiliation, although the others were soon released. The townsmen now tried to pack the jury of twenty-four men, and would not accept as a member anyone from outside the town. Despite a royal mandate that confirmed this right to exclude outsiders in inquest juries, they were eventually forced to accept men from the neighbouring hundreds.

Trials by Jury

The case came to be heard in September 1331 at the abbey's priory at Hertford. Richard of Wallingford was able to ensure that the justiciars (judges) of trailbaston were handsomely wined and dined in advance.[62] He then used strategies that have a modern appearance, challenging every single juror hailing from St Albans. Next he counter-indicted the townsmen—by now another of his servants had died of wounds received on the fateful day—and with such vigour that the townsmen proposed coming to terms. They underestimated his wish for total victory. The deaths had occurred sixteen months earlier, and in the meantime he had been rustling up powerful support—an earl, a viscount, two papal nuncios, and many worthies of the county and beyond. The townsmen had support from one citizen of London and a sergeant-at-law. Their greatest weakness was not that they were less well connected than the abbot, it was that they asked the impossible of him. Once again, they insisted on their right to maintain hand-mills.

The abbot indicted the coroner—John of Muridene, an officer of his own liberty—with having maliciously conspired to indict him and his monks for the deaths. A jury duly found the coroner guilty. The jury likewise found large numbers of the townsmen guilty of the 'divers felonies and transgressions' of which they were accused by the abbot. Eighteen of them admitted guilt and were imprisoned. Forty-two pleaded not guilty and claimed trial by jury. Of these several were found guilty and imprisoned, while others were found to have acted under duress and were released. The prisoners were mostly bailed and later fined. Almost casually, and in only one place (a record of the verdicts), is there mention of a piece of gratuitous cruelty that tells something of the attitude of the townsmen to the black monks. Four named persons, including a dyer and a baker, together with certain others, were found guilty of capturing, imprisoning, and castrating the abbot's chaplain, John of Walden. It is not hard to imagine their reasons, right or wrong.

There were many cases heard on the charge of 'subtraction of suit of multure', that is, of not turning in wheat for grinding at the abbot's mills; and all pleaded guilty. The cases dragged on, but the jury usually decided for the abbot. Thirteen of his tenants claimed the ancient right of using hand-mills. The verdicts are not recorded, but since the townsmen decided to submit to the abbot, we may guess that the outcome was in each case the same. As arranged by the king's council at Westminster, a delegation of four men representing the whole town of St Albans now came to Abbot Richard and agreed to all his demands. They agreed to surrender their deed of liberties, to use his mill, to pay him for the lost use of it, to pay his expenses, and to give a surety of good behaviour.

The abbot and convent naturally thought that at long last victory was theirs. Richard of Wallingford's health was such that he must have suffered great exhaustion, but it is the exhaustion of his opponents that the chronicler underlines. Richard consulted with his right-hand man, Nicholas of Flamstede, and others of the convent, and decided on a gesture of reconciliation. His choice of words to the townsmen was frank. Because, he said, he did not wish to mislead them, or to be tricked by them, they should return to their neighbours to discuss what had been agreed. And if they were all of one mind, they should come back to celebrate with wine and spice-bread. They did so, threw themselves on his generosity, and enjoyed the feast.

Old habits die hard. Within a fortnight, the townsmen had begun to revert to their old ways. When summoned to a meeting of the abbot's council, some dissembled, some looked for an excuse, others pretended to have no knowledge of the affair, some even denied that it had ever been their intention to do as asked, or said that there had been no consensus. Yet others wanted to renegotiate terms. What was an abbot to do? There was nothing in the rule of St Benedict to tell him, although there was much about what he was not to do. Aware that he was sailing very close to the wind, Richard of Wallingford now enlisted the help of a collaborator, and sent him to one of the ringleaders in the town with letters that proved the townsmen's conspiracy. The man was told that if he did not fall in line with the agreement, 'as his accomplices had done', the lord abbot would persecute him to the death. Thinking that his friends had already submitted, the terrified man did so in writing, and entered into a bond for £40. More to the point: he agreed to use the abbot's mill, and he agreed to persuade his friends to do all that he had done. The action was spectacularly successful, bringing in bonds of £2000, none being for more than £100.

The *coup de grâce* came on 13 May next following, in the year 1332, when the townspeople surrendered their silver seal, royal charter of liberties, common chest—and their hand-mills. There was one concession: for a fee of £48 a year—a hefty sum—they were to be allowed to grind their own barley. The chronicler suggests that they surrendered the tokens of their liberty to the convent. Indeed, another document has it that a deputation of twenty-four burgesses of the town, headed by one Benedict Spichfat, came to the chancery asking that their charter might be cancelled and the seal defaced, 'the silver to go to the decoration of the shrine of St Alban'.[63] The humiliation was complete, for the time being, at least. The timing of the agreement concerning surrender to the convent was decided by the astronomer-abbot: it was exactly two years to the day since the murder of Walter of Amundesham, 'the abbot's marshal and shield-bearer'.

The Men of Redbourn

As we have already seen, there was an important shrine in the St Albans cell at Redbourn, which the monks used as a spartan retreat, but the place also had a sizeable number of bondsmen who owed feudal duty to the abbot of St Albans, and the affairs of the larger town did not leave them unmoved. In their case, they refused to pay tallage, a tax levied by the lord abbot more or less at will. In this instance it was meant to pay for administrative reforms. As tithes were tenths, tallage was often a fifteenth. Theirs was not an uncommon protest, and in 1340 the system would be overhauled, but for the time being Abbot Richard found himself with another rebellion on his hands. Challenged in their court, the men of Redbourn at first said that custom required them to pay the toll only when new abbots were elected. When this plea did not work, they decided to legitimise their claim to exemption with a charter from the time of Edward the Confessor, who died in 1066, the year of the Norman Conquest. The charter bore the name of Nigel Niger, presumably meant to be the man who gave Redbourn to the abbey—Æthelwine Niger, or Swart. This anachronistic mixture of names was typical of the language of the entire document, which was found to be a nonsensical mixture of English, French, and a little Latin. The monks of St Albans had a genuine charter of their own from the same period, but even without it they would have had no difficulty in confounding the unschooled forgers. The abbot excommunicated them, and took steps to prevent the same thing happening in future.

The angry but vanquished Redbourn men were forced to swear that they would pay their tallage, but the oath meant little to them. When the chamberlain's beadle went to collect it he was badly beaten, whereupon the abbot despatched a posse of bailiffs. They apparently captured only one poor rustic, the rest having fled, and the man was brought to St Albans where he was roughly handled and put in fetters. Eventually sixty-one people paid their fifteenths, of whom eight were women. Four were excused on the grounds of poverty—not a symptom of monkish compassion, but of the impossibility of squeezing water from a stone.

The men of Redbourn did not forget these insults, but their next rebellious move did not come until the second half of the century, in the time of Abbot Thomas, when they demanded a charter of hunting and fishing rights. When it was refused they broke down the dike surrounding the prior's meadow, so flooding it. By comparison with the popular uprisings in the St Albans district in the later century, this was all very small beer—but even had Abbot Richard and Abbot Thomas been able to foretell the uprising of 1381, it would have been cold comfort. For both of them, Redbourn spelt trouble.

Isabella's Mill

No excuse of poverty and hunger could have been found for the person against whom Richard of Wallingford next took legal action: she was the most famous woman in the land, the queen mother, Isabella. If there was ever a time when he would have turned a blind eye to her failure to pay rent for a mill, that time was past, now that she was disgraced and exiled from London.

The mill in question was at Little Langley. It had been bought from Ralph Cheynduit, in the time of Henry II, and leased to his family for 20 shillings per annum until the lease was purchased, together with the manor, by Queen Eleanor. (The grandmother of Edward III, and queen of Edward I, she was also half-sister of Alfonso X of León and Castile, patron of the astronomers whose tables Richard of Wallingford had at last acquired.) Eleanor's bailiff had routinely withheld the rent. While not directly connected with this fact, it so happens that Eleanor had learned at first hand about running disputes over the milling of grain at St Albans. In Abbot Roger of Norton's time a large body of women, protesting that they were denied use of their hand-mills by him, and waiting for her when she visited the town, were outwitted by his device of leading the innocent queen by another route. When she discovered his trick, Eleanor was plainly not amused, and Abbot Roger had to beg forgiveness—although he defeated the town in the courts on that occasion.[64]

The case for recovering the rent from what was plainly a sizeable mill was one which Abbot Richard thought he might at last win. He began to petition parliament in 1331, and was eventually successful. The king himself ordered an enquiry, which was carried out with great thoroughness, and in 1334 it was arranged that rents be collected from Isabella his mother and set against the debts the abbot owed to the crown at the Exchequer. It is unlikely that Isabella lost much sleep over the case, of which she may well have been ignorant, but it prompts a question of an entirely different sort. There are five surviving manuscript copies of a work of astrology that was written by an unnamed abbot of St Albans for an unnamed queen. It seems likely to have been written by Richard of Wallingford, but for which queen?

The work is no longer one on meteorological astrology, but on the type of astrology that was routinely frowned upon by orthodox academic theologians, dealing as it did with nativities (birth horoscopes and their interpretation) and other questions with relevance to human life. This little treatise was originally written in the spaces above an ecclesiastical calendar, a type of work that would have included days of the month, saints days, Easter tables, the positions of the Sun in the zodiac, and other astronomical information. It bears some resemblance in style to Richard's astrological work *Exafrenon*, and like that work was eventually translated into Middle English.[65] No other abbot of the time had the right expertise to produce any such thing. But for which queen was the so-called *Canon over the Calendar* written?

In the first two years of Richard's abbacy Isabella's star was in the ascendant; but those years were a time during which Philippa, Edward's very young queen, might also have received a gift from the new abbot. She and the king spent much time at the palace at King's Langley, eight miles from St Albans. In works of this kind we often encounter royal horoscopes. Might the calendar have greeted the birth of Philippa's first son, Edward, the Black Prince, who was born at Woodstock on 15 June 1330? In view of the affair of the Little Langley mill, Philippa might appear the more likely recipient, but more than that it is impossible to say.

Mills, Malt and Mast

The barn, the fold, the forge, and the mill were cardinal points in the medieval economy, and it should come as no surprise to find the lord abbot paying them so much heed. Richard repaired the abbey's mills at Park, Codicote, Luton, and the Moor.[66] He built the Stankfield mill at St Albans, from its foundations up. (The word 'stank' means a pond, even a fish-pond, and in this case strongly suggests a mill dam for an overshot or breast-wheel.) He is said to have built a malt mill in the town. Only in modern times has it become usual to have malt germinate in a continuously mobile environment—usually in revolving drums, or stirred by paddles in tanks. The traditional method was simply to spread the moistened grain on the floor and to turn it regularly with a shovel. Richard's dedicated malt-mill might be thought to hint at experiment and innovation; but, as to the details, we can only speculate.

It is conceivable that the malt-mill has a bearing on some remarks made by the chronicler about an agricultural discovery made by Richard of Wallingford about yields of grain. We are less likely to pass quickly over these remarks if we recall how disastrous were failures of the annual harvest in the fourteenth century. They were not to be much remedied by buying overseas: there were some grain imports from the Baltic, but grown in regions likely to have suffered in the same years as England. It was usual to store excess grain from good harvests in anticipation of bad, but two bad harvests in succession spelled disaster. A century of benign weather had lulled people into a state of unreadiness. There was serious famine in 1315, 1316, 1317, 1319, and 1321—years of summers with heavy rain and unusually cold winters. Grain yields were as little as two bushels of yield from one of seed. Disease—rusts, smuts, mildews, and moulds—made matters worse. In times of famine, not only did yields of grain fall dramatically, but there were epidemics in herds and flocks. There were even shortages of the salt needed to cure meats and fish, since salt was largely obtained by the evaporation of sea water. Bread riots brought the social classes into conflict, petty crime increased, people poached of necessity, and in some cases ate grass like cattle.[67] There was clearly much more than the profit motive in the thoughts of an abbot anxious to improve the yields of his manors.

We are told that on a croft at the hospital of St German's, which had been infertile for four previous years, Richard had ordered wheat to be sown on land on which 'mast and *fow*' had previously been spread. The second word very probably refers to the material taken out of ditches when they are cleaned ('fowed'). We know that he had the mill dam at Park cleaned out, as well as the dam of his abbey mill; and, if they were not too old, the fowings would have made good compost. Richard is said to have found in the St German's case that sowing seven bushels of wheat yielded ten quarters (eighty bushels) of good wheat in the autumn—although this was admittedly a good year in the Chilterns generally, as our chronicler adds.[68] A good yield on an ordinary modern farm would be more than a hundred bushels from the same seed, but of course the result depends on the variety sown, and Richard's results were plainly exceptional for the time, or they would not have been reported. And if his stern morality denies him sainthood in the twenty-first century, it cannot be denied that he was commendably organic.

Windmills

There are other references to mills on tenanted abbey lands which might somehow reflect back on affairs in Richard's time. When a certain John Aignel died in 1362—not the Sir John Aignel of cup-bearing memory—his estate included a water-mill called 'Tolpade' at Cashio. (The local people would have been obliged to pay a toll of wheat for its use, which might explain the first half of the name, while the rest could have been 'path', 'toad', or even 'paid'.) The same man also left a windmill (*molendinum ventriticum*) at Wigginton, near Tring. Like the water-mill, it must have been built long before, judging by the fact that both were in a ruinous condition at his death.

The mention of a windmill is much rarer than that of water-mills at this period. The windmill was introduced into England in the twelfth century, just possibly under the influence of crusaders returning from the wars in the Middle East, where they had seen such devices. More probably, the idea for windmills arrived from France, having come from Muslim Spain, where such things were not uncommon. What little is known of the earliest northern windmills comes chiefly from illustrations in manuscripts, in carvings, and on stained glass, rather than from written descriptions. To be

20. A post-mill, after an illustration in the Luttrell Psalter, dating from shortly after Richard of Wallingford's time. Note that the post around which the mill turns is made from the bole of a tree.

effective, a mill needed to have the wind blow more or less into its sails. To allow for changes in wind direction, the body of the commonest type of early mill was made to swivel around a very substantial upright post, such as a turned tree trunk—hence the name 'post-mill'. In such a mill there is a long beam (the 'tailpole' or 'tiller beam') projecting from the rear of the swivelling body. It is this that allows the miller to turn the entire body, with its sails, into the wind—exhausting and potentially dangerous work that was often done by a horse or ox. Early archaeological evidence is rare, but it so happens that the foundations of a fourteenth-century post-mill were found long ago in the village of Sandon in Hertfordshire, about twenty-five miles from St Albans. The fact that so much weight had to be carried around the main post meant that the earliest post-mills were relatively small, but the foundation beams at Sandon were quite substantial, each about fifteen feet in length.

The post-mill idea seems to be a purely European invention, which was eventually introduced to the Middle East by members of the Third Crusade. While the first English mention of a post-mill comes from Yorkshire in 1185, within ten years they had become common enough for the pope to levy a tithe on them. Eventually another mill design was developed, in which the machinery was housed within a fixed masonry or timber tower, above which a rotating cap carried the roof and the sails on their axle ('windshaft'), as well as the first wheel of the mechanical train (the 'brake wheel'). This now more familiar arrangement was already in use by the fourteenth century, but John Aignel's windmill is unlikely to have been so advanced. The most we can say about it is that if it was erected before Richard of Wallingford's death, he would certainly have wanted to probe its inner workings; and to have taxed it.

Unflagging Aspirations

By the time the dispute with Isabella's estate came to a head in 1334, Richard of Wallingford had survived a visitation (prompted by the letter from the monk Richard of Ildesle), and was supposedly sharing his powers with Nicholas of Flamstede, as his coadjutor. He had won a great victory over the town. He had reduced the townspeople once again to a state of servitude that they considered a violation of both history and morality. (Their plight left the St Albans chronicler unmoved. He knew that after Richard's death ill-feeling between town and monastery had surfaced again and again.) Most of the monks were impressed by their abbot's legal performances, and many were beginning to forget his stern treatment of them. There must have been some heart-searching—although less than there should have been—about the unholy pleasure that the monks derived from the ceremony in the abbey church at which the townsmen came to deliver up their hand-mills and their civic chest with its three keys. The abbot gave them a meal, and they gave him presents. These were gifts to God rather than to man, he said. It was a sound convention and no doubt represented his frame of mind, although on the same occasion he decided that their millstones would make an excellent pavement—God's gift to the abbey, as it were.

Did he suppose that they, or their children, or their grandchildren, would forget? In the revolt of 1381, long after his death, the rebellious mob broke into the abbey cloisters and took up the millstones from the floor of the doorway to the abbot's parlour. The stones were taken outside and ceremoniously broken into small pieces. To each person a piece was given. They were, as the chronicler says, distributed like those pieces of consecrated bread which were broken and distributed on Sundays in the parish churches, so that the people, seeing them, would know that they had been avenged against the monastery in their cause. It scarcely needed to be said that if the millstones represented the body of Christ on that symbolic occasion, Abbot Richard was not far removed from Pontius Pilate in the people's estimation—although of course Walsingham could not say as much.

At the time of the dispute with Isabella, Richard was barely forty, and yet, for all his vitality, it was plain to everyone that he was slowly dying. He was so affected by 'the detestable disease of the leper' that he could scarcely speak, but he refused assistance and addressed the town again in terms of reconciliation. It is said that they responded to him warmly, to the extent that he bore himself so bravely in adversity. We are told, however, that there were some in the monastery who were jealous of his success, and of the fact that he achieved it with a minimum of help. The abbot's council is named, and judging by previous reports it is more than likely that Prior Nicholas of Flamstede was the member of council who murmured loudest.

Richard had become familiar with the law, so much so, we are told, that he was more learned than his own council. The wonder is that he found time to attend to so many other things, as his condition slowly worsened. While throwing himself heart and soul into the eradication of hand-mills on a point of law, he was promoting new mill projects on a larger scale than those he had inherited. He read newly available astronomical texts. The routes by which they reached him are entirely unknown to us, but there was a constant interchange of scholars between St Albans and Oxford, and likewise between Oxford and Paris and other universities, by which he might have received materials stemming from Muslim Spain. He was writing on questions of theoretical astronomy at the same time as he was pursuing his adversaries through the courts. And still he found time to design and commence building the grandiose astronomical clock by which posterity would most easily remember him.

Part Three

Time and the Man

I cannot tell you the hour precisely. It is easier to get agreement among philosophers than among clocks.

> Seneca, *Apocolocyntosis*, 2.2

11

Builders and Clockmakers

GREAT AMBITION is often best admired from a distance. It is hard to decide whether Richard of Wallingford's was driven by some indwelling spirit native to his character, or by an awareness that his years were numbered. Whatever its cause, his passion to obtain what was his by right as abbot was a thorn in the side of both convent and town. We sense more than a trace of personal vanity in all of his contentious actions, but he lived in a harsher world than ours, and was as scrupulous as most of his calling and rank. Like many who suffer hardship, he had a stern regard for duty. He knew that he had a duty to build, rebuild, and repair the fabric of the abbey, and that this would require massive financial outlay. The abbey's finances were in poor shape, after decades of poor management and accounting, but under his watchful eye they began to make a hesitant recovery. He needed money, for he had more in mind than the abbey buildings. He had no difficulty in persuading himself that what he wanted, the finest astronomical clock ever conceived, would be to the glory of God and his abbey. The first call on the monastery revenues, however, was for the restoration of the fallen buildings.

The Builder

The cloisters built by Paul of Caen had suffered badly with the collapse of columns on the south side of the nave in 1323. The cloister walk nearest the church had come down with them. Richard began the restoration, but so slowly that it was not complete until seven years after his death. He knew the importance of a good roof. There is more than a suspicion that the magnificent lead covering of the roof that frames his portrait (Fig. 1) makes allusion to the lead he put on the abbot's chamber. The chronicler tells us that by the end of his first two years in office he had rebuilt or roofed nearly all of the houses of the monastery. He repaired the abbot's house in London, its hall, chambers and kitchen, its stables and its chapel. He would have been proud of the fact that this was recorded with a note to the effect that it all cost fifteen shillings less than had been estimated.

He was responsible for a much grander project, one that required not only buildings but a certain vision. The great gate that is now to be seen—in a somewhat spuriously restored form—was built in 1363, long after his death. Much later in its history it was used as the town grammar school, but the great gate stood more or less on the site of an almonry that had been erected earlier by Richard of Wallingford, and it was a building with a similar purpose. The almonry had a short life, for it was destroyed in a violent storm, but its plan tells us something about Richard's priorities. The almoner was the monk responsible for distributing alms to the poor of the town, but his residence also housed schoolboys and their masters. It had a hall, chapel, chambers, kitchen and cellar, and other houses necessary to its purpose, which plainly went beyond the usual one of distributing alms. Scholarship had helped to elevate Richard to high office, and he wished to give the same advantage to others. It is worth comparing his educational initiatives with those being taken at this same period—with markedly less enthusiasm—by the southern Benedictines in Oxford.

In another act of charity, Richard made improvements at the abbey's hospital of St German's, within the town of St Albans. He threw all his energies into improving the paving, the surrounding walls and gates. (It was there that he experimented with his compost, to improve the yield of grain.) He was a scholar who knew that there was a real world outside books, a monk who knew that contemplation alone would not clean out or repair the abbey mills. He erected new buildings at the outlying manors and priories of the monastery. The chronicler does not name them, in the interests of brevity, but he does add a wistful comment of his own. Each abbot, he says, is obliged to care for the works of his predecessors. Richard of Wallingford, by starting so many new projects, was burdening those who followed him in office with responsibilities that would be hard to support. He does not say that things would have gone differently had Richard lived longer.

The burden did not end here, since building in stone and mortar was not where Richard's heart was. 'He built that noble horologe in the church, with great industry and at great expense', we are told, but the chronicle was written with the advantage of hindsight. The description was not that offered by everyone at the time. The abbot refused to abandon the project when some of the monks, 'wise in their own estimation', made it out to be the height of folly. He insisted that his original intention had been to make a clock at smaller cost, bearing in mind the sorry condition of the church fabric—a matter then under constant discussion. What then was the abbot's excuse for his lavish spending? In his absence, he said, the sums allocated to the enterprise had been raised at the instigation of certain of the monks, an increase due in part to the greed of the craftsmen. Would it not have been unbecoming, even shaming, for him not to have finished what had been so begun? The question was of course purely rhetorical. It is clear that at the root of his mortification is the thought that he was mismanaging the finances of his abbey.

His unconvincing explanation tells us much about the circumstances under which the clock was built. It is the oldest clock of which we have detailed knowledge, and yet there were professional craftsmen upon whom he could call to build it. The impression he gives, that he could leave the monastery for a while, only to return to find that he had been given something that he had not commissioned, must nevertheless be taken with a grain of salt. The design was his. We know that he had worked hard and long, calculating the details of his masterpiece. The most likely reason for charging the craftsmen with avarice is that they spent more time on the work than he had estimated, for the simple reason that they were working in what to them was strange territory. In other words, they were craftsmen, not astronomers or mathematicians who had spent half a lifetime in the Oxford schools.

His critics were not only from within the monastery. When, in an unspecified year, Edward III visited the abbey, he gently criticised Abbot Richard for embarking on such a sumptuous work at a time when so much rebuilding was needed, following the disasters in the time of Abbot Hugh. To this Richard replied, with all due reverence, that there would be abbots after him who would be able to find craftsmen to restore the fabric of the monastery, but that after his death there would be none capable of completing the clock. The chronicler adds the comment that he spoke truly, 'for in that art there was none like him after his death, and when he was alive he had no equal'.

Roger and Laurence of Stoke

The drift of these remarks is that Richard of Wallingford was making a new contribution to an older tradition. Even in the *Gesta abbatum* we have evidence for an established craft, and it a curious fact that the name of Stoke seems to link the St Albans clock with East Anglia. To start at the end: we are told that Laurence of Stoke ('Laurentius de Stokes') together with one of his fellow monks, William Walsham, completed certain details of the clock after Richard's death, indeed during the long abbacy of Thomas de la Mare (1349-96). The reference is to the upper dial and wheel of fortune, and we are left in no doubt that they were following the plan of Abbot

Richard, 'master of all these things', although mention is also made of the subtlety of the two monks' carving, which surpassed that of all other craftsmen in the district. Laurence was no stripling monk, but one senior enough to travel to the papal court in 1349, in the party that went to have Abbot Thomas's appointment confirmed, and he is there described as *horologiarius*, clockmaker.

The name of Stoke is met with once more in an account of a dispute over the oblations to a certain cross in the cemetery of St Peter's parish in St Albans, a cross erected by one Roger of Stoke, who is again described as *horologiarius*. The date at which it was set up is not given, but the dispute was in the time of Abbot Michael, Richard's successor, perhaps in 1341. Roger of Stoke was evidently a doubly remarkable person in the eyes of the community, for he is said to have erected the cross, which was carved with extraordinary skill, for his future burial—he was therefore not a monk—and all in six days, abstaining from bread and water while he did so. The mention of six days—the time it took God to create the world—should perhaps alert us to what is coming next: miracles began to happen at the cross, hence the offerings at it, and a dispute between the abbey infirmarer and the vicar of the parish as to who should have them.

It is clear that clockmakers were special people, but the name of Roger of Stoke is important to us yet again, because it is found twice in the period 1322-23 in the obedientiary rolls of Norwich Cathedral Priory. This tells us much about the scarcity of skilled clockmakers. They were evidently called upon to travel sizeable distances. In the first mention we find reference to a payment of 7 shillings to Roger for his attendance on an existing clock and for the carriage of his garments (or cloths) and tools. A year later he is linked with the carriage from London of a large plate for the dial. Then, in the same accounts, his name and that of Laurence occur both separately and together. It is obvious that only two men are concerned, working together in both centres, and that Laurence was the younger. It is quite possible that Laurence entered the Benedictine order as a young man after working on the Norwich clock, and that he was the son of Roger, a layman. It seems almost certain that both of them worked for Richard of Wallingford in his lifetime. It could be that both were called Stoke only because they came from a place of that name, for surnames were not yet completely settled. Stoke, however, is too common a place-name for us to hazard a guess as to Roger's and Laurence's origins. St Albans and Norwich are not close—by road they are about 107 miles apart—so the Stoke that falls roughly midway between them need not be favoured above other places with that name in East Anglia. Stoke Holy Cross, for instance, only five miles south of Norwich, seems rather more plausible.

The Norwich accounts will be of value to us when we address the important historical question of what was the crucial moment at which a purely mechanical clock first came into service. It will be as well to begin by considering a few of the main issues at stake. Mechanical timekeeping was surely the single most important practical innovation of the entire middle ages, both scientifically and socially. It compares for its social impact with the invention of movable type, which can in any case hardly be counted as belonging to the middle ages. Reading glasses, curiously enough, date from almost exactly the same time as the clock. While they brought about many subtle social changes, not everyone had impaired vision. The magnetic compass was for the time being without significance for navigation, despite frequent claims to the contrary. Gunpowder would eventually revolutionise the way people made war, but war was not a daily activity. Good timekeeping was an ever-present need—and when it had not been a need, it was perceived as such, as soon as it became feasible.

21. The Tower of the Winds, erected in Athens early in the first century BC by the Macedonian astronomer Andronikos of Kyrrhos. The building's eight sides face the points of the compass and are decorated with a frieze of figures in relief, representing the winds. Below them, on the sides facing the Sun, were sundials. The tower was surmounted by a weathervane in the form of a bronze Triton. Inside the building there was a water-clock, using techniques devised earlier by Ctesibios and Philo, with a dial of astrolabe type. Water was piped from a spring on the side of the Acropolis. The reservoir for the clock is all that survives of it (illustrated here on the right).

12

Horologe and History

THE ENGLISH RECORDS testifying to the arrival of a mechanical means of measuring out time appear to be the earliest of their kind known from any country—although some interpret them so as to avoid this conclusion. There is every reason to think that they point back to a time shortly after the crucial contrivance was invented. This, the so-called escapement, is the device in a clock that at short and regular intervals of time releases the train of wheels which ultimately indicate the time. That train of wheels was originally powered in a very simple way by a falling weight, a weight attached to a chord or chain wrapped round a barrel, on the axle of one of the wheels in the train. (Only at a much later period did springs provide the motive power.) Left unchecked, the driving weight would have fallen quickly to the floor. The escapement allowed it to fall in a slow and controlled manner, alternately releasing and braking the moving train.

The precise nature of the first mechanical escapement—for which honour there are two rival contenders—will be left to the following chapter. Before coming to that question we shall try to assess in more general terms what external evidence there is for the arrival on the European scene of some sort of working escapement, and for the motives which are likely to have called it forth. But first we must draw a number of distinctions, for this entire slice of history is all too often distorted by an unhappy use of quite ordinary words.

Time and the Hour

The word 'timepiece', for example, is imprecise, but it is so obviously imprecise that it is usually innocuous. It is not usual to use it of a sundial, but in all other respects it corresponds to the Latin word *horologium*, a notoriously ambiguous word. When the meaning of the latter is not clear from early records, it is wise to play safe and use the old English word 'horologe'. The word 'clock' is another ambiguous word, especially in its earliest uses. To those who heard Richard of Wallingford's machinery ringing out the hours, and so helping to regulate their lives, it was indeed a clock, for the word simply derives from the northern Latin word *clocca*, meaning a bell. When Richard of Wallingford described his design for the appropriate mechanism, he said it was 'to ring a *clok*' (*pro sonitu unius clok*), so the word here clearly means 'bell for ringing the hours'. Many early fully mechanical timepieces had no dial at all, and simply recorded the hours with a bell, which explains the transfer of the name 'clock' to the timepiece as a whole. The means for sounding a bell at regular intervals did not have to be purely mechanical, however, and there are many water-driven bellringing devices known to history, even from antiquity.

Some of those who have written on the history of mechanical timekeeping have treated it as no more than an episode in the history of timekeeping generally, submerging it in a catalogue that might run from Babylonian water-clocks and Greek sundials, via King Alfred's candle clock, to the caesium-133 atomic chronometer. This is not entirely unreasonable, unless it goes with the assumption that the motives that brought about innovation were unchanging, which is quite obviously untrue. Others have written with a barely concealed ulterior motive, aiming to show that, like most of the great events of history, this one was determined by a certain favoured historical process. Some historians see the world in terms of economic forces, for example, and others favour social or religious pressures. Some like the idea of ready-made knowledge imported from the East, others find explanations in terms of changing intellectual taste. Yet others are convinced that significant historical change is usually the result of accident—the kitchen boy who lets the cooking spit slip, and notices that it has become an automaton, for instance. Perhaps the most miraculous property of the mechanical clock is that it has something to offer to all of these enthusiasts.

Consider the outlook of a man like Richard of Wallingford, who as an expert astronomer was concerned to measure time accurately. For this he had instruments enough, but they were of use only as long as the heavens were visible. He measured time as we do, dividing the day into twenty-four equal hours, and most of the numerous astronomical tables he used did likewise, but the common people of his day did not use the same convention. As the ancient Egyptians and Babylonians had done before them, they divided night into twelve hours and daylight likewise. This convention resulted in hours that varied in length with the season of the year, giving long daylight hours in summer and short in winter. For the regular clergy, such as those in the monastery at St Albans, and the secular clergy too, the division of time into hours was an important if not a precise art, but it was one that broadly followed the convention of the common people, until the astronomers' alternative ('equal hours') was gradually forced upon all of society by timepieces that beat out time regularly, without reference to the movement of the Sun.

Quite apart from the two chief conventions as to the lengths of hours, there was another needed, as to the time from which the hours were to be reckoned. The ancient Egyptians, for example, commonly counted the hours from dawn. 'Italian hours' were often taken to be hours—at first unequal, later equal—counted from the end of twilight after sunset. Broadly speaking, those like the Jews and Muslims who have used a lunar calendar have begun their time reckoning from sunset. Writing in Paris or Montpellier in 1271, Robertus Anglicus collected together some traditional ascriptions. Many Latins, he said, begin the day in the most natural way, from sunrise, some however from the first sign of dawn, and the astronomers from noon. (Astronomers keeping records of visual observations made at night still do this, to avoid a change of date during the night.) 'The Chaldaeans', according to Robert, begin the day from midnight and 'the Jews' from sunset. He added some support from scripture, and his account needs much qualification, but we need not pursue it any further. The important point is that to a Christian cleric of the thirteenth century these things mattered greatly, since they were connected with the cycle of prayer times and the customs of other religions.

The Jews before the Christian era had established the custom of praying three times a day, at the third, sixth and ninth hours, and the early Christians extended that scheme, adding prayers at midnight (when Paul and Silas sang in prison) and at the beginning of day and night. It was St Benedict himself who added a seventh hour of prayer, compline, so completing the rule for the 'canonical hours', the times of prayer, followed by the church thereafter. It is true that the rule was subject to much local variation. Travellers could find the hour of matins changing, as they moved from place to place; lauds was occasionally combined with it; sext and none could be joined, as could vespers with compline. This could be confusing, but everywhere there was a need felt for a means of deciding when the canonical hours occurred.

Water-Clocks

A simple clepsydra—which was typically nothing more than a vessel discharging water slowly, with levels inside the vessel marked in units of time—was rarely reliable. Putting a float on the surface of the water, and activating a dial by a rope tied to it, did not in itself improve matters. The more complex the timepiece, the more care that is likely to have been put into it, but always there would have been difficulty in allowing for variation in the rate of flow. Once it is graduated, however, a simple water-driven device records intervals of time having nothing to do with the seasons. Water might leave the container at a variable rate, but this is a fault of design or manufacture, and has nothing to do with the unequal hours of the ancient and medieval world, the hours that vary with the seasons provided by the Sun. Most simple devices for keeping time—the candle clock is a favourite in story books—tell equal hours. In India, from at least as early as the fourth century and for ritual purposes even to the present day, water-clocks have been made in the form of bowls, floating on water but perforated so that they gradually sink. They are graduated in units decided by the time it takes to utter sixty long syllables, and sixties of those sixties, hopefully all of equal duration. In the western middle ages, reciting penitential psalms was a technique for measuring time which also presumably led to more or less equal hours—although perhaps not quite equal as between a frosty night and a warm one. Despite all such evidence, as we shall see shortly, ways of indicating seasonal hours reasonably accurately were already known in the ancient world. This was done achieved through the use of complicated, float-controlled, astronomical dials.

Given the importance of the prayer cycle, the monasteries had a need for an alarm system, if only to stir the person who roused the others from their sleep. The oldest detailed medieval account of the construction of any such thing shows that it was water-driven. It is to be found in an incomplete manuscript from the tenth or eleventh century, written in the Benedictine monastery of Santa Maria de Ripoll, at the foot of the Pyrenees in Catalonia. The alarm was a simple affair, a weight-operated striking mechanism, with a rope and weight turning an axle. This operated a flail, ringing several little bells hanging from a rod. There was perhaps nothing more by way of a dial than was needed to reset the alarm, a task that would have been undertaken by the sacristan. (The rules of the Cistercian, Cluniac, and Benedictine orders all gave the task to him or to his assistant.) It is likely that there was an element of guesswork in the setting, when it came to deciding on the (unequal) canonical hours of prayer. From an account of the eleventh-century customs of the monastery of St Victor in Paris, we know that a very crude rule for lengthening and shortening the hours was in use there.[69]

Ingenious though the Ripoll alarm mechanism was—and it should perhaps be seen as an extension to a long Muslim tradition of making water-driven automata—it would not have been difficult to make, and there is reason to think that such devices became fairly common. A commonly

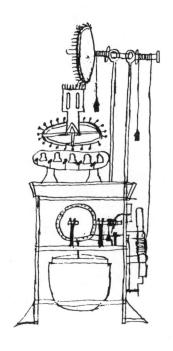

22. Drawing of a water-clock of unknown date, on a blank leaf unrelated to the late thirteenth-century manuscript in the Archivo de la Corona de Aragón where it is found. The drive was possibly achieved through a sinking float, counterpoised by the upper right-hand falling weight (note the crank to raise it periodically, at lower right). The wheel above the water vessel, probably meant to show twenty-four divisions, turned by the falling float, presumably released the falling weight on the left every hour, so allowing the circle of bells to be flailed in an obvious way. The appearance of the two upper wheels hints at simple wooden discs with metal pins inserted.

reproduced thirteenth-century illuminated illustration from a French moralised Bible, for example, while it is hard to interpret, shows a water-driven device somehow sounding a carousel of small bells. It is possible that it had nothing to do with timekeeping at all, but it does seem to have had some sort of water-operated movement, and the illustration includes a conventional Sun-symbol, which hints at a timekeeping function. Indeed, a much later manuscript than that with the Ripoll alarm (but now in the same Barcelona library) has a drawing on a blank leaf which seems to picture a water-driven mechanism of the type in the Bible illustration. Since it is much more intelligible, it is reproduced here (Fig. 22). Whatever the character of these simple devices, we may be sure that peals of tuned bells, even the jangling of small and untuned bells, would have given a certain simple pleasure. Such devices might even have had more uses than one, for then as now, noise could be its own justification. As early as the rule of John of Hertford, abbot of St Albans between 1235 and 1260, it was ordained that whenever a new abbot was installed it should be to the accompaniment of the striking of bells, the sounding of shawms ('which we call mules') with the horologe, the lighting of tapers around the altar, and the uncovering of the throne. Evidently even a water-clock could be introduced into the high ritual of the church, but presumably only on account

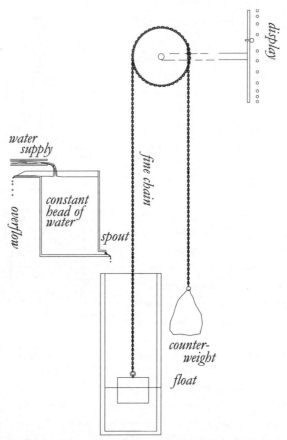

23. The water-powered drive of an anaphoric clock of Vitruvian type. The display is shown in profile, so as not to obscure the drive. (For the clock dial see the next figure.) The method of keeping a constant head of water, giving a constant rate of rise in the float chamber, follows a common conjecture. A drilled polished agate would have made for a constant stream into the float chamber.

of the sound it could make. The water-clock in question at St Albans was emphatically not a small alarm intended only for the sacristan's ears.[70]

On a much higher scientific plane was the device now generally known as an 'anaphoric clock'. This had been used in the Greek and Roman worlds, as well as in Islam, and it carried a dial of the same general type as that designed by Richard of Wallingford for St Albans. That part of the display was in effect a large astrolabe—some would see the very beginnings of the astrolabe idea in the development of the dial of an anaphoric clock. The classical account of the clock's construction was given by the famous Roman architect Vitruvius, writing in the first century BC. What he called his *anaphoricum* was the second of three water-clocks he described as being for winter use, the implication being that in summer one needed no such thing, except by night.[71] He described it as a device with a circular disc picturing the constellations, and rotating once in the course of a day. The disc was turned by means of a bronze chain coiled round its axle. One end of the chain was to be attached to a drum, floating on the surface of water in a cylinder that was slowly filled from an external source (see Fig. 23). At the other end of the chain there was a counterweight. Such a device had been used in public clocks in several places before he described it. He referred to the most famous example of all, that in the Tower of the Winds in Athens (Fig. 21).[72] Fragments of two later Roman anaphoric clocks were found in the nineteenth century, one at Salzburg and the other at Grand (in the Vosges, in north-east France). Since both probably date from the first or second centuries, and since they were found in relatively remote parts of the Roman empire, it seems reasonable to assume that the anaphoric clock was widely diffused in the ancient world.

One of the qualities of these anaphoric clocks will be worth remembering when we consider the St Albans dial, in the next chapter. When briefly introducing that ubiquitous medieval instrument the small portable astrolabe, we saw the significance of its two discs (the pierced rete and the plate under it). The first contains the star map moving round the sky, while the lines on the second show the local fixed framework of grid-lines which we can imagine superimposed on the sky overhead—the horizon, the meridian line, and others. It is not at all necessary that the moving fretwork, the rete, should represent the star sphere and the plate the local reference frame. While this became the almost universal preference on portable astrolabes, the roles of the rete and plate were easily reversed, as in the Vitruvian account and in the anaphoric clocks from which fragments survive. (A conjectural reconstruction of the appearance of the clock at Grand is shown in Fig. 24.) That arrangement (stars behind, horizon and hour-lines to the fore) is not unknown in the middle ages, but the St Albans dial is the only example of which I am aware in which that arrangement is followed but with the star map in projection from the south. For present purposes it is not necessary to look deeper into the question of the many possible variants. The main difference between the two cases mentioned is that the signs of the zodiac circulate in opposite senses. The common order, reading anti-clockwise, is Aries, Taurus, Gemini, Cancer, Leo, Virgo, Scorpio, Sagittarius, Capricorn, Aquarius, Pisces. The clock at St

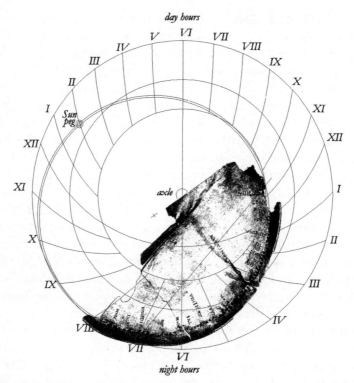

24. The Grand (Vosges) fragment, as illustrated in the Hoffmann sale catalogue of 1886, here shown as it might have been fitted into the display. This should be considered in conjunction with the previous figure. The Sun marker, here placed in an arbitrary position, is to be read against the fixed grid of hour-lines. Here it shows the second hour of the day, on the system of seasonal (unequal) hours. Some of the surviving plate was later lost, but the larger part is now in a museum in Saint-Germain-en-Laye. Note that there is no hole for an axle for the moving solar plate, which was presumably cased in some sort of disc, perhaps of wood. The plate is not accurately made. The centre of the ring of holes, here indicated with a faint cross, should be appreciably further away from the axle, representing the pole of the heavens.

Albans resembled the Salzburg plate in following this convention. The clock at Bourges and the Grand plate reverse the order. Since both ancient traditions survive, perhaps neither was ever entirely lost in the intervening millennium, although reinvention cannot be ruled out.[73]

There is no marker for the Sun on the zodiac ring of an astrolabe. Scholars were often told to mark its position with ink, or to hold a rule at its place.

On the Vitruvian anaphoric clock dials there was a series of (ideally) 365 holes around the ecliptic circle, into which a ball on a peg could be plugged to mark the daily positions of the Sun. The idea was that the peg would be moved manually.[74] The better medieval clock dials of astrolabe type, whatever the convention of plate and rete, also showed the Sun, but in the ideal mechanism it was moved mechanically rather than manually. No one in the middle ages solved the problem of moving the Sun correctly more elegantly than Richard of Wallingford.

The First Cluster of Records

The first English payment with reference to a horologe (*horologium*) that might have been truly mechanical is recorded in a manuscript roll that originally came from the room of the prior at Norwich. It is dated 1273, and is followed by another payment from 1290 for repair or adjustment (*emendatio*) of the horologe. Nothing is known about it, beyond the fact that, when a large new horologe was built in 1321, mention was made of payment for a rope or chord (*corda*) for the old horologe. In the earlier case we are obviously not dealing with a sundial, although just possibly with a water-clock. (As explained earlier, the simplest arrangement would have required a chord to be tied to a float, in a vessel from which the water was gradually drained. Wrapped round an axle, on the end of which was perhaps a pointer, the chord would have been tied to a counterpoise.) The 1321 horologe was quite certainly an escapement-controlled clock. The same word (*horologium*) was used for the old and the new in the same document, without further distinction. The conclusion is often drawn that the old was likewise fully mechanical, but it is unwise to pronounce judgement on the basis of such incomplete evidence.

Another written record that probably refers to a mechanical clock is one in the annals of Dunstable Priory, a house of Augustinian canons in Bedfordshire. The record is brief: 'In the same year [1283] we made a horologe, which was set up above the rood-screen (*pulpitum*)'. The reference is to the stone screen separating the choir from the nave of the church. It carried a crucifix in some form, which in this case was accompanied by painted images of John and the mother of Jesus—they were repainted in 1293. In some churches the rood-screen was large enough to carry an organ loft, so it could easily have accommodated a large clock case—as did that at Bourges in France in the fifteenth century. It is interesting to see that the canons made the horologe themselves, but that they did not think it important to sing the praises of any one member of the community for his genius. The strongest argument against the idea that this was merely a water-clock, however, is its placement. The canons would have not put a small clock in such a place, for it had to be visible to all. It could be argued that even a water-clock could have had a large dial, but then there is the question of the channelling of water to the top of the rood-screen, and the regular replenishment of a supply cistern. This record seems more likely than not to have been to a mechanical clock.

That from Dunstable is one of a cluster of similar records referring to *horologia* and appearing suddenly in English ecclesiastical annals of the last two decades of the thirteenth century. There are records, some of them admittedly obscure, from Exeter (1284), London (Old St Paul's, 1286), Westminster Hall (1288), Merton College, Oxford (1288), Norwich (1290), Ely (1291), and Canterbury Cathedral (1292). There follow records for Salisbury (1306), and outside England for San Eustorgio, Milan (1309), Cambrai (1308 or 1318), and one or two more dubious instances. There are two famous passages from Dante's *Paradiso*, but opinions as to their date are as divided as those on their interpretation. They were written after 1308, perhaps after 1316, and certainly before 1321, the year of the poet's death. Almost as surely they refer to a mechanical clock.[75] Taken singly, the records are usually cryptic, but taking them as a group they invite the conclusion that a new form of complicated mechanical timepiece had suddenly become available. We are told, for instance, that at the London church of St Paul's, Bartholomew the clockmaker (*orologiarius*) drew 281 loaves of bread over a period of three quarters and eight days, early in 1286, and more later in the same year, after he was joined by William of Pikewell. A device requiring labour on this scale was probably too complex to be anything other than a mechanical clock. The only real alternative to this interpretation is to suppose that there was a sudden surge of enthusiasm for building much more elaborate water-driven devices than had been in use hitherto—and for this there is no independent evidence.

Perpetual Motion

There was evidently something new happening by the 1280s, or even by the 1270s, and in view of this fact we are fortunate to have an independent fragment of evidence from a very different source, which effectively tells us that the necessary controlling device (escapement) had not been invented by 1271, but that it was then being actively sought. This comes from the astronomical commentary by the scholar Robertus Anglicus on the *Sphere* of Sacrobosco. In the same context as that in which he informed us about different ways of counting the hours of the day, Robert remarked that makers of horologes (*artifices horologiarum*, probably professional water-clock makers) were trying to make a wheel that would complete one revolution for every revolution of the celestial sphere, that is, would turn once in a sidereal day. He carefully explained the slight difference between time as reckoned by the Sun and time as reckoned by the stars, after observing that those artificers had not then managed to perfect their work. He knew the principle of a weight-drive, for he mentioned in very general terms a carefully made and uniformly balanced wheel with a lead weight hung by a chord wrapped round its axle. He was not talking about the rather similar arrangement of the counterpoise in a water clock. The outstanding problem was clearly that of finding a controlling mechanism for a falling weight.

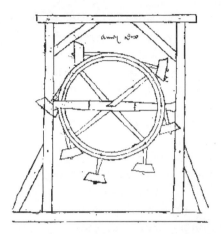

25. Villard de Honnecourt's design for a perpetual motion machine.

Historically important though this passage is, it is so only because it appears to be the *last* of its kind, suggesting that the invention of the escapement came afterwards. The invention might conceivably have been made already, although one imagines that news of such a thing would have travelled fast, and Robert seems to have been generally well informed. The fact that we know of several earlier statements to the same effect, and none afterwards, suggests that the invention came soon after Robert's.

Earlier references can tell us much about the ambitions and motives that brought about a complete transformation of time-telling methods. There is one such statement in the portfolio of drawings by that great thirteenth-century French source of out of the way architectural information, Villard de Honnecourt. Compiling his famous portfolio, at an unknown date, but not far in time from 1230, he included a design for a perpetual motion machine, in which a wheel has seven mallets pivoted around its rim (Fig. 25). The idea was that as the wheel turns they should flip over in such a way that there are always four on one side of the wheel and three on the other, so that the imbalance keeps it turning. Under the drawing is a caption (in French) telling us that 'masters have often striven to make a wheel turn of its own accord' and that 'here is a way to do it with an uneven number of mallets and with quicksilver'.

It is occasionally assumed that the mallets were meant to be filled with mercury, but it is more likely that the reference was to another unworkable suggestion, found in Islamic sources, using mercury in the compartments of a drum, mounted on a horizontal axle. The idea was that such a drum should have an odd number of compartments, each partially filled with mercury, and that this would guarantee—as it was vainly believed—that there would always be more on one side than the other, so that it would turn endlessly. This machine is not to be confused with a device controlled by mercury that was apparently designed in the mid-1270s, and which does indeed work. The latter is described and illustrated in the books of Alfonso X of Leon and Castile, and is ascribed to the most productive of his astronomers, Isaac ben Sid (Fig. 26). This too had a drum, but now one which was divided into twelve compartments, the panels separating them being pierced with small holes. It was turned slowly under the action of a weight on a rope wrapped round the axle, and as it did so the mercury leaked from compartment to compartment, the viscosity of the mercury providing most of the force counter to the driving force. Improving upon the tradition of the old anaphoric clocks, the system turned an astrolabe dial. (The illustration is a modified version of that in the manuscript, but is true to the style of common thirteenth-century astrolabes.)[76]

Those who sought perpetual motion in the middle ages were not seeking energy free of charge. They were simply trying to create a machine that was self-moving, in the sense that it did not require constant human interference. Even those unworkable machines, such as Villard's, which were imagined to work without winding or other human interference, were aimed at producing a movable display, not at cheating nature. The point is well illustrated by the title of one of the Alfonsine books, 'On Water Wheels, Mills and Presses that Move by Themselves'. They moved with the help of water, of course, and not exactly 'by themselves'. The Alfonsine mercury device needed winding from time to time, to raise the falling weight. It had in common with Villard's idea the use of gravity, which we automatically associate with the first mechanical clocks. The search for self-moving mechanical devices continued fitfully throughout the middle ages. Three or four different devices were sketched in a fifteenth-century treatise from the northern Veneto, for example, by a man who hints that he had built them and that they did not work.[77] The situation becomes especially interesting when we see how writers such as Roger Bacon tried to bring scientific theory to bear on the problem. When scholars of the thirteenth-century and later did so, moreover, gravity was not the only working principle they entertained.

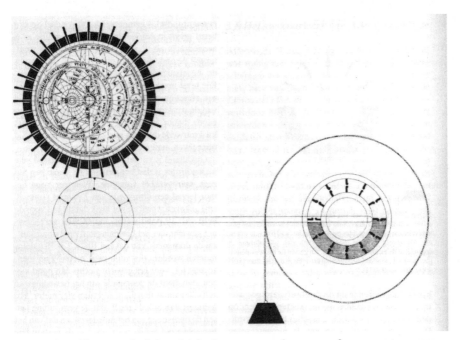

26. The Alfonsine clock in which the seepage of mercury from one compartment to the next allows the counterweight to fall slowly. The figure was redrawn by Manuel Rico y Sinobas for his edition of the text (1863) and the mechanism is reasonably true to the original.

Bacon's is one of the few names now widely associated in the popular imagination with the empirical sciences in the middle ages. He is remembered more often than not for his visions of the future, but far more important, historically speaking, is the way in which he tried to bring theory and practice together. In a work 'on secret works of art and nature and the nullity of magic', written perhaps in 1248 and probably addressed to William of Auvergne, bishop of Paris, Bacon attacked magic on the grounds that science and the technical arts can perform far greater wonders. After a chapter concerning marvellous machines, such as submarines, automobiles, and flying machines, he comes down to earth with more realistic things—curved mirrors and lenses, Greek fire, gunpowder, and the magnet. And then, before turning to such matters as the prolongation of life and alchemical secrets, he named something that he considered would be more valuable than anything he had already discussed: a self-moving astronomical sphere.[78]

He was at pains to show that, while it was not then within the power of mathematicians to produce the required result, an expert and faithful experimenter (*experimentator*) might succeed. Such a person would make a model of the heavens from suitable material and by suitable artifice, in such a way that it turned with the daily motion by its very nature. The idea he was putting across was that the experimenter first studies the physics of the heavens, and so learns how to produce or simulate their motions here below. In a passage from his famous *Opus maius*, dating from 1267, he more or less repeats that idea of simulation, but as before he clothes his account in mystery. What did he believe was the deep-seated physical cause that could make a sphere turn with the daily motion of the stars?

It seems almost certain that his answer was magnetism, and that he had a particular 'expert and faithful experimenter' in mind, a contemporary scholar whom he is known to have greatly admired, but about whose biography we know very little. It seems clear from remarks made by Bacon in his *Opus minus* (1268) that the man was Pierre de Maricourt, from Picardy, more often known as Petrus Peregrinus, Peter Pilgrim. Bacon speaks of 'one of the greatest of secrets in the experimental sciences, or indeed anywhere'. This, he said, concerns 'a body or instrument which is to move with the motion of the heavens, and transcend all instruments of astronomy', and it must be made from a magnet. It is not known whether Bacon had met Pierre de Maricourt in Paris. Was he unusual in sharing in the secret, or was it a talking point among scholars in that city? The second seems more likely, in view of the fact that William of Auvergne, writing in the period 1231-36, had explained the motion of the celestial spheres in terms of magnetic forces.[79] At all events, it was in 1269, within a year or two of Bacon's completion of the *Opus maius*, that Pierre proposed a solution to the self-moving wheel in a remarkable treatise on the magnet (the natural magnet, or lodestone) and its applications. He was convinced from his observations that the poles of a magnet point north and south because they derive their virtue from the poles of the heavens; and that each part of the magnet (he made spherical lodestones) likewise corresponds to a different part of the heavens. It was a short step from this principle to the idea that a magnetic sphere, pivoted perfectly so that its axis (defined by its magnetic poles) is in the line of the celestial axis (joining the north and south poles of the heavens), will turn with the daily motion of the heavens. The idea was that it would be held fast to the heavens, so to speak, by a force of attraction. In short, it would have the makings of a simple astronomical timepiece.[80]

Pierre de Maricourt was writing whilst encamped with the duke of Anjou's army, then besieging Lucera. It is hard to believe that he would then have had many opportunities to experiment, but in a final chapter of his work he makes a remark which shows that there is more to the activity on which he is reporting than mere armchair musing. He has seen people exhausting themselves, he says, in vainly trying to make a wheel—notice that he does not say a sphere—move round with perpetual motion. He therefore offers a design for a timepiece driven by a magnetic motor, which he imagines will provide perpetual

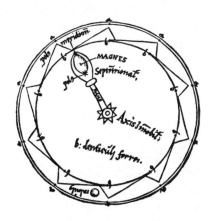

27. Pierre de Maricourt's design for a magnetic motor that was to provide perpetual motion.

motion. His motor comprises an egg-shaped lodestone on the end of a silver spindle, all within a wheel-like silver casing (Fig. 27). The spindle moves round like the pointer on a dial, attached to a gear at the centre and carrying the lodestone at its tip. The lodestone moves just clear of the inner edge of the casing, where there are iron nails (or teeth) placed slantwise. As one of its poles passes a slanting nail, its attraction for the nail gradually increases to a maximum as their separation diminishes, after which its momentum carries it on to the next slanting nail, where the process is repeated. The details are unimportant, and of course the highly imaginative design would have been as unworkable as Villard's. If he ever tried to make his motor, he was no doubt wise enough to keep quiet about the result. Not that he was without a cast-iron defence against all who found that his methods failed to work. 'Ascribe your failure to your lack of skill rather than to a failing of nature', he had said on an earlier occasion.

In view of Pierre de Maricourt's reference to the 'many persons' he had seen trying vainly for a solution to the self-moving wheel, it is quite possible that it had become something of a local literary convention when Robertus Anglicus echoed it two years later. It is the later date, however, which bears most directly on the question of the non-existence of an adequate escapement, assuming that Robert was well informed. Just as interesting, however, is the fervour of all concerned. As Bacon explained, in his letter of 1248, the reason for it was quite simply that such a self-moving thing would make 'all the instruments of astronomy superfluous, both special and ordinary'. Adding that the treasure of a king would not be able to compare with it, he wrote as though possessing such an astronomical clock would be like possessing the secret of the workings of the universe. The mechanicians were playing for high stakes.

Mechanisms and Motives

It is a remarkable fact that the first Norwich reference to what might have been a mechanical clock came just two years after the statement by Robertus Anglicus; but even if we reject that Norwich fragment, we still have half a dozen likely English references from within twenty years of it. It seems that when Richard of Wallingford was made abbot of St Albans, the mechanical clock was perhaps half a century old—allowing ample time for the mechanism to have evolved significantly, and its social function too. Church clocks, even within monasteries, were to some extent open to a wider public—the sound of their bells certainly so. The character of society, however, was not uniform across the length and breadth of Europe. Italy is especially interesting, for the character of its many competing and rich city states quickly gave rise to a new style of public clock there. This second historical phase produced a string of Italian records, starting perhaps with Orvieto—a sizeable clock was repaired in 1307-8, so was not new then. We do not know its nature, or that of the Ragusa (Dubrovnik) clock mentioned in 1322 as requiring a clock keeper, but it is likely that they were fully mechanical. In 1336 Milan had a public clock, to which we shall need to return, since it struck the hours in a way somewhat similar to that of the St Albans clock. From this time on, records of public clocks multiply: by 1353 the list includes Parma, Padua, Monza, Vicenza, Trieste, Genoa and Florence, while between 1351 and 1353 in London an Italian clockmaker directed the building of a clock for the Great Tower of Windsor Castle. The place occupied by such tower clocks in the public consciousness has unfortunately led some historians to write, not only as though they were first on the scene, but as though the demand for the commodity 'public time' was what led to the invention of the mechanism which provided it. Neither idea is acceptable, on the basis of present evidence.

Astronomical Motives

The fact that Richard of Wallingford's clock is the first of which we know the details should clearly not be taken to mean that it was wholly original.

He died in 1336, and by then there had been time enough for traditions to have been established for making the iron gears needed to produce the daily motion and its subdivisions—not the gears most often used by the millwright, which had been mostly of wood, but in styles loosely related to them. As for the controlling mechanism, the escapement, there might have been more than one working design. The form of the escapement, as opposed to its proportions, is touched upon so lightly in Richard's manuscripts that we may be sure that he was not the inventor of the underlying principle, and that his reader was expected to know what he was talking about. We cannot be sure about the sort of reader he had in mind, but presumably he was addressing himself either to the professional clock-builder with a fair degree of education, or to the virtuoso scholar. It is the astronomical complexity of the St Albans mechanism that virtually guarantees our claim that nothing before it was quite like it in those complex respects, for the number of scholars in Europe capable of designing such a thing was very small indeed, and it is scarcely conceivable that the personal circumstances of any of them would have led to the building of a comparable machine.

What can we learn from the first fragmentary church records, and scholarly statements of ambition by such men as Roger Bacon and Pierre de Maricourt before them? Those records have been used to justify some surprisingly divergent opinions about the origins of mechanical timekeeping. Like any other great idea, mechanical timekeeping has a convoluted ancestry, influenced from many directions. Those who, in describing it, begin by introducing the problem of timing the hours of prayer—excusable though that may be in the biography of an abbot—can expect to be accused of loading the dice unfairly in favour of an ecclesiastical origin. Our examples from the university world, to which the abbot also belonged, and from which there is ample scholarly evidence of a desire to represent the daily rotation of the heavens, should mitigate the charge; but did others not have a need to settle times, a need which they could not satisfy with sundials, water-clocks, counting sheep, singing metrical psalms, or other existing methods? What about city-dwellers, and others who lived a similarly well-ordered public life? What of the astronomers, when they were going about their ordinary astronomical business? They certainly wanted to record times as precisely as possible, and while they were quite well equipped to determine them without the help of clocks, they would have been open to any easier way that offered itself. Mechanical clocks were generally thought reliable enough to be used for simple astronomical tasks only towards the end of the sixteenth century. As it happens, astronomers had already made occasional use of water-clocks for astronomical observation in the ancient world, and Roger Bacon, in his *Opus maius*, shows that he was aware of the fact that some were still doing so. He compares ancient time-measuring devices to the water-clocks that were being made in his own time, by which short intervals of time—such as the time it takes the Moon to cross the horizon—could be worked out with greater accuracy than was to be had from instruments like the quadrant and astrolabe. Clocks were therefore, even then, not the prerogative of church, court, and city.

The astronomers had other motives, however, for seeking an accurate driving mechanism. They were interested in more than the hours and minutes of the day, in more than the daily rotation of the heavens. As was Richard of Wallingford, the better astronomers were anxious to streamline their techniques for calculating planetary positions, and this led to the design of suitable computing devices, equatoria, of which the albion was an outstanding example.[81] To take matters one stage further, and create a material model of the heavens, with the Sun, Moon and planets moving visibly round a central Earth, was something greatly to be desired. Anyone representing the positions of the heavenly bodies in this way was bound to be sacrificing accuracy, by comparison with what could be had from careful calculation; but precise positions were not the chief aim of those who built such planetary displays.

When Bacon alluded to the desirability of a three-dimensional spherical model, in his letter 'on secret works', his was little more than a castle in the air. That it did not indicate the outlook of the specialised astronomer can be seen from his rhetorical suggestion that a machine would render other instruments unnecessary. Bacon's remarks point to a desire to encapsulate God's cosmic scheme in a machine, in a broad, qualitative sense; and that is an idea which was much more easily grasped by ordinary people than the wish of the astronomers to have numerical exactness in their models.

A geared astrolabe dial alone, such as was found on anaphoric clocks, provided a moving image of the stars. It could be moved automatically or simply cranked by hand. Adding the Sun to the dial could be done by plugging a ball representing the Sun into a different hole along the ecliptic every day. Automating this is difficult, since it moves round the ecliptic circle at variable speed. Finding suitable gearwork not only for the Sun but for the Moon and planets too was a far more difficult problem that had been tackled by astronomers since ancient times. It was not only complicated by the need to make them all turn around the same axle, the axis of the world. Even giving them their own separate dials leaves the extraordinarily difficult problem of making them move at their various and variable speeds. Here were challenges in plenty to stimulate astronomical minds.

Most of our early evidence for such planetaria is for two-dimensional displays, plates or wheels embodying an astrolabe, with perhaps the Sun and Moon and just occasionally the planets moving round the zodiac on it.[82] The Alfonsine mercury device had nothing more than a simple astrolabe dial. There are, however, earlier records of planetaria moved by trains of gear wheels. Archimedes, for example, the greatest mathematician of Greek antiquity, is credited with having made some such model, and another was ascribed to Posidonius. What is without doubt the most technologically advanced artefact surviving from the ancient world, a geared device brought up from the sea bed off the Greek island of Antikythera at the beginning of the twentieth century, and still not completely understood, can be loosely described as a geared celestial model from which readings could be taken.

However ingenious their wheel-work, and however accurate the relative placement of the planets on the wheels, there was one disadvantage shared by all early planetaria of any complexity: they had to be cranked round by hand. They lacked a drive that would make them show the state of the heavens at all times automatically. Bacon tells us that this was still the case in his time, the mid-thirteenth century. Here was a desideratum that might well have supplied a technically-minded astronomer with a motive to devise an escapement-controlled timepiece. Islam has a long history of water-driven automata, into which cosmic elements were occasionally introduced, and the Alfonsine mercury clock stands in this tradition, but we have no firm evidence for a fully fledged water-driven planetarium. The ambassadors of al-Ashrāf, sultan of Damascus, presented a gift of a costly bejewelled planetarium of some sort to the Holy Roman Emperor Frederick II, when he was in Apulia in 1232, but there is no reason to think that it was self-moving. In the 1960s I found a Latin text, seeming to stem from the thirteenth century, and describing in some detail 'a device of certain wonderful wheels', a geared planetarium but with no indication of any driving mechanism (Fig. 28).[83] Since the text was closely connected with northern Italy it was tempting to see a link between it and the sultan's gift to Frederick. It is unfortunate that we do not know whether the highly ingenious scheme outlined in the text was ever made, but it is extremely important because it shows that there were individuals capable of calculating some of the necessary gear ratios. This they did, with great skill.

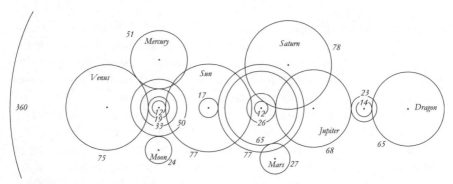

28. One of several possible arrangements for the device described in a northern Italian text of the thirteenth century ('Fiat columna ...'). The circles represent toothed gear wheels. Circles that touch indicate meshed gears. The arrangement shown can account for the mean motions of the Sun, Moon and planets, each of which could have had its own separate dial. There are ways in which epicyclic motions and the motion of an astrolabe dial may be added conjecturally, through the large wheel of 360 teeth. It is conceivable that the actual configuration was quite different, using the same gear ratios but with gears mounted on concentric tubes, allowing everything to circulate around a central Earth—described in the text as a parchment-covered ball. This would have required high technological expertise, but cannot be dismissed out of hand. Richard of Wallingford's clock used tubes of the required sort, but fewer.

Astronomical Motives Questioned

What does all this evidence suggest, as far as the history of mechanical timekeeping is concerned? In the 1950s Derek Price expressed the view, often and forcibly, that the wish to drive planetaria was paramount, and that the mechanical clock was essentially a degenerate form of the driven planetarium.[84] In 1983, in an ambitious book on the long history of time-telling, David Landes objected strongly to Price's thesis. Putting aside Landes' objection to the notion of degeneration, which is not an essential part of Price's argument, and his assumption that Price was saying that these makers of driven planetaria did not care about time measurement for its own sake, we find an argument along the following lines: simple mechanisms precede complex, and therefore clocks preceded driven planetaria, which are much more complex than they.[85]

This is an unacceptable argument, since it is beyond all reasonable doubt that geared planetary displays are older than the mechanical escapement-controlled drive. It is also an argument that tacitly rules out the idea of the simple bell-ringing clock as a spin-off from the other technology. There are, in the history of technology, many instances of simple devices being taken from complex ones—saucepans from the nose-cones of rockets—and they warn us against any argument that is based on increasing complexity with time. This is not to say that the conclusion of such an argument will necessarily be wrong, of course.

It would not be hard to invent other motives for striving to produce an escapement-controlled drive, but modern views of time—for instance as an entity that begs to be continually monitored—can all too easily mislead us. It is not that abstract questions on time were never addressed. In Oxford in the thirteenth and fourteenth centuries there were strenuous debates, for instance, on whether time was a physical or intellectual entity, on how the time continuum can be related to numbers that are discrete, on how the human soul is related to time, and so on. Richard of Wallingford is bound to have been aware of these discussions, and even at some stage to have taken part in them, but it is hard to believe that there was any connection between such abstractions and the severely mechanical devices we are discussing, for the people concerned were from very different callings. Influence in the other direction seems much more likely. Clock metaphors in literary works are not uncommon, for example, but that fact does not provide any motivation of the sort we are seeking. To take one example: the French scholar and bishop Nicole Oresme considered the possibility that, at the creation of the world, God put special forces in the heavens, by which they moved continually thereafter like *un horloge*. (The text, *Livre du ciel*, is in French, not Latin.) Our question, however, runs in the opposite direction: was it a wish to create an image of God's created celestial world, turning with its daily motion and so forth, which prompted the discovery of the invention that made the image possible?

The short answer is that we do not know. But why should we want to know? The reply will often be that we want to fit these events into our wider view of historical cause and effect. If the controversy discussed here proves anything, it is that there are many historical routes that lead to, and throw light on, the situation in which we find Richard of Wallingford. It is misleading to pretend that we can identify a single line through history on which a single 'Eureka moment' lies. (The regulation of prayer times ... simple water-driven alarm ... mechanical escapement ... clock ... planetary clock. Here is just one example.) The astronomical clock was a complex machine, compounded out of parts, each of them devised for reasons peculiar to it. There was the motive for ringing the hours in the monastic day, and there were the motives for creating various sorts of astronomical display. Someone, somewhere, somehow found an escapement. This was a necessary but certainly not a sufficient condition of the remainder, and we shall never know who, where, or how—even though we might make an informed guess. The argument in the past has been much too abstract. It seems highly probable that the mechanical escapement was found by someone working with a bell-ringing mechanism. That person would simply have realised that much the same oscillating device could be used for another purpose, that is, as an escapement. We shall return to this possibility.

To What Was the Mechanical Escapement First Applied?

The monastic and scholarly worlds were largely distinct from the established class of dedicated artisans to which our attention was drawn by the very mention of men described as *artifex* (artificer) or *horologiarius* (horologe maker). Was the escapement principle found in the course of an attempt to improve the reliability of existing water-driven timekeepers, for supply to those who used them already—within the church, at court, in the law-courts, or other places? Early records of the activity of such artisans, after the apparent breakthrough was made, have so far been found only in connection with the church, but the early church records are strangely laconic, giving no sign of local pride. The reason is that they come from terse financial accounts, to which sentiment is foreign.

Some of the earliest payments are for renovation, and tell us little. The Merton chaplain 'Master G.'—he was invariably named in this way—was given 4s. 4d. in 1288 for an *opus horologii*. This 'work of a horologe' could mean almost any time-telling device. It is doubtful whether it means 'work *on* a clock', for *opus* is usually the thing created. The relatively small sum suggests perhaps no more than a sundial, or just possibly an addition to an existing mechanism. A mechanical escapement, added to an existing water-clock, perhaps?

The Westminster Hall record rings true, although our best authority is the great seventeenth-century legal historian John Selden. He reports that in 1288 justice Ralph of Hengham was fined 800 marks (£533 6s. 8d.) for halving a poor defendant's fine. This much is borne out by a legal record dating from 1484, but Selden went on to add that Hengham's fine—a very substantial sum, more than equal to the task—was earmarked for a clock in a clock tower at Westminster Hall, to be heard in the courts of law. An Elizabethan judge, John Southcote, had much the same story, and said that the clock was still in place in his time. (He died in 1585.) While this is not conclusive, it seems unlikely that Westminster, the centre of political power, would have been far behind other institutions in the matter of an important new invention.

The custodian of the horologe at Ely (1291) was paid 3s. annually for looking after something, which cannot have been a trifling object, for many labourers would gladly have done a month's hard work for that sum. The Canterbury record for the 'great horologe in the church' (1292) tells us that it cost £30, which gives us a much better idea of the magnitude of the enterprise, especially when we make a comparison with the 1322 accounts for the horologe from Norwich. They list payments made to Roger and Laurence of Stoke for over two years, and to a third *horologiarius* Robert for four terms. They also list irregular payments to smiths, carpenters, masons, plasterers, and painters; and payments for iron, copper, bells, ropes, gold leaf and other materials. All this came to £52 9s. 6½d., a sum comparable with the annual income of a small priory and almost exactly three-quarters as much again as the cost of the Canterbury horologe. In all cases there would have been much free labour on hand from the religious community concerned, which introduces an element of uncertainty into the grandeur of the end result.

But what exactly was that? Bell, display, or both? And what kind of display? That the horologe at Dunstable Priory was on the rood-screen suggests that it had something to display, whether or not it was of an astronomical character. At Exeter, Roger of Ropford was a bell-founder who 'looked after the horologe', but he also looked after the organ, which clearly had no bell, so we can say only that there was some sort of mechanism. (One does not speak of looking after a sundial. There are many historical references to men who made both clocks and organs, but it is well to be aware of the fact that before the early fourteenth century the word 'organ', *organum*, could be used of almost any musical instrument, or even of a small engine of some sort.) Ely had a true bell-ringing clock after 1371, but that does not tell us whether the Ely horologe of 1291 indicated the hours by bell or dial or both. The Norwich assembly of 1322 is the first recorded instance of something with a full complement of all that we could expect—an astronomical dial, automata (with fifty-nine images and a procession of monks), and bells of some sort, probably small. The clock itself was anything but small. To take just one detail from the financial records: the iron plate for the dial weighed 87 lbs and was probably five or six feet across. The St Albans clock was on the same grand scale, and it agrees with later evidence from many other centres—Strasbourg, Durham, Beauvais, and Rheims, for example. The very fact that the Norwich clock had so much large—and therefore heavy—hardware to move suggests a mechanical drive, not a water drive.

None of the evidence from church accounts settles the question of whether bell or display came first to the English churches. There is one line of reasoning which should certainly be dismissed immediately. It is that astronomers were not interested in the drive, since the rate at which an astronomical display unfolds is of no great moment to its dramatic content. This is to ignore the motives expressed by Robertus Anglicus and Pierre de Maricourt, not to mention the painstaking calculating undertaken by Richard of Wallingford and others before him. But still the question 'Bell or display?' remains strictly unanswerable. The St Albans clock need not have reflected a common ambition. The first working escapement would have been quickly expropriated by both parties. Simple does not always precede complex. Automated bell-ringing preceded the invention, and in later centuries many church clocks simply indicated the hours by a bell, and had no public dial whatsoever. (There would have been only a small dial on the clock frame for setting purposes.) It was in the richer centres, such as Norwich, St Paul's, and St Albans, that truly elaborate displays made their appearance, often with automata in addition to celestial display and clock. But that ultimately provides no answer. These places were rich enough to want and afford all they could buy.

As for non-astronomical automata, they too were nothing new. Like automated bells and planetaria, they too antedate the mechanical escapement. There are famous examples recorded in texts from antiquity and the Islamic world—model doves, moving peacocks, human figures, and other devices moved by water, steam and falling weights. In the tenth century, the Lombard diplomat Liuprand of Cremona visited the court of the Byzantine emperor in Constantinople and reported on the marvellous rising and falling throne of Solomon he had seen there, with automata in the form of singing birds, and lions that roared with open mouths as they beat their tails on the ground. By the thirteenth century, Europe was trying its hand at the same sort of thing. We have already heard Roger Bacon on the subject of the marvellous machines that lay in the future. His seemingly prophetic words were prompted by a new enthusiasm for marvels resembling those from the East. It comes as no surprise to find that he and Albertus Magnus both acquired shady and unmerited reputations for having constructed, respectively, a talking head and an iron man. Entirely authentic were the automata created in the palace at Hesdin (in the Pas de Calais) for the count of Artois in the late thirteenth century. Among other things, these devices simulated thunder, and covered the onlookers with water or flour, according to sex. Automata of a more devout nature—model saints appearing at windows, the Virgin and child, and other figures—were added to many early church horologes, where they still have the power to attract onlookers. They were not always confined to the display inside the church. Surmounting the bell-tower of San Gottardo in Milan (1330-36) there is an angel turned by clockwork. St Paul's in London (1344), and later the cathedrals of Canterbury and Chartres, had similar angels. That general idea was not new, for in the previous century Villard de Honnecourt had drawn a system of ropes and pulleys for turning an angel manually—perhaps secretly to inspire awe in the populace. In every century since then, clockwork-operated jacks (jacquemarts, mannikins) have been placed on the outside of buildings—galloping horsemen, bell ringers, and trumpeters, for example. Sometimes there has been a clock dial and sometimes not. Reference to the subtlety of the carving done at St Albans by Laurence of Stoke and William Walsham, in connection with the finishing of the work, need not have referred only to the clock case. If Norwich could have a fine display of moving figures, why not St Albans? There is, to be sure, no firm evidence that Richard of Wallingford planned any such thing. All the firm evidence we have points to an exercise more severely academic and puritanical—and 'without equal'. It would be a great mistake, however, to underestimate the human desire for theatre, of whatever kind, even if the desire was not Richard's.

29. Richard of Wallingford, with pastoral staff and abbot's mitre, pointing to his clock, high in the south transept of the abbey. He is depicted as suffering from leprosy in this manuscript illumination. From MS Cotton Nero D.7, fol. 20r. *(British Library)*

13

The St Albans Clock

AS OFTEN HAPPENS when academic spirits turn to administration, Richard of Wallingford found it to his liking. He was of that rarer breed, however, which refuses to allow its mind to lie fallow. As far as is known, none of the monks was capable of sharing his intellectual life in any depth. The chronicler could speak only fleetingly of this side of his character. 'He compiled many books on diverse sciences and arts, together with several astronomical and geometrical instruments that were unknown before his time.' A gloss added to the manuscript mentions the abbot's books on astronomy, geometry and other special sciences, 'in which he excelled above all others in his time', and then mentioned the albion and the meaning of the word, 'all by one'. Such things were not for an untrained bystander to judge. The writer knew enough to mention the liberal arts, the curriculum shared by all university men, but he felt compelled to refer also to the more visible mechanical arts, in which the abbot is said to have outstripped the greatest masters in knowledge and understanding. A great abbot could be builder and engineer, as well as scholar, lawyer, and administrator. The two small symbolic late medieval portraits we have of him show him in these different roles (Figs 1 and 29).

How effectively was this man, stricken with leprosy, able to pursue the 'liberal and mechanical arts' in the years left to him? We have seen that he revised and improved his chief mathematical treatise, *Quadripartitum*. He became familiar with, and used, the best astronomical tables of the time, those first prepared under the patronage of Alfonso X in the 1270s. They had been modified in certain respects by Parisian astronomers a few years before Richard was made abbot, but it would be another decade or two before they were in general use across the western academic world. The image of him in his study (Fig. 1), taken from the chief manuscript of the *Gesta abbatum*, shows him dividing a circular brass disc, something he must often have done during his Oxford years. Was the image anything more than a token of his former self? The mitre on the floor and the pastoral staff symbolise his rank, but was the artist—who almost certainly never saw him—painting nothing but symbols? The workbench, with set square and ruler, the dividers he holds in his right hand, and the desk in which an astronomical quadrant hangs, next to a cupboard for books, are all things which members of the convent would have seen for themselves. If the artist's imagination was not enough, then perhaps some of the older monks prompted the artist, or perhaps he was copying from another drawing, nearer to the scene itself. Whatever the truth of the matter, Richard of Wallingford did not betray his scientific past.

The Treatise

At some stage or other, Richard began to write a thoroughly academic treatise on the art of designing an astronomical clock. He began by setting out the theory of calculating the numbers of teeth, sizes, and relative angular speeds of wheels needed for what he called 'an astronomical horologe showing the movements of the planets'. In typical medieval fashion he illustrated the theory with elementary examples, but before long he was introducing very precise astronomical data into his examples, using the best available astronomical tables of the time, the *Alfonsine Tables*. Most of those who have occasion to mention his place in the history of the clock are inclined to ignore completely the highly original nature of this part of the work. It is as though one were to write a history of television and mention only boxes and screens. There are a few scattered precedents for his formal theory of wheel ratios, the chief of them from the first century of the Christian era, when the Alexandrian mathematician and inventor Heron wrote on the relation between the times of rotation of touching discs and meshing wheels. Richard of Wallingford, however, attacked these problems in a new and original way, owing much to the techniques he had learned as an astronomer.

To take a crude example: given a wheel that is made to go round once in a day, it requires a minimum of intelligence to produce another that goes round once in an hour. (One might, for instance, simply mesh a wheel of 240 teeth with a wheel of ten teeth.) Using the day wheel to drive another that goes round once every 27 days 19 hours 17 minutes 42 seconds (this would be useful for representing an aspect of the Moon's motion) is a problem of very different order of difficulty. A deprived modern child—forced to dispense with a pocket calculator—might reduce the two periods of time to seconds and look for a highest common factor. Another, instructed by a suitably ancient teacher, might use continued fractions. Richard of Wallingford devised a method requiring what he called 'tables of proportion' for his wheels. Their use was 'by inspection', but this should not give the impression that his techniques were lacking in mathematical ground rules. Far from it. He gives complicated rules, for example, to cover the cases where you do not find what you are seeking in the tables. And here we must repeat a point made earlier, about the enormous difficulties the medieval mathematician encountered when enunciating mathematical results without the concise algebraic notations available to us now. Twenty or thirty lines of Latin were often needed to express an idea that we can express in a single formula.

After sixteen propositions, the theoretical part of the planned treatise comes to an end. The practical part was plainly meant to follow on, but the abbot clockmaker did not live long enough to put this part of his house in order, which is why it is necessary to look into the state of the manuscripts that have passed down to us.

The Manuscripts

The St Albans clock has been routinely mentioned in most histories of the clock. It was known from the testimony of John Leland, whom we have already met in his capacity as Henry VIII's antiquary, that Richard of Wallingford had written down certain rules relating to it. It was written, said Leland, 'lest by a mistake of the monks such an extraordinary machine should become worthless, or by ignorance of its structure it should fall silent'. No such work was known to the assiduous St Albans historian of the 1790s, Peter Newcome. In 1956, the late Derek Price suggested that nine propositions in a manuscript in Gonville and Caius College, Cambridge, might be related to it. (They turn out to be a part of the theoretical section described above, and a St Albans connection is confirmed by certain theological material in that volume, although dating only from the late fifteenth century.) Another manuscript, later and even scrappier, is now in the Royal Library in Brussels, with just two and a half chapters of that same first part. For most purposes these fragments can be ignored, for there is another manuscript, one from which almost all of our evidence for the clock and its workings now derives. It is in the Bodleian Library in Oxford, MS 1796 in the largely scientific collection of books assembled by the seventeenth-century antiquary Elias Ashmole. I first recognised it for what it was in 1965, and a glance at its internal arrangement will make it easy to see why it had gone unnoticed for so long.

The volume is an unusually chubby book, small and thick and containing over 200 parchment leaves bound within leather-covered wooden boards. It contains more than a dozen distinct texts, all of them astronomical or astrological. More significant is the fact that it contains the texts of all of the scientific works commonly ascribed to Richard of Wallingford, with the exception of the long mathematical *Quadripartitum*. It was probably copied within fifteen years of Richard's death by a St Albans scribe, for in the margins of five different pages there are inscriptions recording the fact that the book belongs to the abbey. Indeed on four pages the inscription is more specific, telling us that the volume belonged to the subsacrist, brother John Loukyn. This too is noteworthy, since the sacrist, whose task it was to look after the sacred vessels and vestments, was also in charge of bells and bellringing, and so of clocks. (It is from 'sacristan' that we derive our word 'sexton'.) It is quite possible that this volume is what Leland had in mind when he mentioned a work written by Richard of Wallingford that was meant to guard against loss of information on the clock with the passage of time.

John Loukyn can probably be identified with a lay brother of that name, recorded in the *Gesta* in the late fourteenth century. He was probably not the first owner of the volume. A lay brother is unlikely to have had much use for the theoretical astronomy it contains, and the volume was beyond all doubt copied by someone who had been close to Richard of Wallingford, or at least to his legacy of writings. The texts are mostly fragmentary—fair copies, perhaps, of leaves he had copied for his own use. The astrology could have fallen within almost any monk's range of interests, but the other texts reveal an interest in theoretical astronomy and instrumentation. They include, for example, a work on the astrolabe ascribed (wrongly) to Massahalla, a work on the new quadrant, which I think was by the Merton astronomer John Maudith, and four pages on measuring the altitudes of stars. The last makes mention of a gnomon, 63 feet high. (Was it a feature of the church roof, the edge of the eaves, perhaps, that Richard used to measure altitudes accurately?) The most telling argument for the abbot's connection with the contents of the volume is that—in addition to the edited and well-polished theoretical chapters on clock design—it contains a jumbled and unedited set of drafts dealing with the design of the actual clock, as well as a few additions by an incompetent but ardent disciple.

The four relevant sections were not copied consecutively, but at separate places in the volume. Then, as if that were not enough, at some later stage in history the leaves were rebound in the wrong order, splitting up the contents still further. (They remain misbound, but there are clues enough for us mentally to rebind the leaves, once the basic fact has been recognised.) By the time we reach the third set of leaves on the clock, it becomes clear that we are dealing with two different drafts describing the same device, of-

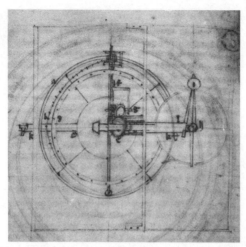

30. The variable-velocity drive in the St Albans clock, from MS Ashmole 1796, fol. 167v. The drawing uses a mixture of perspectives.

ten running parallel but occasionally introducing supplementary information. The third sequence of chapters actually begins to repeat itself, and the fourth does likewise, although here it is easy to be misled, when the text deals first with the design and then the assembly of the designed items. The overwhelming conclusion to be drawn from all this is that we are faced with a copy of parchments left in an incomplete and disordered state at Richard of Wallingford's death, and that if we disregard a very short and feeble attempt to imitate his style, the Ashmole manuscript presents us with the literary core of the work he was still engaged on at that time. It is a remarkable volume, despite its disorderly state, for it contains a small number of detailed mechanical drawings that would be hard to parallel from any earlier period. (Figs 30 and 31 reproduce the two most important.)

The Escapement

We are very fortunate to have these parallel drafts, for they show how Richard of Wallingford gradually improved his technique. By revealing his thought processes to us, the drafts also allow us to dismiss any suggestion that he was simply collecting together the ideas of his predecessors. There are two respects, however, in which his very brevity leads us to the conclusion that he was taking over one particular established device, and that is when he gives measurements for the component parts of the escapement and its bell-ringing counterpart—the two work in similar ways—without

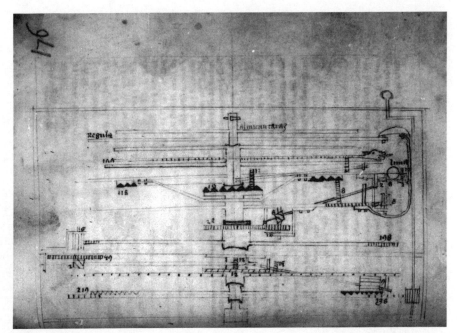

31. The astronomical trains of the St Albans clock. It would be difficult to find a more elaborate machine drawing earlier than this. It shows two of the four compartments of the vast clock frame. (From MS Ashmole 1796, fol. 176r.)

feeling the need to provide any very explicit description of them. In short, he leaves us with our old historical problem of deciding on their origins.

Despite the St Albans evidence, it is still widely supposed that the first mechanical escapement was what has long been known in English as the 'verge and foliot'. Relatively few are prepared to advance any argument for adhering to that old idea. Those who try to establish the history of the device on the basis of the very dubious etymologies of the two or three relevant words are unlikely to make any progress. A 'verge' is simply a rod, and all agree that the rod in question is the spindle—almost always vertical at this period of history—that turns first one way then the other. The word 'foliot' is often said to relate to the foolish dancing back and forth of the beam, or whatever was carried above the rod; but, whatever its meaning, no one suggests that it relates to that part of the device which pushes the spindle and beam back and forth, which is where the familiar escapement differs from Richard of Wallingford's. The phrase 'verge and foliot' can

32. The commonest form of early escapement, generally now known as the verge and foliot. The crown wheel is at the end of a train of wheels driven ultimately by a falling weight. As shown here, one of its teeth is pressing on the upper pallet (on the verge, the vertical rod) and forcing the foliot to turn, the left-hand weight swinging towards us. Eventually the lower pallet will be caught by an opposing tooth. The foliot will be brought to a standstill and its motion will eventually be reversed.

just as easily be applied to the one device as the other, although that fact is of no great consequence.[86]

The common escapement was made along the lines illustrated in Fig. 32. There is certainly no harm in describing it as a 'verge and foliot', although this terminology was given wide currency in English writings only after 1899, when F. J. Britten published his highly influential *Old Clocks and Watches*. He there drew attention to the fact that the word *foliot* was used for the governing mechanism in a clock in a fourteenth-century poem by Jean Froissart, *L'Orloge amoureus*. It is not without interest that Froissart, the renowned Brabant poet and historian, did not consider a mechanical allegory to be out of place in a group of love poems.[87] The clock plainly counted as something of a wonder, and it is now considered likely that

Froissart learned of its workings when passing through Paris in 1368, from Henri de Vick, who was then constructing a clock for the royal palace of Charles V. Unaware of the St Albans alternative, Britten assumed that the escapement was in the form illustrated in my Fig. 32, but Froissart's reference to the clockmaker setting the foliot, the spindle, and 'likewise all the pins', fits far better with the idea of an escapement of the St Albans type, as we shall see. The common sort usually has teeth that are never likely to have been called pins—although some small and crude wooden clocks did later have pins in their place.

The better-known sort of escapement survived well into the seventeenth century, and even beyond, but it was eventually superseded by a succession of pendulum-controlled escapements. The first known depictions of the common escapement are in the manuscripts describing the so-called *astrarium* of Giovanni de' Dondi (Fig. 33), where the foliot cross-bar is trivially replaced by a crown. Assuming that this depiction represented Dondi's actual machine, his reasons for using a crown need not have been entirely aesthetic. It is conceivable that he was making use of a redundant wheel from an earlier escapement—perhaps based on one that had been imperfectly made. He would have needed to add weights to it, to adjust its time of swing. No hint of the fact is given, but it would have spoiled the appearance, of course.

The method of functioning of what eventually became the common early type of escapement (with opposed pallets) is easily understood with the help of a diagram. As drawn in Fig. 32, the crown wheel is to be imagined turning anti-clockwise, at the end of a train of wheels ultimately driven by a weight suspended from a rope wrapped round a barrel, in the familiar way. The verge has two metal plates (pallets) attached to it. As the upper pallet is pushed by the crown wheel, the cross-beam (foliot) rotates, bringing the left-hand end nearer to the viewer. There comes a point at which the tooth of the wheel slips past the upper pallet entirely, but at that moment the lower pallet is caught by a tooth of the crown wheel below, so bringing the verge and its foliot to a standstill before reversing their motion. Eventually the upper pallet is caught by another tooth on the crown wheel, and after this complete cycle the whole process is repeated, for as long as the driving train supplies the necessary power. The rate at which the foliot turns is markedly dependent on its moment of inertia and on the driving force. The moment of inertia of the foliot can be changed by adjusting the positions of the weights at its ends. In these respects the device is quite unlike a pendulum, which swings with a natural period that depends (ideally) on the pendulum itself, and not on the driving force.

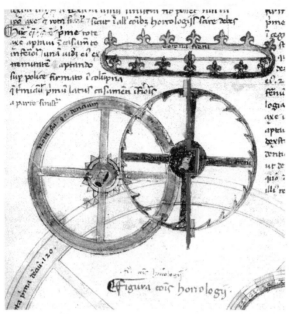

33. The earliest known drawing of what became the common verge and foliot escapement—the oscillating crown excepted. From the principal manuscript of Giovanni de' Dondi's treatise on the *astrarium* in MS 631, Biblioteca Civica, Padua, fol. 13r.

The oscillating cross-bar on Richard of Wallingford's clock would have been essentially no different from that on the device just described, and for the sake of clarity is not shown in our drawing of the overall arrangement of his escapement. This is shown in Fig. 34, in which the two wheels, fixed firmly to the same axle, turn together, being again driven round at the end of a train of gears, the driving force being provided by the ordinary falling weight arrangement. We cannot say exactly what form the cross-bar took. From a mechanical point of view its precise form matters little, as long as it is symmetrical around the axis of the verge. Richard of Wallingford refers to a *quadratura plumborum*, a 'quadrature of lead weights', presumably something of rectangular shape carrying lead weights. (*Quadratura* was a word used in the middle ages for rectangular stone building blocks, but that is of little help.)

The regular reversals in the motion of the verge and the lead weights it carried back and forth were achieved by two sets of pins acting on a block of metal of roughly semicircular form, and fixed to the verge. Richard of Wallingford calls that block a *semicirculus*, 'semicircle' and the verge a

34. The main components of the St Albans escapement (*strob*). The uppermost bar (*quadratura plumborum*) is here drawn without its adjustable lead weights, to avoid obscuring the rest of the figure. The horizontal bar between the two pin-wheels is fixed to the clock frame. Its only purpose is to support the verge (shaft, *hasta strob*) from below. The engaging of the pins with the semicircular double pallet (*semicirculus*) alternates between the two wheels, as shown in the next figure and as explained in the text.

hasta strob, 'strob shaft'. A sequence of the actions of the semicirculus is shown in Fig. 35. For each drawing of it, the pins in its neighbourhood are to be considered to move away from the viewer. The way in which the pins give an oscillatory motion to the verge should be obvious, but it is explained in the caption to the figure. By suitably shaping the block, this type of device can be made to operate very efficiently indeed. It has certain advantages over the more familiar device, the chief of them being that inserting pins accurately and at uniform spacing into a wheel is very much easier

35. An illustration of the oscillatory motion of the verge caused by the pins on the twin escape wheels acting on alternate pallets of the *semicirculus*. One should imagine the pins to be moving away from the observer. The furthest frame of the sequence shows the verge being turned counter-clockwise (looking down) as the pin on the right-hand wheel pushes the pallet away. In the third frame that pin has just slipped off the pallet, and a pin on the left-hand wheel is about to catch the opposite pallet. Pushing this pallet away, it brings the turning verge to a standstill and then changes its direction to clockwise. This continues until in the sixth frame that pin too slips off its pallet. The next pin on the other wheel now takes over, and the verge begins to turn counter-clockwise. The *semicirculus* would not have been an exact semicircle. The centre of rotation may be placed slightly forward of its centre; or, alternatively, its pallets may be slightly blunted.

than cutting a set of teeth on a crown wheel so that all are of exactly the same form and size. On the semicirculus, the only curves to be matched are those of its two halves, its pallets, as we might call them. Another advantage is the smoothness of its action, which for some strange reason is often flatly denied. Of course the skill of the maker is a material factor.

Dondi's astrarium, which uses the common device with pallets at opposite ends of a diameter of the wheel, was a visually magnificent planetarium, driven by clockwork. The mechanism was housed in a seven-sided brass frame, with each planet moving round a scale on a separate face, and other dials and scales below. It is often bracketed with the St Albans clock, but it was only begun in 1365, nearly thirty years after Richard of Wallingford's death, and—like the basic manuscript describing it—was not completed until 1380. It is a sad fact that we seem to have only one slender piece of evidence for the type of escapement in use anywhere during that interval of over forty years—from the poem by Froissart mentioned earlier—although Dondi writes as though his escapement would have been familiar to his readers. Did his version, which was to become so common, also antedate Richard of Wallingford's? Dondi's assumption of familiarity tells us absolutely nothing of relevance to this question; and in any case, Richard of Wallingford made a similar assumption of familiarity. It is hard to believe that the two inventions were independent, for they share so much. If they are not, what reason can we offer for dating one before the other?

The Order of Invention

While I can see no irrefutable argument either way, it seems to me likely that what became the common type of escapement was inspired by the St Albans type, rather than the reverse. Since both make use of the forced oscillation back and forth of a bar, either could have been inspired by a pre-existing bell-ringing mechanism, that is, one in which a falling weight creates a rhythmic striking of a hammer on a bell. As we shall see shortly, Richard of Wallingford describes a bell-ringing mechanism which is closely related to his escapement, while a belfry clock in Nuremberg from a later period rang a bell with a hammer on the cross-bar (foliot) of a mechanism of the commoner type. (In both cases, of course, the striking is separate from the escapement which controls the timekeeping side of the clock.)

In favour of the idea that a simple bell-ringing mechanism of the semicirculus type came first is the fact that it could have been more crudely made. Precision in the pivoting of the strob shaft would have been relatively unimportant, and a firm pivot between the two pin-wheels would not have been strictly necessary. The strob shaft could have ended at the semicircular block and have been suspended from a leather strap (perhaps this explains the word 'strob'), the shaft being steadied not from below but from the side. The pins might have been hammered into a simple wooden

36. The strob mechanism, used not as an escapement but to ring a bell repeatedly.

drum, rather than inserted into two separate wheels (Fig. 36). It is easy to imagine that such a crude bell-ringing device, with a falling-weight drive left to run unchecked, suggested itself as the first mechanical escapement.

It is quite true that if the other type of mechanism (with opposed pallets) came first, as a bell-ringing device, that too could have suggested itself for use as an escapement, but there are other reasons for accepting the priority of the St Albans device. Its action seems inherently more likely to suggest itself. This is not simply a case of vague intuition: people find it easier to envisage the working of the semicirculus in the absence of a model, and easier to understand in the presence of one.

Some of those who consider that St Albans double-pallet escapement to have been an independent invention, or an offshoot from the now more familiar form, occasionally try to argue from its failure to survive, at least in large numbers. This is no argument at all. If the double pallet escapement came first, but the familiar form proved to be the preferred form, then the former would have failed to survive on that account. The common device with opposed pallets would probably have been preferred on the grounds

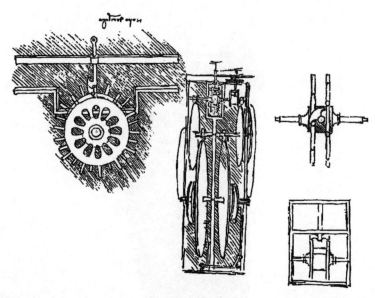

37. Drawings by Leonardo da Vinci of an escapement of the St Albans type. After Codex Atlanticus, now in the Biblioteca Ambrosiana, Milan.

that it is more economical in metal. This has nothing to do with mechanical merit, for many an excellent invention has been the victim of economic pressure. In any case, we know that the double pallet did survive, for along with the escapement with opposed pallets it was described briefly and drawn many times by none other than Leonardo da Vinci. There is a work of the late fourteenth or early fifteenth century describing an escapement with a double wheel which is in some way related to it.[88] And we recall Froissart's reference to pins rather than teeth, in his poem *L'Orloge amoureus*. His words are not at all suggestive of a clever provincial variant of some standard form.

We find different forms of the double-pallet device at various points in those volumes of Leonardo's bundled papers now known as the Codex Atlanticus and Codex Madrid.[89] Most of the relevant pages probably date from the last decade of the fifteenth century, when Leonardo was working for Ludovico Sforza, duke of Milan. Various attempts have been made to identify some suitably famous clock as his source. He reproduces a whole variety of designs, and not only one, which if anything seems to lessen the likelihood that he was depicting actual machines. The most relevant sketches are in the Codex Atlanticus (Fig. 37), but scattered through the pages of the Codex Madrid I there are ten mechanisms in which it is easy to see connections with one escapement or the other—with double pallet or opposed pallets—and they are equally divided between the two.[90] As for the broad principle of design, however, the way in which we are to fill in the gap between Milan and St Albans is entirely unknown to us. The Leonardo evidence does not prove that the device with double pallet came first, but it is an embarrassment to those who wish to relegate it to the shadows. It has even been said that a link between St Albans and Milan need not be sought, since there is no textual evidence for the reconstruction I offered for the St Albans escapement in the late 1960s. Since I had made a working model based on the Ashmole manuscript before learning of the Leonardo drawings, with some of which it conformed very closely, and since I do not lay claim to second sight, I am not persuaded by that particular argument.[91] Indeed, if Richard of Wallingford's and Leonardo da Vinci's double-pallet escapements were intrinsically different, we should be in the even more embarrassing position of having the relations between three different types to explain.

The St Albans Striking Mechanism

At two points in Richard of Wallingford's drafts for a treatise on his horologe he provides information on the striking mechanism of his clock, giving numbers of teeth on the relevant wheels, and some measurements. We can be in no doubt that the oscillatory action needed is achieved in essentially the same way as that of the escapement, but again some things are taken for granted, and these we can surely assume were well established in his day. Without going into the intricacies of later striking work, one or two general principles are worth mentioning, since they remained remarkably constant, apart from the choice of twelve or twenty-four strokes in the clock's 'hour-striking'. (This is a technical term, as will be explained.) There was a wheel in the main St Albans clock which turned once every twenty-four hours and which would have had twenty-four pins around its side, to trip the striking mechanism once every hour. Clockmakers refer to this as 'letting off the striking work'. It is usually done by having the pins move a pivoted lever, the other end of which in general prevents the striking work from moving—for example by falling into a slot in one of the wheels. When this 'locking lever' is moved, it disengages for a time from whatever device it is locking (in later clocks usually a slotted disc), allowing the hammer mechanism to strike the bell a suitable number of times before the lever falls back into a locking position once more. In the simplest case the bell will be struck just once before the lever locks again. At St Albans the mechanism was more advanced.

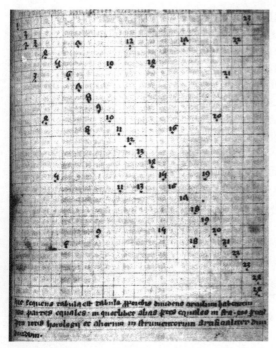

38. Richard of Wallingford's table ('figure') for the spacing of the pegs on the barrel in his mechanism for hour-striking. From MS Ashmole 1796, fol. 181r.

Richard of Wallingford mentions the use of a spring to shut off the strike, but the passage in which he explains his striking mechanism was at first hard to understand. Part of the difficulty was due to the fact that the manuscript is misbound at the crucial point. Richard tells us to look for a 'figure' on the next page, and there seemed to be none that would answer to what was needed, anywhere in the manuscript volume. The 'figure' turned out to be a table of numbers, nearly two hundred pages distant. It embodies one of the manuscript's greatest surprises, a device which we very probably owe to the abbot himself, since the table is set out in a way that would have come naturally to an astronomer, but not to an ordinary clockmaker.

It seems intuitively obvious that the first mechanical clocks which struck the hours on a bell would have produced either a single stroke every hour, if the bell was large, or a short spell of ringing a group of small bells, perhaps tuned bells, such as are occasionally illustrated in association with water-clocks. The custom of striking a large bell with the number of strokes indicating the hour—eleven at eleven o'clock, and so forth—is now familiar to everyone. It obviously requires a new mechanism, over and above those for letting off the strike and moving the bell-hammer (or tipping the bell), to count out the right number of strokes. For very many years, following a statement by a certain fifteenth-century monk of Malmesbury, it was thought that hour-striking, as this is called, originated in the year 1373. Eventually historians had their attention drawn to a statement in a chronicle of Galvano Fiamma—first published in 1723—which seemed to sug-

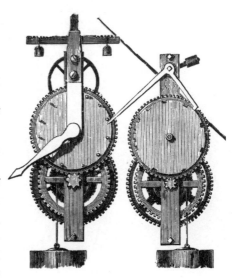

39. A nineteenth-century drawing (supposed to be of the fourteenth-century striking clock built by Henri de Vick for the French king). It is used here to show how, at the hour, a pin in the wheel carrying an hour hand pushes on the L-shaped lever and lifts a pallet out of a slot on the striking mechanism on the right. The falling weight of the strike drives the mechanism sounding the clock bell a number of times proportional to the interval between slots. The slots are so arranged that the number of strokes corresponds to the hour. (This is so-called 'hour-striking', here on a twelve hour system.)

gest that a clock installed in 1336 in the tower of San Gottardo in Milan had hour-striking. Again this was on the 24-hour system alluded to by the monk of Malmesbury, rather than the system of 12 + 12 hours with which we are all familiar today. But what is absolutely clear from the St Albans manuscript is that the clock described by Richard of Wallingford—who died in the year of the Milan installation—already had hour-striking.

The mechanism 'for sounding a bell' (*pro sonitu unius clok*) according to the number of the hour, as described by Richard of Wallingford, is interesting not only because we know of nothing older, but because—with one possible exception—it seems to be of an otherwise unknown type. Whereas the lever that locks and unlocks the mechanism for striking the bell usually does so by engaging with slots on a circular plate (Fig. 39), Richard of Wallingford uses a barrel with pegs inserted into it at suitable intervals. The 'figure' alluded to earlier is simply a table giving the spacing of those pegs. One peg releases the strike, that is allows the bell-ringing 'strob' mechanism to operate, while the next peg stops it, after a number of strokes proportional to the spacing of the pegs (Fig. 40). The number of strokes needed for twenty-four hours (1, 2, 3, ... 24) is 300. If the barrel were to have had all the pegs in a single ring, great precision would have been needed. Richard of Wallingford avoided this need by putting the

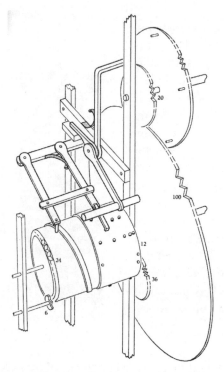

40. The St Albans mechanism for hour-striking. The counting out of the strokes is here proportional to the interval between pegs on the barrel. How the barrel is scanned is conjectural. Another possibility is that the lifting pin followed a spiral groove in which the pegs were set. The general principle is not in doubt.

pegs on a spiral track around the barrel. He probably arranged for a second barrel to operate some sort of tracking mechanism. Ours at least is one possible solution to this tracking problem that works well. In some respects it is conjectural, but even if it is not exactly like the original, there can be little doubt about the general technique of scanning a drum, which is ingenious and apparently original. It is encapsulated in the 'figure' for the placement of the pegs (Fig. 38), which can be imagined as a plan to be wrapped round the barrel.

Later tune-playing carillons were often activated by tripping the hammers on a series of tuned bells in much the way that the pins on the drum in a musical box pluck the teeth of the comb to make the sounds. A chiming clock, one capable of playing a snatch of a tune on the hour, often worked in this way, using an additional chiming-barrel (not to be confused with that for the strike). Despite its apparent resemblance to them, the St Albans clock-striking mechanism was very different: it was a counting device, of great sophistication, and much more intricate than carillon-type devices for playing sequences of notes.

Perhaps the only early hour-striking mechanism of which any trace survives, and which uses an arrangement at all comparable to the one here explained, is yet again to be found in a drawing by Leonardo da Vinci in Codex Madrid I. This shows an inherently simple arrangement, despite first appearances to the contrary (Fig. 41). Imagine that a large screw is enclosed in a tube, and that a rope fixed to one end of the screw and coiled round it comes out of a hole in the side (not the end) of the tube. Pulling the thread will shift the screw along the tube. Suppose now that part of the tube is cut away, so that the end of a pivoted lever can press into the screw thread at just one point; and that the screw thread is slotted at intervals. Such slots will be the equivalent of Richard of Wallingford's pegs. By falling into a slot the lever will be capable of locking the screw, and when pushed out, the lever will allow the screw to turn through a suitable angle, that is, until it encounters another slot. The screw will be turned in the usual way, by a freely falling weight attached to the rope.

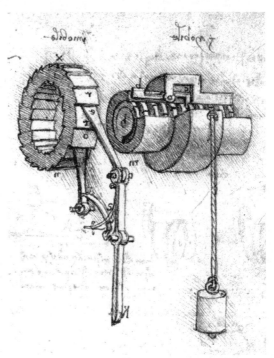

41. A drawing by Leonardo da Vinci of a mechanism for hour-striking. This clearly follows the same general principles as those of Richard of Wallingford's clock. (From Codex Madrid I, fol. 12r.)

Leonardo's counting mechanism has certain mechanical drawbacks, chief of which are friction on the rope and the difficulty of returning the mechanism to its initial state at the end of twenty-four hours, but for present purposes it is the use of a spiral, so reminiscent of the St Albans device,

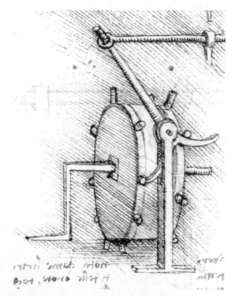

42. A drawing by Leonardo da Vinci of a mechanism of the general St Albans type for producing a shuttling motion, such as might have been used to ring a bell. (From Codex Madrid I, fol. 7r.)

that is especially interesting. Strangely enough, the briefly worded explanation Leonardo offers for his device relates only to a separate part of the mechanism, for releasing the locking lever once every hour, that is, for lifting it out of a slot.[92] It is not necessary to discuss this here, or to consider the oscillatory striking device Leonardo had in mind. Whatever it was, it is neither described nor drawn on the page in question; but there is, in the same volume, one 'escapement' in particular which would have served excellently, for it was clearly meant to provide an oscillatory straight-line motion (see Fig. 42.) It is of the St Albans type.

The similarities between the Leonardo spiral and Richard of Wallingford's description, together with the presence of a double-wheel and double-pallet escapement in many of Leonardo's drawings, all go to reinforce our earlier conclusions about the diffusion of the escapement of St Albans type. Richard's clock incorporated designs that were never common, but neither were they peculiar to some shady backwater. We now have two fine mechanisms which seem to have spread to Italy, if they did not come from there, and which still did not prevail. It is hard to believe think that any sceptic will insist on independent invention; but if the examples mentioned here are not independent, then the all too common argument that the St Albans designs were isolated inventions is plainly false.

Developments in Italy

Hour-striking was of enormous social significance, when the bell could be heard by the inhabitants of large towns, for it allowed them to keep track of the equal hours of the day, and so brought the rigid ordering of the European labouring classes one stage nearer. Had the autocratic abbot of St Albans lived to see this dramatic social change, he would doubtless have approved of any part he might have had in it, but of course the sound of the St Albans clock bell would have had a very limited range. In Italy, the organisation of urban society ran along very different lines, and some have been tempted to use the Italian case and invert the argument, suggesting that the invention of the mechanical clocks was an answer to a demand for closer social control. Much has been made in the past of the San Gottardo reference, which has been taken to imply that hour-striking was the invention of an unknown Milanese craftsman—or at least that there is no better candidate for the title—and that it was an answer to the pressing needs of the citizenry of Italian towns.

There is more at stake here than a trivial dispute over priority. Claiming priority for the Italian device has been used to give to Italy most of the credit for the revolution in mechanical timekeeping that unfolded in the years around the end of the thirteenth century. The argument put forward by one of the best of modern writers on these matters, Gerhard Dohrn-van Rossum, begins with a discussion of a change in the public consciousness. There is no doubt that in the course of the fourteenth century ordinary people began to distinguish between the hour of the day as judged in traditional ways and regular mechanical time, clock time. This change, he suggests, required public clocks—although he defines 'public' generously, and allows it to refer to any large group of people, even a monastic community. He then produces a predominantly Italian list of public clocks, civic tower clocks, where 'public' is therefore now taken in a different, and narrower, sense.

The Italian list is certainly impressive for the latter half of the fourteenth century, but not for the period before Richard of Wallingford's death. A repair to an *ariologium* in Orvieto (1307-8), and mention of difficulties of keeping Italian hours there, might or might not point to a mechanical clock. A Modena (1309?) reference might be only to a bell, albeit with a civic function. Parma (1317) had a bell to time the hours for labourers; if originally it was not operated by a clock, it does seem to have been so operated by 1336. Ragusa (1322), modern Dubrovnik, employed a clock-keeper. Valenciennes (France, 1325) signalled the opening and closing of markets by an *orloge*—it has other names, and probably referred to a mechanical clock, just as the other vague references might have done. If so, then it is probable that in the psychology of ordinary people time was beginning to be regarded in a new way, as something proceeding at a regular pace; and that it was doing so in those northern Italian more rapidly than elsewhere in Europe, especially in the later century. However, in view of the oddity of the Italian start to the day, which changes with sunset, even this idea should be treated with care.[93] Old habits died hard. The clock on the tower of the Palazzo Vecchio at Florence was repaired in 1512, and not until then was it altered so as to indicate the hours in our present way, twelve before and twelve after midday. The 'new' way was described at the time as the French style.

Dohrn-van Rossum's argument, designed to put Italian technology at centre-stage, now turns in a new direction. He justifies his decision to exclude from his list (taken up to 1360) clocks with automata and astronomical indications, on the grounds that we nowhere find 'time-indications' associated with the latter—by which he presumably means a bell sounding the hours.[94] He states baldly that 'early striking clocks are so far attested only in the urban sphere', and he is far from being alone in this. He quotes Jacques Le Goff and several others to the same effect, and another writer, Werner Sombart, who even goes so far as to suggest that the absence of church striking clocks was the result of active resistance. This is nonsensical. The St Albans clock had this, and so, surely, did many earlier English and French devices in churches, albeit on a smaller, more intimate, scale. It may be worth noting that when Richard of Wallingford invited his subdued villeins to partake of spice-bread, they were to return at either the first or the third hour. It would be interesting to discover how they knew the hour.[95]

The conclusion drawn by Dohrn-van Rossum from his list of 'public clocks' is that they 'and the modern system of hour-reckoning originated in the Italian cities'. If this is only a statement about mass consciousness then it is harmless enough, for it would be a quibble to point out that equal hour reckoning was known to every university educated man of the middle ages, or that reckoning from sunset is not exactly modern. He goes on further, however, to claim that 'the technology, presumably invented in Italy, was at first exported to other countries only by Italian technicians'. It may well be the case that Italian clock builders quickly rose to pre-eminence, for there was a powerful craft tradition there, and we know that there were Italian tower-clock builders working in Avignon, Perpignan, and England after Richard of Wallingford's time. He is able to point out that Bohemia, Austria and Poland for a time used Italian hour-reckoning, which goes to strengthen the point about the power of that developing Italian tradition, but that does not legitimise his 'presumably invented in Italy'. The mistake is to speak of 'the technology' as though it was something peculiar to public clocks in his narrower sense. It was not. The technology of the bell-ringing clock was by and large a more *primitive* form of that used for clocks with a visual display. The only new problem to be solved, if it had not been solved earlier, was that of tripping the hammer of a large bell.

This last point brings us back at last to the tower clock of San Gottardo in Milan, and Galvano Fiamma's mention of hour-striking. Galvano was a Dominican friar in the service of Azzo Visconti, duke of Milan, and it is unfortunate that his account is unduly chauvinistic, as well as brief and careless. Under the year 1336 he relates that there were many bells in the tower and an admirable clock, '[admirable] because it is a very large *tyntinnabulum*, [and] it strikes one bell (*campana*)...' with one stroke at the first hour, two at the second, and so forth up to twenty-four. The clock was, he said, absolutely indispensable for people in all stations of life.[96]

This is an important text, and we should try to make some sense of it. Taking the words at face value, the Latin most readily suggests that a very large *tyntinnabulum* (normally a small bell) struck the main bell. This is bizarre. It is scarcely less strange to do as some have done, and take *tyntinnabulum*, a word denoting a cow-bell or something of that size, to mean a bell clapper. Does it simply mean that the clock is remarkable because the noise it makes, the tintinnabulation, is very great? Whether or not this was the intention, it does not need to be the *tyntinnabulum* which strikes the one bell, since that which (Latin *quod*) strikes the bell could have been meant to be the *horologium*, the machine taken as a whole.

However we read Galvano's opening passage, there is no doubt about the ringing of a bell with a number of strokes corresponding to the hour. But this is not the end of our questions. Are we to assume that the system of Italian hours (counted from sunset, or a short time thereafter) was still in use? If so, since the time of sunset changes with the cycle of the seasons, either a mechanism for adjusting the start of the day was called for, or such a change was effected manually. Perhaps the clock was readjusted by hand every sunset. There is no doubt that throughout Christendom—possibly earlier, but certainly by the fourteenth century—a bell was commonly rung at sunset, the so-called *Angelus*, and that all who heard it were exhorted to say three Hail Marys (Ave Marias). There would have been nothing new in having a bell-ringer on duty at that hour, and he might have done the setting.

As a subsidiary question: was the first (single) stroke in the sequence of twenty-four hours given at (or around) the time of sunset? If so, this would not have been the usual system of Italian hours. If the first stroke was placed conventionally at the *end* of the first of those twenty-four hours, then sunset would have been marked by twenty-four strokes—an inescapable Angelus indeed.

Fiamma's remarks are a very slender foundation on which to build any historical edifice, and they do nothing to counter the claim that the earliest hard evidence for hour-striking is that from Richard of Wallingford's treatise, where the many details of the mechanism are painstakingly set out. How likely is it that there was a connection between the English clockmakers and those in northern Italy? We do not have to suppose that the influence travelled from England to Italy, for the two earliest known clocks with hour-striking could have had a common ancestor anywhere in Europe. We cannot say with any certainty that the idea travelled southwards, but the quasi-astronomical style of the 'figure' of the counting device tends to favour the idea that it did so, having originated with the astronomer-abbot of St Albans. The mid-fourteenth century was a period of feverish horological activity across the face of Europe, and technological ideas would have travelled fast, both then and later. (Other records show hour-striking to have also been in use in Padua in 1344. We recall that there was hour-striking in Glastonbury in 1373, and it is possible that it had even reached Moscow by 1404.) Leonardo da Vinci's drawings of the St Albans type of double-wheel escapement, and of his counting mechanism for the strike making use of a scanned spiral, were neither of them of the type finally adopted throughout most of Europe. This fact merely serves to strengthen our feeling that there was a real link.

Richard of Wallingford as Engineer

The St Albans clock was of iron. While the abbot would have directed the labours of the smiths, and have viewed their work with the informed eye of a blacksmith's son rather than wielded a hammer himself, we need only look at the wonderfully intricate thirteenth-century iron grating that still stands in the south aisle of the presbytery to appreciate the quality of the work he could command (Fig. 43). Since there was already a class of artisan going under the name of *horologiarius*, there is no reason for thinking that Richard of Wallingford needed to call on the services of millwrights for this particular task—although, as we know, he made much use of them for the abbey mills.

The millwright's was a much more securely established craft. Medieval millwrights were men who needed many skills, to build both the mill housing and the machinery, and they became masters in the design of trains of gears for transmitting power. There are very few points of resemblance between the trains of gears in a mill and those in a clock, however, and those are superficial. In a clock, the ratios between the numbers of teeth on the wheels need to be carefully calculated, whereas precise ratios

43. A remarkable example of thirteenth-century ironwork, in the south aisle of the St Albans presbytery. It now fronts the chantry chapel of Humfrey, duke of Gloucester (1391-1447).

are of little consequence in a mill. Even in the seventeenth century, mill gears were still usually crude wooden affairs—for instance using wooden pegs to mesh into a lantern gear, as shown in Fig. 44. No millwright needed to create gears of such refinement as the contrate wheel with spiral grooves which Richard of Wallingford designed to produce a smooth and continuous drive (Fig. 45). (A contrate gear is one with teeth at right angles to its plane.) No millwright would have had any use whatsoever for the transported train of gears by which the Moon's motion was adjusted with extraordinary accuracy, wheels which will be explained more fully in due course.[97] We are simply not in the world of millwrights, although it is conceivable that they were brought in to help—for instance to make wooden patterns for the ironwork, or to create the wooden barrels.

It is right to admire the extraordinary mathematical accuracy of the wheel ratios in Richard's clock, but in doing so we should not overlook Richard's command of engineering principles of a much higher order than those needed by millwrights. This is not to disparage their work: Richard might have benefited from their accumulated expertise, especially in the impor-

44. A lantern gear meshing with a pegged wheel, both of wood, and typical of many different sorts of medieval and early modern mill machinery. A detail from Georgius Agricola, *De re metallica* (Basel, 1556).

tant area of reducing the sliding friction in bearings. It seems to me likely that the St Albans mechanism used brass or bronze bearings to carry its wrought iron axles, although there is no firm evidence for this.[98] We know very little about the best engineering practices of the middle ages, but the spiral contrate wheel deserves more than a passing reference. We cannot be sure of the nature of the pinion which engaged with it, but to engage with spirals its teeth must at the very least have been cut slantwise. For a reasonably thin gear, this would have been enough, but if the pinion was cut from a thick chunk of metal, its teeth might conceivably have been spiroid. It is very tempting to describe the gear and pinion system as a member of the modern general class of bevel gears.[99] We need not go to such lengths, however, to appreciate the simple fact that their chief advantages are much the same: they make for a smooth drive of high efficiency between non-parallel shafts, especially shafts at right angles.

Richard of Wallingford used helical gears (pinions) elsewhere in his clock, for the same reason as they are used today, to transmit power between non-intersecting shafts inclined at an angle. It is interesting to see that he used them only where the power to be transmitted was small, as in the turning of a Moon globe. From experience he would have discovered that their sliding motion introduces more friction than was justifiable in the central trains of the clock.

45. The spiral contrate wheel in Richard of Wallingford's clock. This was meshed with a pinion in a form about which we can only conjecture. If short, it might have resembled a simple cog with slantwise cut teeth. If long, the teeth might have been more obviously spiroid.

The Building of the Clock

From the two styles of gearing which requre the cutting of spiral teeth, we have clear fourteenth-century evidence for techniques unknown from an earlier date. They remind us of how little we do in fact know about past engineering. There are some early clocks that are known to history only by mention, or by a few sentences, as in the case of the clocks at Dunstable, Exeter, St Paul's in London, Ely, and Canterbury—to take nothing later than the thirteenth century. The earliest medieval English smith known by name, whose works and contract survive, was Thomas of Leighton, who flourished in the 1290s and who may have been somehow connected with nearby Dunstable Priory. His skill is evident from the grille he made for the tomb of Queen Eleanor in Westminster Abbey—witness its die-stamped foliage—but whether he was ever commissioned to make clockwork is only a matter for speculation.[100] There are, however, clocks known through moderately detailed records of payments made for the materials and labour needed to build them. Such was the clock with large astronomical dial on which Roger and Laurence of Stoke and a certain Robert worked for the Norwich Cathedral Priory, for which payments were made to blacksmiths, carpenters, masons and plasterers, ropemakers and bell casters, a woodcarver and painters, and workers in brass and copper. In a very few instances early records have survived which, while not describing the clock in detail, give some idea of its working and of the human industry demanded of those responsible for it. One such collection of records tells of the building of a clock at Perpignan, capital of the Pyrenean province of Roussillon, at a time when it was ruled by a branch of the family of the kings of Aragon.[101]

Those Perpignan records confirm the impression already given by the St Albans chronicler as to the high cost of any such project, and the elevated status of the master clockmaker. They also hint at the excitement of all who witnessed the enterprise, from without as well as within. The records were kept by the steward, Ramon Sans, who in 1356 was commissioned by King Pere IV to recruit and pay workers and purchase materials. The project was considerably simpler than its St Albans counterpart, and was completed in nine months, but still it cost 1667 livres, the equivalent of perhaps £400 sterling—a sum greater than the entire annual income of most monasteries, and about half that of St Albans.[102] The master clockmaker was Antonio Bovelli, *plombarius* to Pope Innocent VI in Avignon. Bovelli's title suggests that he had worked in lead, laying the pipes and roofing of the papal buildings, but his responsibilities must have covered metal-working and engineering more generally. He brought ten clockmakers with him from Avignon. Like the specialist tools he needed, he paid for them out of his own considerable fees, which amounted to more than a sixth of the total cost.

The scale of operations at St Albans would have been roughly comparable with that at Perpignan, although the abbey clock had not been designed for public service and called for no clock tower. Bovelli was assigned rooms in Perpignan castle. He fenced off much of the courtyard, where he built furnaces and had smithies constructed on site—anvils, bellows and all. Raw iron was purchased, in bar form and in blooms, and was hammered to copy the shapes of wooden patterns created by carpenters under Bovelli's guidance. The clock frame weighed nearly a ton, and the iron for it cost over 34 livres. Raised by a crane into the room prepared for the clock in the tower, the parts were assembled there. The driving weights of about half the weight of the clock were cast in lead and hammered into frames, a much simpler matter than the casting, raising, and hanging of the bell, which weighed about three tons.

Such was the public interest in the venture that Bovelli had to organise his carpenters to build scaffolding to regiment the crowds. When the bell was finally cast, the king, his court, and all who had worked at the project—the cooks perhaps excepted—rejoiced with a feast. Coaxing the clock into working as Bovelli intended was a less appealing process, and one that seems to have dragged on without ever having been satisfactorily achieved. It seems to have been a constant cause of concern, and within thirty years the bell was being sounded by paid ringers, equipped with hour-glasses. We have already seen reason to think that its St Albans precursor was much more successful, but few of those who marvelled at that would have given much thought to its accuracy, or even have had a clear idea of what that entailed. Accuracy was not the quality they looked for in the clock. To them it was more than a mere timepiece.

46. A general view of the reproduction of the workings of the St Albans clock, made for the Time Museum, Rockford.

14

Machina Mundi

> This world indeed is a great horologe to itself,
> and is continually numbering out its own age.
>
> John Smith (d. 1652), *Select Discourses,*
> ed. J. Worthington (London, 1660), p. 142

THE ROMAN POET Lucretius, in his great poem on the nature of the universe, *De rerum natura*, was fond of likening the world to a machine, an entity with structure. Indeed, he thought it was a machine that would eventually disintegrate, in a way that would prove hard to fit into Christian thinking, but his image, and the phrase *machina mundi*, 'the machine of the world', caught the medieval imagination and was often repeated.[103] The great Oxford scholar Robert Grosseteste used it no fewer than three times in the first six sentences of his elementary work of astronomy, 'On the Sphere'. Sacrobosco chose to end his still more popular treatise on the same subject with a sentence in which the same phrase occurs. He claimed to be quoting Dionysius the Areopagite—the man converted by St Paul—on the unnatural eclipse of the Sun during Christ's Passion: 'Either the God of Nature suffers, or the machine of the universe is dissolved'. It is evident that the very notion of a machine carried with it the idea of a potent but unseen rationality. It is no accident that with the advent of the mechanical clock the conviction grew. The audible and visual qualities of the endlessly beating escapement provided poets and other writers with images of orderliness and temperance which meshed comfortably with conventional images of an orderly and temperate universe. What had begun as an abstract metaphor could now be invested with more specific meanings, by taking the different component parts of a clock into account. This was done even in the writings of one of the most abstruse theologians of the day, Thomas Bradwardine.

Bradwardine's finest work bore the title *De causa Dei contra Pelagium* ('In Defence of God against Pelagius', Pelagius being a fourth-century monk who had famously insisted that spiritual salvation depended on personal human effort). The book was not completed until 1344, nearly a decade after Richard of Wallingford's death, but the two men are likely to have known one another. Even if they did not, the work tells us much about the place of the clock in the fourteenth-century imagination. Bradwardine uses the image of God the divine clockmaker, whom he likens to the weight which drives round the train of wheels in our little clock, each with its ordered and definite motion. He goes further, referring to Hermes Trismegistus, and hinting at a perpetual motion provided by magnetic forces—shades of Pierre de Maricourt—'or else in some other more ingenious way'. Bradwardine was of course not proposing mechanical solutions of his own but making a point about the freedom or otherwise of the human will. Likening individual men to the *machina mundi*, he said that everything philosophers had said earlier about the world's relations with its First Mover—following Aristotle's doctrine—could be immediately transferred to the relations between man and God. At the end of the century Geoffrey Chaucer, in his *Nun's Priest's Tale*, likened the cock Chauntecleer to an abbey clock. Since he praised Bradwardine in the same poem in connection with the question of human freedom and predestination, it is not unlikely that he too had Bradwardine's use of the clock metaphor in mind.

It is not necessary to pursue Bradwardine's argument here, but we might well ask how it was that the clock and the universe could be spoken of in the same breath. It would have been difficult for any St Albans monk to have marvelled at his monastery's astronomical clock without allowing his thoughts to turn in the direction of the theology of Creation. That this state of affairs existed must have owed much to a long tradition of making planetary models and related computing devices, of kinds discussed here more fully in Part Four. In this tradition, Richard of Wallingford earned himself a central place. A clock did not have to include astronomical elements, but the fact that it included them so often in the first century of the mechanical escapement's history tells us much about their makers' priorities. This is not a question of our wishing to choose astronomy above monasticism as an answer to the clock's true origins. It is simply that every clerk who crossed the threshold of a university believed implicitly that the universe studied by the astronomers was the world created by God.

The Clock as Instrument

As we have already seen, turning an astrolabe dial with the right motion—it rotates in about four minutes less than an average solar day—was something that had been done on anaphoric clocks with a water drive in the ancient world, but it is impossible to say how well known such an arrangement was in the fourteenth century, even to scholars well versed in astronomy. Richard of Wallingford was very probably aware of it: it might have been a familiar feature of some water-driven devices in his time, and we should not forget his experiences on the long journey to Avignon. He was no ordinary onlooker, but an expert astronomer, without peer in England, and intimately acquainted with the techniques for calculating the positions of the Sun, Moon and planets in the zodiac. He would have grasped the principles behind any machine he witnessed almost immediately. He is unlikely to have been the first to attempt to link the recently invented mechanical clock drive to an astronomical display—the Norwich material hints at such a thing—even though it is extremely unlikely that there was any earlier mechanism that could have compared with his in complexity.

The evidence for what went before it, as we know, is incomplete. When it comes to deciding on priority of invention, even explicit medieval testimony is often unreliable. Take John of Whetehamstede, for example, a man born about a century after Richard of Wallingford and himself abbot of St Albans. Whetehamstede wrote an encyclopedic work called *Granarium* ('The Granary', a play on his own name) in which he included a list of inventors of astronomical instruments. Mentioning the albion in connection with Abbot Richard, its designer, John adds the names of other instruments ascribed to him, but decides to reserve judgement in view of an absence of written evidence. At this point in the manuscript he adds a later marginal note, naming other astronomers in connection with some of the instruments on his list, and remarking simply that there are some who ascribe 'the rectangulus and astronomical horologe to Richard, abbot of St Albans'. The statement is not without value: it tells us that the St Albans clock was regarded as more than a mere timepiece. It was something to be classified with the most important astronomical instruments in the medieval repertoire.

In what sense was the clock regarded as an 'astronomical instrument'? It was a piece of hardware and it was astronomical, but is that all? Much has been written during the last century about the rise of experimental method in the middle ages, and many a writer has been led astray by the frequently used Latin word *experimentum*, which often meant little more than 'relevant personal experience'. It scarcely ever related to a controlled experiment by which a theory could be tested, or by which crucial measurements could be made. In much the same way, the word 'instrument' is easily misunderstood. Like crucial experiments, instruments devised with the sole purpose of testing a theory, instruments that may even be discarded after use, were virtually unknown in the middle ages. There were instruments used—as now—to take measurements, either to improve a theory (for instance, when a planet's position is measured, to provide some important parameter or other) or to provide some specific information (such as the local latitude, or the time of day). We all too easily forget a third class of instrument, however, by which something is demonstrated, exhibited, or displayed. Celestial and terrestrial globes and armillary spheres are examples. Some would claim that even the common astrolabe was rarely more than this in the middle ages. This is the category in which John of Whetehamstede's gloss places the St Albans machine, and with it other scientific instruments which had a more familiar place in the educational processes of the time. The astronomical clock was well described as an instrument in this sense. It was a product of the university schools, but it embodied more than a touch of theatre as well.

What did this 'instrument' show to the onlooker, in the first half-century of its existence? An astrolabe dial was the basic element in its display—indeed, an astronomical clock at this period could be fairly defined by that single characteristic. Even long afterwards, very few showed the Sun, Moon, or planets. There was, however, the thirteenth-century northern Italian 'device of certain remarkable wheels' which—while it was apparently not driven by a mechanical clock—did have Sun, Moon, and a full complement of planets in its display.[104] We cannot say whether the Norwich clock showed the planets, or even whether it showed the Sun and Moon. We do not know who designed its dial-work, or the level of his astronomical expertise.

The St Albans clock showed the Sun and Moon, but what about the planets? In the sixteenth century John Leland mentioned various aspects of the St Albans dials, but said nothing of the planets. John Bale, who followed Leland in his survey of English writers, added nothing on this point. John Pits, however, who followed in their footsteps, spoke as though the planets were there; but perhaps he was doing no more than run through a list of all the things a good seventeenth-century astronomical clock was expected to have. We should not place much weight on his remark, but there is no need to pass from a rejection of Pits to doubting Richard of Wallingford's intention to show the movements of the planets.[105] The theoretical part of his treatise sets out the method of calculating the numbers of teeth on planetary wheels, and there are tables drawn up for cutting planetary wheels of 206, 255, 199, 224, and 235 teeth (Fig. 47). The tables give the angular spacing of the teeth (or groups of teeth) quoted to degrees, minutes, seconds, thirds, and fourths—a sexagesimal fourth being a degree divided into 60 x 60 x 60 x 60 parts! (This practice is to our eyes nonsensical, but most medieval astronomers found the temptation to continue long division irresistible.) Emmanuel Poulle's suggestion that these might not have been due to Richard of Wallingford can be safely dismissed, for the simple reason that they are intimately associated with a table for a wheel of 331 teeth for the Sun. This is as good as the abbot's signature, as we shall see.

The incomplete and disordered designs he left at his death for the building of the machine, as far as we have them, make no provision for a planetary display. It would not have been difficult, mechanically speaking, to have arranged for a system in which each planet

47. A table for cutting the teeth of planetary wheels in the St Albans clock. (MS Ashmole 1796, fol. 116r.)

had a separate dial. Arranging for all of the planets to move at their correct speeds around the zodiac ring (which is itself moving, and off-centre) on the dial of an astrolabe clock would have been extremely difficult. Such an arrangement would have required numerous concentric tubes, for example, such as are found in the remarkable astronomical clock of Philip Imser, now in the Technisches Museum in Vienna, but that work dates only from 1555.[106] We know that the St Albans clock had tubes for the Sun and Moon, but on balance it seems unlikely that a full set was ever made for the planets. We may be quite sure that Richard of Wallingford wanted a planetary display. If they were there at all, the planets would most probably have been shown on separate dials, clustered around the main dial.

48. A typical medieval illustration of the theme of Fortune turning her wheel, and so producing changes in human circumstances in the course of life. (Copied from the lost twelfth-century *Hortus deliciarum*, a work we met in connection with Fig. 16 above.)

Tides and Fortune

Why Richard of Wallingford's cosmic machine was truly extraordinary, and why it remained so for more than two centuries, was that it embodied gearwork of outstanding theoretical accuracy and mechanical ingenuity. The ordinary passer-by, to whom this was not at all evident, would have been far more easily impressed by a wheel of fortune on the clock face, and a separate dial showing the tides at London Bridge.

Both of these would have been relatively simple wheels to organise. Let us assume for the time being that the Sun and Moon have been taken care of. The connection between the tides and the Moon was by this time well known, at least in outline. There are, for example, at least ten tidal diagrams in medieval works stemming from the writings of the Venerable Bede, the great seventh-century Jarrow scholar, who addressed the question of the relationship between the tides and the motions of the Moon.

He had a good understanding of the need to distinguish between astronomical and meteorological influences on tidal patterns, although later scholars copied his data without realising that tidal extremes at different places, even on the same coastline, do not generally coincide. We know of a thirteenth-century manuscript which was no doubt once in Richard of Wallingford's chamber and which has a simple scientific procedure for determining the times of flood tide. It belonged to the abbey of St Albans and contains some astronomy and geography by John of Wallingford, monk and historian, who died in 1258.[107] There the time of flood tide at London bridge is calculated using the following rule: the time is 3h 48m on the first day of the lunar month, and advances by 48 minutes every day, so restoring the time to 3h 00m on day 30. While this is only an approximate solution to the problem, it was certainly good enough for monks from the abbey who wanted guidance on the time to disembark from what was in fact their nearest port. On the clock, therefore, all that was needed was a wheel that turned through a full circle (twenty-four hours) in thirty days, and this would have been child's play to arrange. Richard of Wallingford might of course have provided an improved set of tidal parameters, using the more accurate lunar data provided by his main Moon dial. There is in fact a subsidiary wheel on one of his manuscript diagrams which would have served such an improved tidal theory very well.

The wheel of fortune mentioned in connection with the clock, not only by Leland but by the fourteenth-century St Albans chronicler, presents us with more difficult problems, mechanical and conceptual. Medieval literature abounds in references to the Roman goddess Fortuna, a capricious deity who raises and lowers the hopes and prosperity of men and women. Fortuna introduced an element of chance into an otherwise rigidly ordered universe; and without chance, many felt that they had no hope. We meet the goddess in art and poetry, as well as in historical and astrological writings, and the basic medieval imagery does not vary much. She may stand on her wheel, even turn with it, or turn it with a handle. Occasionally it is toothed, and driven by another gear. The wheel might even be seen broken on the ground, but usually it is intact. Most of these ideas come from old Roman images, as does the central idea that Fortuna changes human affairs through her powers. Often a ruling prince sits atop the wheel, and a deposed prince is then shown falling off at its lower levels (Fig. 48). Just occasionally, Fortune is represented in literature as the servant of God, performing God's will; but that image is not easily reconciled with Christian belief. The image of Fortuna turning her wheel in the church of a Christian monastery might be thought incongruous, and there are other difficult questions to answer. At what speed does the wheel turn? There is no tradition of any precise rate at which the common literary figure was supposed to rotate.

Is there any better explanation of what the St Albans wheel of fortune might have been? An answer is perhaps to be sought in the technicalities of scholastic astrology, which was repeatedly defended against its critics in standard ways. The usual Christian defence of astrology was to argue that the influences of the heavenly bodies 'incline but do not compel'. Now there was in astrology a well known doctrine of 'the place (or part) of fortune'. It is quite conceivable that the wheel mentioned in the records supplied the means of working that out. One of the chief purposes of the concept of a place of fortune was to predict the length of life of a person whose birth horoscope was known. The rules for calculating it were complex: the easiest method was to subtract the Sun's longitude from the Moon's and add the longitude of the ascendent (the point of the zodiac ascending over the horizon at the moment of interest). We know that the St Albans clock was equipped to provide all of the necessary bits of information separately, but devising wheelwork to show the answer on a single wheel would not have been straightforward. References by Leland and others to the clock's complicated diagrams incline but do not compel us to think that there was more to be seen on it than could be easily understood by the common onlooker. Perhaps the wheel of fortune was one such thing, meant chiefly for the expert astrologer.

On Reading an Astrolabe Dial

When we come to the Sun and Moon, we no longer need to speculate, for the surviving text reflects the painstaking work that went into the design. Before we can appreciate even a summary of Richard of Wallingford's achievement, however, we must come to grips with the nature of the astrolabe face of this type of clock. The finer points of the theory of the astrolabe are not needed. We are not dealing with a small instrument adjusted by hand, as was an ordinary astrolabe, or with the numerous calculations that could be made on the small instrument. All the evidence is that the clock at St Albans was set up aloft, on a gallery below the high window in the south transept, and that it was large—of the order of six feet across—and so was designed to be viewed from a distance. That it was not a small thing can be judged from several aspects of its design, and the same conclusion is perhaps to be drawn from the fact that the housing of the mechanism, repaired during the abbacy of John of Whetehamstede (1420-40), was then known as 'le clokchambre'.[108]

49. The dial of the St Albans clock, as reconstructed for the Time Museum, Rockford, Illinois. The general pattern is that of a common astrolabe, except that the roles of rete and plate are reversed, so that here the zodiac (ecliptic) and star map are on the plate, behind the fixed rete of hour-lines, horizon, lines of equal altitude (almucantars), and lines of azimuth.

First and foremost we should mentally separate the movement of the Sun from that of the stars. The grid at the front of the clock (Fig. 49) can be thought of as no more than a fixed mesh over the St Albans sky. Its chief lines are the central vertical, namely the observer's meridian line, and the horizon, the circular arc with a small '6' at each end. (That grid is usually placed on the *plate* of an ordinary astrolabe, which is behind the star map, not in front of it, as here.)[109] The star map turns clockwise, as does the sky to an observer in the northern hemisphere looking towards the southern part of the horizon. The polar axis of the sky coincides with the axis of rotation at the centre of the dial. As for stars that rise and set, they rise over the eastern part of the horizon (on the viewer's left), culminate on the meridian, and set over the western horizon (on the right).

The Sun has its place in the sky, and is in this respect like the stars, but its place changes with time. It moves along a path through the stars (the ecliptic, through the middle of the band of sky we call the zodiac), completing its circuit in a year. Its annual movement is contrary to the clockwise direction of its daily rotation, so the daily movement of the stars as a whole is slightly faster than that of the Sun.[110] It is the Sun, however, which provides us with our hours and minutes of time, twenty-four hours being the time it takes the Sun to return to the meridian.

It has to be said that this is only a crude account of solar timekeeping. Astronomers drew a careful distinction between the true Sun (the actual Sun, which does not move quite uniformly) and the mean Sun (an idealised body, which ideally tells us clock time). We shall ignore the difference until we come to the following section of this chapter.

Richard of Wallingford needed to arrange for the motions of both the star map and the Sun. He seems not to have arranged for his model Sun to stay in the zodiac band, but simply put it on a pointer, turning full cycle in one day. Such a pointer can be thought of as the hour hand of the clock, an hour hand that rotates only once in twenty-four hours, not twice. It is not strictly an hour hand, however, since it follows the true Sun, not the mean. The place of the true Sun in the heavens is where the rod crosses the ecliptic circle on the dial.

Needless to say, the workings of the entire machine—the turning of the star map and the Sun pointer, letting off the strike every hour, showing the Moon's motion, the tidal dial, and all the rest—depended on the steady beat of the escapement, which was used to create a 24-hour motion (in mean time) to which all other motions were then related. The wheels providing that daily motion constitute what is called the 'going train', as distinct from the 'striking train', with its separate falling-weight power supply. The choice of the numbers of teeth in the wheels of the going train had to be matched to the beat. It seems that the favoured value for this beat was nearly 11 seconds, which to the ears of people accustomed to the tick of a domestic pendulum clock is surprisingly ponderous. Its advantage was that the clock would have required rewinding less often than one with a fast beat.

There was an enormous amount of information to be had from the astrolabe dial as regards stars, Sun and Moon, for those educated people who had been taught the use of an astrolabe. Even the ordinary monk, without the advantages of a university education, would have required very little instruction to be able to use the dial to good purpose. By day, he would have been able to read off the position of the true Sun in height and direction (altitude and azimuth). By day or night, he would have been able to read off the hour in seasonal (unequal) hours, using the arcs below the horizon which mark them.[111] It is impossible to say whether the clock was ever illuminated by night, but if not, the point of much of it would have been lost. Like the Sun, the Moon can be regarded as a moving point on the star map, although in the case of the Moon its motion carries it round the zodiac once every month. The informed monk would have been able to judge the Moon's position relative to the Sun's at all times, in particular when the two were together in the zodiac (at new moon), at right angles (at first and last quarter), or at opposite points of the zodiac (at full moon). Such information was of value when planning a journey by night, but it was also considered to be of value when deciding whether it was safe to bleed a patient—and bleeding was routinely undergone by all of the monks as part of a regimen of health.

As for the Moon's phases, our monk was not required to work them out from the astrolabe dial. Richard of Wallingford had so arranged matters that the Moon was represented by a ball, half black and half white, which rotated behind a window so as to show the correct phase at all times. As though this was not enough, he had introduced a mechanism whereby the Moon was covered to the appropriate degree of the ecliptic on the rare occasion of a lunar eclipse. These contrivances would perhaps not strike us as spectacular, but as far as we know they were then both unprecedented and extraordinarily challenging to the designer.

The Sun's Variable Motion

The explanation of the relationship between the Sun and the stars offered so far has been oversimplified in several respects, in particular in regard to the Sun's annual motion through the zodiac. It had been known for well over two millennia that the Sun does not move at constant speed against the background of stars. One classic Greek geometrical solution to the problem of accounting for its actual motion was to suppose that it does in fact move at constant speed round a circle, but that we are not at the centre of that circle, which is why its angular speed appears to change over the course of a year. Ptolemy handed on this explanation, together with his much more complex lunar and planetary schemes, and later astronomers were generally content to change only the parameters of the solar motion. We shall touch on these questions briefly, in connection with the albion, in Part Four.[112] In the early seventeenth century, Johannes Kepler showed that, like all the other planets, the Earth moves round the Sun in an *elliptical* orbit, and he presented the laws by which the speeds of all of them along

50. Richard of Wallingford's masterpiece of gear design, his oval contrate wheel of 331 teeth, engaging with a pinion of eight teeth. See the next figure.

their paths vary. Despite this and later revisions, the old assumption of a Sun moving round us on a suitably chosen eccentric circular path, and at constant speed, gives results corresponding quite well to what we actually observe; but in any case it would be wrong to judge the St Albans astronomical clock by any but the accepted theories of the day. The question, then, is how did Richard of Wallingford make the Sun pointer move at a variable speed so that its ecliptic position (on the star map) was correct in the light of the astronomy of his time?

How he did so was by a combination of mathematical and mechanical genius. He used a carefully calculated oval gear in contrate form, and arranged for it to engage with a pinion of eight teeth having an axle lying along the radius of the oval wheel. The overall appearance of the oval wheel is shown in Figs 50 and 51. The train of gears connecting the small pinion to the rest of the clock is by no means unimportant, but could no doubt have been supplied by a few other intelligent astronomers of the day. The arrangement of the oval gear and radial pinion required intellectual origi-

51. A detail of the previous figure.

nality of an entirely different order. To understand why, we must recognise the complexity of the problem they solve.

It is easy enough to see intuitively that a non-circular wheel will move the pinion at variable speed—or vice-versa—but how is the form of the wheel to be chosen so that exactly the right motion will ensue throughout the year? This would have been difficult enough, even if nothing more had been necessary than to represent the Sun's motion around the ecliptic, since that is not centred on the clock dial. (The Sun's motion is along the ecliptic, through the middle of the zodiac—the band of sky wide enough to accommodate the Moon and planets. The standard Ptolemaic theory gives the Sun's motion around the ecliptic, not around the poles of the heavens.) Richard actually needed a variable-speed drive twice over: not only is the ecliptic off-centre, but the Sun moves around it in any case at a variable speed. Now it should be intuitively obvious that, with evenly spaced teeth on the oval wheel, the nearer they are to the centre the faster they turn the pinion; but how the distance from the centre is to be decided, in relation to the astronomer's model giving the solar motions, is far from simple.

Richard solved the problem of the doubly variable motion, and did so to a high degree of accuracy. He did so without explaining his method, in what was an unfinished part of his treatise. (In fact he left us two slightly different versions, even here.) If John Leland—or indeed almost any other person before or after him—had looked into the workings of the St Albans

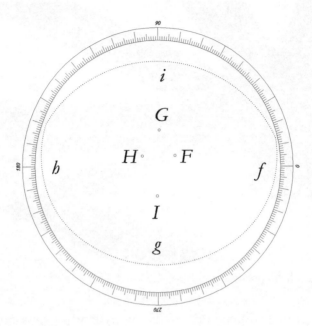

52. The geometrical scheme for laying out the oval wheel, as explained. The arc g is of a certain given radius and has it centre at G; and so for the other three arcs. The centres (G etc) are placed with the help of the surrounding circular scale.

clock, it is doubtful whether he would have given the oval gear a second glance, and yet it embodied a mathematical solution to the problem here outlined that is perhaps without equal in the entire history of medieval or Renaissance rational mechanics. Those who question this judgement should try to produce an ideal theory of their own, establishing the form the oval wheel should take. Its smooth mechanical action is also of course an important aspect of its design, but it was in deciding on its form that Richard of Wallingford's mathematical training came into its own. The form for which he eventually settled is shown in Fig. 52, an oval created out of four circular arcs. Each arc has a precisely calculated radius and centre, and these are quoted in ways that show the strong influence of standard astronomical procedures. Distances were given on a standard scale of sixty units and their sixtieth parts. Directions to centres were quoted using what is to us a protractor, but which was then regarded as an ordinary astronomical circular scale, with angular measure specified by zodiacal sign, degrees, and minutes of arc. On our version of the figure for laying out the shape of the wheel we have placed 331 small dots to mark the placing of the teeth.

Merely by considering them we begin to recognise the technological difficulties of creating that contrate iron wheel, its diameter of the order of the height of a man. But no less a cause for admiration is another of the wheel's mathematical properties. The figure of 331 is chosen, not randomly, but to produce in combination with other wheels in its train a precise (average) annual motion. (It was not Richard of Wallingford's first solution to his problem, for he seems to have also toyed with the idea of an oval wheel of 219 teeth with a pinion of nine teeth.) When we calculate the circumference of the wheel and divide it by 331, however, we find to our astonishment that the spacing of the teeth is exactly one unit on the scale of sixty by which radial distances are measured. To have so elegantly satisfied those two requirements, that for variable motion through shape, and that for average motion through the total number of teeth, was truly a *tour de force*.

The various parts of the display that pivot around the centre of the dial at their different rates were carried on tubes around the central axle. The oval wheel and its pinion were the key to the link between the axle carrying the Sun (the outer tube, 'the Sun's tube', in the highly stylised and incomplete figure) and that which carries the star plate (the inner tube, 'the diurnal tube', which has another rod running through its centre). The word used for each of these tubes was *caliga*, a Latin word for a Roman military boot, a word used in the middle ages for leg armour. Much of the vocabulary of Richard of Wallingford's treatise gives evidence of the need to create new words for new concepts; but since other words eventually took their place, his do not figure in our dictionaries.

If only to drive home some of the distinctions drawn in this section and the last, we may note that the difference between mean solar time (clock time, as rung hourly by the bell) and true solar time (as shown by the Sun's image, fixed to the Sun's tube) is known as the 'equation of time'. The more elaborate sundials from a later period often tabulated it, so allowing people to set their watches by using it as a correction to the sundial reading. Since it exceeds a quarter of an hour at certain times of year, it is not trivial. To those who were prepared to learn how to read its dial, Richard of Wallingford's clock showed this quantity with near perfection.

Hammering out the vast and complex oval wheel to its rough form, then filing it down to its precise size, and marking out and filing its teeth, must have kept the St Albans clock-builders occupied for many a long week. No doubt they had some assistance from members of the convent for the more mundane drudgery. Fashioning the tubes, however, must have been even more difficult, and here a blacksmith's skills would have been taxed. Simple and multiple tubes were eventually not uncommon, but they were usually incorporated into clocks on a much smaller scale. How would they have been made? Perhaps they were forged from thin metal strips laid around a cold rod as core. They could then have been filed down, or sanded smooth on a lathe. Casting seems less likely, but we have no reliable evidence one way or another for anything beyond the fact that the clock which was designed to have them did indeed exist.

The Moon and Dragon

The St Albans clock was an affair of great complexity, and this is not the place to trace its workings in detail, but there are other respects in which it makes use of design characteristics without clear precedent, and we should mention a few of them, even if only briefly.

The wheels for the Moon's motion and those for the Sun's are of course interrelated, for they are both part of a grand system driven by the going train. The going train would not have been particularly accurate, by modern standards, but that is not how we should judge the computation of the gear system. If we accept the rate of the main clock, then we can say that the Sun's mean motion is only about seven parts in a million in error. (Again, Richard of Wallingford was using the best contemporary data, from the *Alfonsine Tables*.) As the text implies, this error will be passed on to the Moon's motion; but again, the quality of the result was by any reasonable standards quite remarkable.

As stated earlier, the Moon was represented on the dial by means of a half-blackened ball, rotated to show its correct phases. This required that one of the large concentric wheels carried around with it a subsidiary train of gears. There was more to this transported gear train, however, than a control for the Moon alone, and here we must introduce a far more remarkable refinement, and an item of standard astronomical theory on which it depends.

Richard of Wallingford seems to have been fascinated by Ptolemy's theory of eclipses. He equipped his albion with a means of calculating them quickly. This part of the albion was abstracted in the fifteenth and sixteenth centuries by several German astronomers, who shared with Richard not only an interest in the basic astronomy of eclipses but a belief in their astrological significance.[113] When Lewis of Caerleon was imprisoned in the Tower of London by Richard III, perhaps between 1482 and 1485, he whiled away the time calculating eclipse tables using the methods set out in of Richard of Wallingford albion treatise. These tables were still thought important enough to be copied, together with the canons for their use, at the end of the sixteenth century.[114] There is no doubt that eclipses were considered to be important, for whatever reason. An eclipse of the Moon occurs when it enters the Earth's shadow, as cast by the Sun. It only occurs when the three bodies are close enough to a perfect alignment for this to happen, which will be when the Moon and Sun are on opposite sides of the Earth—'opposition', namely at full moon. If the Moon and Sun were always in the same plane, this would happen at every full moon, but in fact the plane of the Moon's path is inclined to that of the Sun (the ecliptic) at roughly five degrees. Any two inclined planes have one straight line in common. Only if the Sun and Moon are fairly close to that line at a time of full moon will it be possible for an eclipse to take place. If they both reach the line simultaneously the eclipse will be total. (We may call this key line the line of the Moon's nodes.) For other situations the eclipse will be partial, or not occur at all. Ptolemy's *Almagest* set out a detailed method of calculating the conditions under which an eclipse would happen and its 'circumstances' if it did (such as the degree of obscuration and the timing of the eclipse).

Richard of Wallingford managed to build into his clock what amounted to a computing mechanism whereby eclipses of the Moon were correctly represented on the dial whenever they happened in the heavens. It is possible for there to be no lunar eclipses in a given year, although there may be as many as three. When they occur, the rotating ball of the Moon would of course be near to showing its white half, since such eclipses occur near full moon. At the appropriate time, the ball was drawn inwards along a slide so that it passed under a circular obscuring disc (representing the Earth's shadow) until exactly the right degree of obscuration occurred.

In all of this, the proximity of the Sun and Moon to the key line mentioned earlier, the line of lunar nodes, plays a crucial part. The inclined orbits of the Sun and Moon, as seen against the background of stars, have two points of intersection on the star sphere, joined by the line in question. We now describe those points as the rising and falling nodes of the Moon's orbit, the points in which that orbit seems to meet the ecliptic, the Sun's orbit. From ancient to quite recent times astronomers have called them the Head and Tail of the Dragon, *caput draconis* and *cauda draconis*. The Dragon in question is the mythical creature that eats the Moon during an eclipse. The St Albans clock had a plate with a dragon's head and tail at diametrically opposite points, and by watching the slowly changing position of the black and white lunar ball in relation to the Dragon plate, the monks would have been able to anticipate future eclipses. The more intelligent members of the convent might have probed deeper, and have tried to verify the accuracy of the mechanism against appearances in the heavens. The device would then have been an 'astronomical instrument' in two different senses. For all of the monks, however, the clock was a piece of religious theatre, comparable in this respect with a morality play, but now encompassing God's created world, and some of the most recondite knowledge of its workings then to be had—knowledge that a monk needed to study at Oxford to acquire.

15

Legacy

RICHARD OF WALLINGFORD'S LEGACY was as many-sided as his life. What later ages inherited from his mathematical and astronomical writings is not to be measured in such simple terms as the discovery of new phenomena with a telescope, or indeed any other easily comprehensible empirical discovery. His stature within the sciences is to be measured by the originality and fertile nature of his analyses of existing mathematical procedures. Very few of his contemporaries had the slightest inkling of the importance of this, the product of his Oxford period. To most of them, he was a clockmaker. To the townspeople he was more. He was an oppressor. There are modern historians, however, who would add another dimension to the story, although not one concerning Richard of Wallingford alone. In the search for motives behind the invention of the mechanical clock, they are not satisfied with talk of a desire to display the theatre of the universe, or of a wish to perfectly represent the motions of its parts, or to mark out the hours of prayer in the monastic day, or even of the desire to have a town or other community synchronising its affairs through the sound of a public clock. They see social control, even working class suppression, as the chief motive. This is generally presented as an economic thesis: it is not suggested that suppression is necessarily an end in itself—despite the occasions in the abbot's life when it took on that hue. As an economic thesis, it is all too often a case of putting the cart before the horse—the fallacy of *post hoc ergo propter hoc*, as a schoolboy of the time would have said.

Time the Controller

No one can doubt that social control has been aided by the fact that we possess a shared time. We are all of us in some degree slaves to the clock in ways undreamed of in the middle ages, although our regulated experience is not something that was entirely foreign to those who lived in monastic communities. The idea, however, that some medieval clockmaker was driven to invent an escapement in order to bring about social control in the outside world is clearly insupportable. In St Albans, for example, it was not until the fifteenth century that a clock tower with twin bells was built for the use of the townspeople, and even the Italian town clocks of the previous century were not early, in horological terms. In any case, bells were in use for social control long before they were rung by mechanical means. The curfew bell is a case in point, but even that was less menacing than it seems. It had been used to instruct people to cover their fires as a community precaution against fire. This was not, in the first instance, a particularly sinister case of class suppression. (It seems to be a myth that the curfew was introduced into England by William the Conqueror as a means of political repression, although the word itself is from the French *couvre feu*.) Even such simple usage may change. The chief function of a curfew bell in late medieval St Albans seems to have been to close market trading, that is, to control not a subservient work-force but the traders of the town for their mutual interest. The town's Gabriel simply took over the function of a curfew bell that had been rung previously from the abbey tower. Roger of Norton (abbot from 1260 to 1290) had presented his abbey with one such bell. It is even possible that it was rung by a mechanical clock, if our general chronology is right.

The organisation of the townspeople's day was certainly to some extent under the abbot's control, and so it continued to be. No doubt the automation of Gabriel led to greater efficiencies in the running of the town, and was therefore in the interests of successive abbots. Even so, those who see the mechanical clock as a device called into existence by the desire of capitalists to control their work-force will need to redefine their terms to fit more closely the social realities of those communities where mechanical clocks first appeared. Since they will need to adapt their arguments to fit the late thirteenth century, under 'capitalist' they will have to include the monastic hierarchy, and perhaps even give it pride of place. Under 'work-force' they will need to include feudal villeins, those who held their land or property by bond-service for churchmen. And having done these things, the thesis will have lost most of the force they wanted it to have. As the abbey bells had done, after the decline in the power of the church the early morning ringing of Gabriel doubtless helped to keep the work-force in order, but these bells disturbed the slumbers of everyone, regardless of social position. This was never more evident than when, by almost universal agreement, Gabriel's hourly ringing was stopped, after complaints about it came to a head in 1861.

Time's Fell Hand

Richard of Wallingford needed no bell to force the townsmen to surrender their charter in April 1332. He scarcely needed his voice. His physical condition had been visibly declining during the early 1330s, and he was now barely able to speak. Two months after that surrender an order went out from the royal court at Woodstock that in view of his infirmity the itinerant justices should excuse his absence when next they met in Bedford. By 1334 he was in serious straits. A great thunderstorm then set fire to his chamber, and from that day to the end of his life he was in constant pain. He died at last on 23 May 1336—at an age of about forty-four.

Richard of Wallingford's successor was Michael of Mentmore, an Oxford scholar who had been master of the schools in St Albans. They were the first two abbots of the place to hold a university degree, but by no means the last. Richard's tomb in the abbey church still survives: to a person facing the altar it is the second from the left, that with the longest and narrowest stone. The inscription is lost, but it was in French, the language with which his successor seems to have been most at home. In 1786 Richard Gough recorded these words from it: 'Richard gist ici Dieu de sa alme eit merci. Vous ke par ici passes Pater e Ave pur l'alme prierunt ... jours de pardun averunt...' ('Richard is buried here, God have mercy on his soul. You who pass this way, say an Our Father and a Hail Mary for his soul and you will have [a number of] days of pardon'). The system of indulgences was then everywhere taken for granted. Even those who said their Paters and Aves when hearing the morning and evening Angelus were promised indulgence at a rate of exchange ruled upon by various popes, earlier and later. Damage to the brass, even in Gough's time, prevents us from saying just how many days of remission of punishment for sin Abbot Michael thought the memory of his predecessor to be worth.

The Man

We recall that the Fourth Lateran Council of 1215 had created a legislative body for the Benedictine order, the order of black monks, and that the superiors of their houses in each province were to meet every three years to legislate for reform.[115] Between then and the year of Richard of Wallingford's death—when they were combined—there were two separate chapters in England, those of the provinces of Canterbury and York. Visitations were made of them in 1265 by Cardinal Ottobon, who was sent to England as papal legate by Clement V. We know that Richard of Wallingford wrote a treatise of some sort on their statutes, although it is now lost, but we do have a list of his own statutes and those of his successor, thirty-five in all. Designed to legislate for every important contingency, there were things a mere abbot could not control. Every would-be reformer was liable to find himself thwarted by appeals to a conservative Rome that were likely to reverse his decisions. In this respect he steered a safe course, and his concern with obedience to the basic rule of St Benedict was in any case never in doubt. It was something that had become obvious to his monks as soon as he was appointed, since he issued an edict that he would make a visitation and ensure that the rule was obeyed. He is known to have written a commentary on the Benedictine rule too, but like the other treatise it is now lost. To devote so much effort to this kind of writing is generally supposed to betoken an arid and litigious nature, but here it was perhaps more a case of old scholastic habits dying hard. On the credit side of the account, some would say that the rules he drafted for Redbourn show a more humane side to his nature, and an understanding of the need for retreat and regeneration. A more cynical analysis would present those rules as primarily a means of controlling his monks in the most efficient manner possible.

It is wrong to suppose that Richard of Wallingford was nothing more than a rule-directed scholastic, brought by ambition and circumstance into a spiritual world where he did not truly belong. His actions must have been moulded to a great extent by his terrible infirmity, and his overall behaviour as abbot was determined to a great extent by his feudal position; but high office did not leave him without compassion, and leprosy naturally encouraged humility. Some of his personal prayers, mentioning his leprosy, still survive. In one, he honours his parents. In another he offers thanks for his high standing, despite the sins of his youth. In a third he refers again to his good fortune in the form of social elevation from lowly birth to a seat among princes. His forebears had enjoyed good health, he says, and yet none had ever been graced by God with such honour as he, despite his having been smitten with a most foul disease. It never ceased to amaze him, he confesses, that God so inclines the hearts of the great towards him that they can bear to be in his company and show him respect, in spite of his dreadful speech and the deformities of his leprous face and hands. In a fourth prayer he alludes once more to his former sins of the flesh, even drunkenness; and he confesses too to anger and hatred and evil intent towards people close to him. Much of this confession is conventional—in many ways, after all, he was just an ordinary human being—but there is no doubting his honest self-criticism, especially at the place where he speaks of regret for his ardent desire for revenge against those around him, revenge which he 'either took or was sorry to be unable to take'.

He was certainly not a man to be crossed. The chronicler of the *Gesta* had a measure of respect for his severity, putting it down to his infirmity and contrasting it to the laxity of his predecessor—which made his severity all the more necessary. By the time of Richard's death, most of those within the abbey walls would probably have agreed. The townsmen, after all, were now in their rightful place. As a final gesture, he had forced them to pay the expenses of knights of the liberty at parliament, expenses he had previously been expected to pay himself, as lord of the liberty. The dependent priories had been made to feel their dependence. Was that not as the Benedictine rule meant it to be?

'In prosperity circumspect, ... in adversity he was patient and magnanimous. In all things and to all men, in word as in example, he was far-seeing and gracious.'[116] Such praise was never lightly given in the *Gesta*, but then from personal qualities the writer soon turns to material questions, and the financial balance sheet in particular. Looking back over the abbot's achievements, the writer could not agree with all of his priorities, but Richard would not have been disappointed by the final verdict on his rule. The monastery buildings had been almost completely reroofed, the mills were in repair, there was a new almonry for housing poor scholars. More than £1000 had been spent on this mammoth programme, for which the architects were deputies of the royal master mason. The foundations for a new cloister were laid, starting with a foundation stone bearing Richard's name and the year, and under it were placed fragments of relics of saints, including Amphibalus. And the abbot is praised for having spent thirty shillings on this work out of his own pocket, thereby not burdening his baronial domain. The sum was plainly well spent.

During what was a period of general economic decline in England, the abbey manors had prospered during his rule—despite the occasional purchaser of wool defaulting on payment. (One such case gave his successor, Abbot Michael, much trouble.) Richard had never left the running of the farms entirely to others; and, as we have seen, he had even experimented with ways of increasing yields of corn in his fields. Despite all the acclaim, however, the chronicle ends on a mildly censorious note. The abbot had built a beautiful ceiling for the chapel of his chamber, most pleasing to the eye; he had roofed his chamber in lead; and he had engaged in sundry other building works—too many to recount. It would have been better, we are told, if he had spent less on buildings and had paid off more of his abbey's debts. This reprimand follows hard on the heels of the accounts of his having been censured for spending so much money on the clock. New ventures are all very well, 'but an abbot should first make sure that the work of his predecessors is being continued'. He was, after all, only a cog in the Benedictine machine.

Dissolution and Survival

The dissolution of the abbey in the time of Henry VIII went relatively smoothly, by dint of stratagems worthy of the most litigious of its former abbots. Cardinal Wolsey had it 'in his keeping' from 1521 to 1530. He spent time in a house he owned at nearby Tyttenhanger, so that those who claim that he never visited the abbey are probably mistaken, but it was certainly never uppermost in his thoughts. His successor was moved aside in 1538 for Richard Boreman, who had made it clear in advance that he would hand over the abbey to the crown. After this, apart from the church, most of the monastic buildings were granted to Sir Richard Lee (1550). Nicholas Udall, a personal foe of Lee and no great lover of women in general, said that this was the 'reward and hire of a whore', and that Lee's wife had been 'loved too well' by the king.[117] (If true, payment had been much deferred, for Henry had died in 1547.) The church building remained a crown possession until 1553, when Edward VI sold it to the mayor and burgesses of the town to be their parish church in place of St Andrew's chapel. The latter was adjacent to the church to the north, and they had been raising money to rebuild it for more than a century. The Lady chapel was now turned into the town grammar school.

One could say that the descendants of Richard of Wallingford's adversaries had at last revenged themselves on him and his kind, but theirs was a Pyrrhic victory, for the town was simply unable to meet the cost of proper upkeep of the vast building. In the words of an earlier abbot, it swallowed up revenues as the sea swallows rivers, and no doubt the abbey's great clock, 'without equal in Europe', was swept away by the financial tides too. The town had by this time a simple clock of its own, by which social needs could be satisfied. We can only guess at the attitudes of the townsfolk to the abbey clock. It had been created by an abbot who might still have been remembered for his severity. It had been maintained by monks bred in traditions to which they were outsiders. Once the community to which it belonged was dissolved, the great astronomical clock suffered the fate of so many great monastic achievements—death by attrition. The abbot's astronomical and mathematical writings were valued more in southern Germany than in England, and for most Englishmen it was as though he had never existed.

The biographical remarks by the abbot's contemporaries within the monastery dwelt, naturally enough, on matters they could understand. All were familiar with his iron will, but very few had an academic training by which they might gauge his intellect. This could be safely left to kindred spirits in the universities, where he was much appreciated. One of the last sixteenth-century English voices to sing his praises was Henry Savile, a man of comparable genius, who in 1570 put him in company with Richard Swineshead, Roger Bacon, Archimedes and Ptolemy.[118] The abbot would have been honoured by the last two names, and perhaps even by the two Oxonians. A generation or two after Savile, it became the fashion to denigrate medieval learning as a whole. As had happened in the case of Richard's clock, the achievements of medieval science quickly faded from the general consciousness. Neglect is a venial sin; dismissal is something more. Those who search for the origins of modern science and refuse to look back to a time before the likes of Galileo and Copernicus do a disservice even to their own cause. On the other hand, it would be equally wrong to portray the scientific achievements of a man like Richard of Wallingford without some reference to the larger picture. Providing both will be the purpose of the last part of this book.

Part Four

The Springs of Western Science

> It is usual to write poetry on appealing subjects, so producing a simple work. But I must wrestle with numbers and the names of unfamiliar things, with the seasons and diverse states of the world and of the signs, and even with the divisions of the houses.
>
> Manilius, *Astronomica*, 3.29-34

16

The Migration of Ideas

> Lyte Lowys my sone ... suffise to thee these trewe conclusions in Englissh as wel as sufficith to these noble clerkes Grekes these same conclusions in Grek; and to Arabiens in Arabik, and to Jewes in Ebrew, and to Latyn folk in Latyn; whiche Latyn folk had hem first out of othere dyverse langages, and writen hem in her owne tunge, that is to seyn, in Latyn.
>
> Geoffrey Chaucer, *A Treatise on the Astrolabe*, Introduction

ASKED TO NAME the chief glories of the European civilisation of the middle ages, very few historians would give much thought to the sciences. Some would mention the architecture of the cathedrals and great churches, others the founding of the universities. Literature—whether Latin or vernacular—and music, art, theology, even logic, would all have their advocates, but the best hope for the sciences would be to have them slipped in under the wing of technology. The more fundamental sciences are perceived as having drowned in a sea of speculative thought, unable to rescue themselves with precise measurement, or crucial, theory-testing, experiments. Those historians who try to portray medieval science on a larger canvas, claiming to show it as it was, and not as we could wish it to have been, are prone to suffer from more subtle modern prejudices, tacit in the contrasting categories of progress and stagnation, advance and decline, preservation and loss, knowledge and superstition. The line they draw between knowledge and superstition, between what is rational and what is groundless belief, is always especially revealing, for it shows how 'science' is being defined; and the definition is seldom one that would have been recognised in the middle ages.

What if we were able to ask a medieval scholar to tell us how a science should be judged? The answer to our question would change subtly from one historical period to the next. Invariably we should have heard that there is one supreme wisdom, given to us by the authority of holy scripture. Beyond that, we might have been told about the obstacles to truth—misleading authority, unreliable custom, ignorant popular opinion. We might have heard of a contrast between inner, even mystical, inspiration, and the evidence of the external senses. There would have been some difficulty had we tried to force our informants to be more precise, and to distinguish more plainly between what we term a science and knowledge in general, *scientia*. It is in this regard that we should detect the most notable watershed, dividing those who knew their Aristotle from those who did not. The Church Fathers, who held Plato in high esteem, were not especially familiar with Aristotle's writings. In the early middle ages little was known of his work in western Christendom beyond the extremely influential translations made by Boethius of his logical writings, in association with another work of logic, Porphyry's *Isagoge*.[119] In the later middle ages, on the other hand, from the end of the twelfth century onwards, there was almost no branch of university learning untouched by 'the Philosopher'—as Aristotle was then often known. After the arrival of his *Sophistical Refutations*, a work aimed at exposing forms of fallacious reasoning, his theory of language and logic became more widely and intensively studied. One after another, in the later twelfth and early thirteenth centuries, his writings on physics, cosmology, and metaphysics began to arrive on the scene in Latin translation, and his name and reputation were soon firmly established. There was science in plenty before all this, but Aristotle showed scholars what properties it needed to have to qualify as such. The sheer volume of late medieval commentary on Aristotle, and of other writings in an Aristotelian spirit, explains why so many writers are prepared to treat the science of the later middle ages as a footnote to Aristotle. It was this, but it was very much more.

The Latin Tradition

It would be hazardous to claim to have located the first stirrings of a new scientific awareness in the west, since even through the gloom of the early middle ages there are many shafts of light to be seen. From the Roman world, Christian and non-Christian, various simple astronomical traditions had been passed down. There were fragments of Hellenistic planetary theory, and more ancient views on the structure of the world, but the available texts were almost invariably deficient, to the extent that they presented their material without its supporting rationale.[120] The western world was long in coming to realise that a bundle of unrelated facts is not a science, however large the collection. It was rarely appreciated that the most powerful systems of thought draw part of their strength from their theoretical structure, and that deductions could be made within them, and tested against experience. This is not to deny mountains of data their place in history. Pliny's weighty *Natural History* was just such a thing, and its influence on western culture was very great, despite its generally unsystematic quality. At the risk of descending into stereotypes, however, we can say that the exact sciences as retailed by late Roman writers were usually lacking in the deductive qualities of the Greek models from which they were so often derived.

The one moderately clear example of an ordered science from the early middle ages—although one that became progressively more remote from empirical test as time went by—was that by which the calendar was governed. This had a more or less continuous history. The Roman state had used astronomers, first to draft and then to look after the calendar. With growing disorder in the kingdoms which succeeded to Roman rule, responsibility for the calendar was gradually assumed by the church, where it remained until early modern times. The computus, the set of traditional rules by which the priesthood could in principle—if not always in practice—determine the date of Easter and other movable feasts, was taught in monastic schools, and later became standard fare in the arts courses of universities. As a supplementary science, for measuring in hours rather than days, clerics were taught simple techniques for judging the time of day, and—in a religious setting—of prayer. The Venerable Bede, writing at the outset of the eighth century, was a master of such things, and there were others, but their sources were very remote from the ancient astronomical texts that could have brought the subject to life. Of course there were those who thought that the rules were divinely given, and should never be changed. By the tenth century, however, a handful of western scholars began to seek out Muslim astronomy, and in time a select group of western scholars began to look critically at the calendar, and demand that it follow respectable astronomical norms.[121] As we have already seen, Richard of Wallingford wrote a treatise on the subject, although it is now lost.

From Córdoba to Western Monasteries

That western scholars were encouraged to look to eastern cultures for scientific enlightenment in the tenth century and later was probably in some way connected with the experiences of Benedictine monasteries on the western fringe of Christian Europe, and the spread of information from them to the rest of their network. This was a time of widespread disorder in Europe, following a decline in the fortunes of the descendants of Charlemagne. The emirate in Córdoba, on the other hand, was then passing through a period of political and cultural brilliance, even outshining the caliphate of the Abbasids in Baghdad. Original Greek works, and Arabic and Hebrew commentaries on them, were beginning to arrive in Muslim Spain in large numbers, acquired from as far afield as Baghdad, Damascus and Cairo. In the second half of the tenth century, schools of mathematics, astronomy, and other sciences were established, whose scholars were fully capable of improving upon the material acquired from the east. The example set in Córdoba was later copied by Arab rulers elsewhere in the peninsula, for example in Seville, Valencia, Saragossa and Toledo. Astronomy was there far more than a game with texts: new observations were made, and important new instruments of calculation and observation were developed.

It was not long before Córdoba's cultural influences spread to the Christian states in the north of the peninsula, where Latin translations of many scientific works were made. It was characteristic of life in Muslim Spain that there were numerous contacts between religious groups. Potential translators included the many bilingual Jews and Mozarabs—Christians who owed allegiance to Muslim rulers and whose culture was in many respects that of the ruler. As a rare and precious piece of evidence of significant Muslim-Christian interchange, we still have a tenth-century collection of treatises dealing with the atrolabe, other related astronomical instruments, geometry, and water-powered timekeeping. It was produced at the Christian monastery of Santa Maria di Ripoll, the Benedictine house at the foot of the Pyrenees mentioned in Part Three in connection with a monastic water-clock. The texts in this manuscript are mostly translated from Arabic originals, one of the exceptions to the rule being that water-clock. The collection is doubly valuable because it provides us with a potential link between this region and the rest of Europe, where related texts are found from an only slightly later period.[122]

The most renowned Christian scholar to provide us with evidence of intellectual contact between the Christian north of Spain, Catalonia, and Muslim al-Andalus to the south and west was Gerbert of Aurillac. While still a Benedictine novice, he had been put in charge of the count of Barcelona to obtain instruction in the liberal arts, and his guardian saw to it that he was taught mathematics, arithmetic and music by the bishop of Vic, not far from Ripoll. Whether or not he was the first to carry such new learning across the Pyrenees, he was certainly the first scholar of note to do so. Taken to Rome, the young scholar astonished the pope with his knowledge. He ascended the ecclesiastical hierarchy rapidly, eventually himself becoming pope—the first French pope—as Sylvester II. Gerbert continued to correspond with his Catalonian hosts after his return: his example proves what a powerful influence even a single person in his situation could be. His scientific influence was at first through cathedral and monastic schools, starting from Lorraine. He helped to create a Latin vocabulary of the astrolabe, corresponding closely to that of the Arabic texts he knew at one remove. There are other early works on the instrument that stemmed directly or indirectly from his writings, and it is worth observing that two of their authors, Fulbert of Chartres and Walcher of Malvern, were like Gerbert in that they had reputations for learning going far beyond these astronomical exercises. In the middle of the eleventh century we find another Benedictine monk, Hermann the Lame, continuing in the same tradition in the monastery of Reichenau in a remote Austrian valley. For long it was a mystery how such a person could have composed a treatise on the construction and use of the astrolabe. The answer is simply that he too had somehow come into contact with the Ripoll tradition, presumably with the help of the religious order to which he and it belonged.

Such connections are not numerous. In some ways it is astonishing that the evidence for them has survived at all, but before the end of the same century we find another piece of evidence of a Benedictine diffusion of knowledge. The monk Constantine the African had travelled widely in the East and had acquired a remarkable command of languages before he moved to St Benedict's own monastery at Monte Cassino. There he translated around twenty important medical works from the Arabic, with striking consequences. Although slow to be adopted by the medical school of Salerno (in Campania, south of Naples), his translations eventually provided one of the main supports for the school's extraordinary reputation, which in medicine exceeded that of any university before the late middle ages.

From Eastern Islam to England

The reputation of Andalusian and other Iberian learning, especially in astronomy, medicine, and natural philosophy, grew to the point where scholars from remote centres travelled there to learn what they could, languages permitting. There was also movement in the opposite direction, which in a few cases helped to whet the western appetite for eastern learning. One instance was when Petrus Alfonsi visited the court of Henry I of England, around 1110. Petrus was a Spanish Jew (Moshe Sefardi) who had converted to Christianity with no less a person than King Alfonso I of Aragón as his godfather.[123] In England he found that there was already some interest in astronomical matters. He was able to offer advice, for example, to the Lorraine scholar Walcher, who after travelling in Italy had been made prior of Malvern. From a work adapted by Walcher from a text by Petrus (on eclipse calculation) we can judge the sense of intellectual excitement to be had from such exchanges. Walcher's work has many weaknesses, and he was often out of his depth, but he wanted desperately to master the new astronomy. It was now clear to him that he must abandon his old and primitive methods of calculation, and learn the new techniques from the East. In fact the techniques to which Petrus introduced him were those of al-Khwārizmī, a ninth-century Baghdad scholar from central Asia. They were an uncomfortable combination of Greek (Ptolemaic) and Hindu theories, but the inconsistencies were not appreciated, and would hardly have been considered serious had they been seen. New learning was new learning, and the finer points could wait.

Of those in England who reacted to Petrus Alfonsi's revelations the most famous was Adelard of Bath. He travelled to the Arab world via Salerno, where he tells us he discussed medical matters and scientific questions such as magnetism. It is possible, but not certain, that he learned Arabic in Sicily. A work of natural philosophy he dedicated in 1116 to William, bishop of Syracuse, shows no influence from Arabic writings. Once he had returned to England, however, in the 1120s, he translated from the Arabic two astronomical books and three on astrology. He then translated, and partially adapted, the astronomical tables of al-Khwārizmī, in a Cordovan version—116 tables, with thirty-seven short explanatory chapters, canons explaining their use. After this, he added at least one new translation to those made earlier of Euclid's geometry, and more than fifty manuscripts of it are still extant. His knowledge of Arabic was flawed, but through his translations he too played a part in the creation of the medieval Latin astronomical vocabulary, on which many vernaculars have drawn.

Adelard at some stage in his life wrote a work of natural philosophy, his *Natural Questions*, addressed to an unnamed nephew. This work, which avoids all explanation in terms of the supernatural, and which makes occasional allusion to Arabic writings, was no doubt an influence for the good in western science, but unlike the astronomical tables it can be parallelled elsewhere in the west. The same is true of a work he wrote on the astrolabe. The al-Khwārizmī tables and their canons, however, were in a different category of importance, for they were far superior to anything hitherto available—for example, in the computus literature. At last, western scholars could make predictions of solar, lunar and planetary positions with a high degree of accuracy. At last they were able to cast accurate horoscopes—required by the best medical practitioners of the age, quite apart from those who wished to use them for political and personal ends.[124]

The High Tide of Translation

While western scholars were doing what they could to acquire such new materials as these, scientific learning in the Iberian peninsula was enjoying a Golden Age. While we are not primarily concerned with the Islamic tradition as such, it followed in the footsteps of Hellenistic learning in treating the cosmic sciences in two broadly different ways. They may be well illustrated by taking two contrasting Andalusian personalities. Looking to Spain from the direction of philosophy, we see the great twelfth-century Aristotelian commentator Ibn Rushd (Averroes) towering over the scene. Leaving aside, for the time being, the question of how he came to influence and inspire western scholars so forcefully, we may contrast him with a mathematical astronomer from an earlier period whose name—he was Ibn al-Zarqālluh, in Latin Azarchel—was just as well known in the West. Regardless of the relative merits of the two scholars, it was the reputation of such astronomers as Azarchel that brought the first influx of translators into Spain, and not philosophy in the Aristotelian tradition.

In the second half of the eleventh century, an important group of astronomers formed something approximating to a school in Toledo. They included Ibn Sā'id and Azarchel. The latter began his career as a trained artisan. As a maker of instruments and water-clocks, he entered the service of the qadi (a religious judge) of Toledo. He remained there until some time between 1078 and 1080, when the discomfiture of repeated invasion by the Castilians persuaded him to move to Córdoba; and there he lived until his death in 1100. His name became a byword for astronomy, mainly because it was attached to an important set of astronomical tables. They were derived from many sources, and were not in a strict sense his, but he adjusted them intelligently, and wrote canons for them. These 'Toledan Tables' eventually superseded those of al-Khwārizmī—which, as we saw, were still in use in England long after Azarchel's death—and remained in use well into the fourteenth century.

It was to translate works of the same general character as Azarchel's that the greatest of all translators on the Iberian front, Gerard of Cremona, remained in Spain for most of his life. Gerard, who was born around 1114, was seemingly educated in astronomy in Latin schools before going off in search of Ptolemy's *Almagest*, in the mid-1140s. He died in 1187. Daniel of Morley, an English scholar who travelled through France and Spain when Gerard had been there long enough to be called by Daniel 'Gerard of

Toledo', tells of hearing him hold public lectures on an astrological work by Abū Ma'shar; but it was as a prolific and gifted translator that Gerard went down into history.

There were other translators active in Spain. John of Seville, a slightly older contemporary, was possibly a native Mozarab. Born in the south, he spent much of his career in the Christian cities of Limia and Toledo, where he translated astrological works in the second quarter of the century. At much the same period we find another Englishman, Robert of Chester, working with the Slav Hermann the Dalmatian on a translation of the Koran. Robert also translated an astrology by the eastern philosopher al-Kindī, as well as al-Khwārizmī's algebra. There were still others active at the same period and later, but none whose industry could compare with Gerard of Cremona's.

Gerard's taste, as it is revealed in the list of works he translated, was by no means narrow. He translated Aristotle's *Posterior Analytics*, his *Physics*, *On the Heavens*, *On Generation and Corruption*, and *Meteorology*, but he cast his net far beyond the Aristotelian canon. It was not by accident that he selected scientific writings for translation. Gerard found himself in the midst of a markedly scientific culture, and there he stayed for the simple reason that his own interests were primarily scientific. The extent to which he was helped by his pupils is unknown, but over eighty substantial works of translation go under his name. There are about twenty titles in geometry, mathematics, optics, and theoretical mechanics. They include the writings of Euclid, Archimedes, Menelaus, and works first written in Arabic with similar subject-matter. Gerard translated the thirteen books of Ptolemy's greatest work of astronomy, *Almagest,* and a dozen other works of astronomy or astrology. One of these, by al-Farghānī (Alfraganus, in Latin), became a standard university text-book in western universities, where it was also available in an alternative translation by John of Seville. Gerard provided translations of more than twenty works of medicine, notably by Galen, al-Rāsī, and Ibn Sīnā (Avicenna, as he was known in Latin writings). He produced others on magic, alchemy, divination, and geomancy. There are yet other works which may or may not have been his: whatever the truth of the matter, a few of the likeliest of candidates are translations of the works by Apollonius on conic sections, Abū Kāmil on algebra, Azarchel on the use of astronomical tables, and Ibn al-Haytham (Alhazen, to the Latins) on optics. By his extraordinary enterprise, Gerard of Cremona was almost single-handedly changing the direction of western scientific learning. There were other routes to the recovery of a textually purer Greek science, but by their nature they were lacking in the

Andalusian embellishments which gave to many of Gerard's sources their cutting edge.

Approaches to the Greek Aristotle

The earliest Latin translations of Aristotle in the wave of activity that was by Gerard's time beginning to sweep across medieval Europe, and also the more numerous, were made not from Arabic but from Greek texts. Several of them were made in the second quarter of the twelfth century by a Venetian-Greek cleric, now known as James of Venice, who had contacts in Constantinople. Others, a few years later, were due to a certain Henry, a Norman churchman who moved between Sicily and Constantinople. With the Emir Eugenius, he translated works by Ptolemy, Euclid, Aristotle, and Plato, at the imperial court in Palermo. This was a time of much scientific activity in Sicily, enough to draw scholars from as far away as Norman England to study it. Greek texts on astronomy, optics, and mechanics made their way into the Latin world from there. Almost simultaneously, however, as Christian forces were beginning to take back little by little the conquered Iberian peninsula from Islam, translation from the Arabic began to supplement and outstrip that which was done directly from the Greek. Even previously unknown Greek writings were to be found in Arabic translation; and alongside them, of course, there were many original studies in Arabic.[125]

While advanced technical astronomy was as yet a minority interest in the western academic world, natural philosophy had a larger following, and so it was that the greatest of all commentators on Aristotle was seized upon as soon as his work could be put into Latin. This man was Ibn Rushd, in Latin Averroes, known also simply as Commentator. Born in Córdoba in 1126, the son of a qadi, and grandson of a qadi who had been imam of the great mosque in Córdoba, like them he was thoroughly trained in law. After spending much time in Marrakech, he was made qadi of Seville, and finally of Córdoba. The lawyer in him found the analytical style of Aristotle's writing naturally appealing, being quite unlike the mathematical style of Ptolemy's astronomy, which he disliked and mistrusted. The division of human intellects into those who prefer science to have a linguistic structure and those who are happier with mathematical forms is of course not peculiar to the middle ages, but we should remember that mathematical training in the universities was as yet not very advanced.

Averroes wrote not only commentaries on Aristotle but works of philosophy and medicine of his own, and frequently struck a chord in Christian theological circles with his discussions of the relationship between scientific and religious truth. He helped to form Christian opinion in another way, by stressing the opposition between Plato—the darling of Christian philosophy in the earlier middle ages—and Aristotle. Averroes was also a sharp critic of Avicenna. Avicenna still retains the reputation of the leading eastern Islamic philosopher and medical writer of the middle ages. Born in what is now Uzbekhistan in 987, he died in Hamadan (now Iran) in 1037, after a distinguished career. His works happened to arrive on the scene in both Paris and Oxford around 1200, roughly twenty or thirty years before the first strong hints of Averroism in either place. Had Avicenna's influence on western natural philosophy not been tempered by Averroes, the cause of western science would certainly have been set back considerably.

Averroes died in 1198, having suffered a reversal of fortune in his last years, when his doctrines were anathematised and his books burned in his own society. This did nothing to discourage his Christian admirers, although by no means all Christian philosophers were his admirers—to some he was indeed Antichrist. Averroism was one of the most potent of forces for change in the western scientific mentality in the later middle ages, and the Tempier condemnations of 1277 were just one manifestation of unease at the changes the new Aristotelianism was bringing about. The lavish praise heaped on the Greek philosopher by Averroes was at the very least good publicity. Having stated in a preface which he added to Aristotle's *Physics* that its author was the very creator of physics, logic, and metaphysics, since nothing of value had been written on them before him, Averroes went on as follows:

> I say that he brought them to completion, since none of those who followed him has been able to add anything to his writings or find any serious error in them. That all this should be found in a single man is a strange and miraculous thing. The being who is thus privileged merits being called divine, rather than human, and that is why the ancients called him divine.

There is much more of the same strong rhetoric in the other works of Averroes, even to the point where we are invited to use the man's perfection as a reason for praising God. This sort of thing is not in itself enough to explain the high regard in which Aristotle was held by a long succession of natural philosophers in Paris, Oxford, and elsewhere, but it must have had some influence—of course, of a negative sort in the eyes of those who detested Averroes and his followers. (Dante played safe, and put him neither in Hell nor in Paradise, but in Limbo.)[126]

One of the most competent and productive of all medieval translators of Greek philosophical and scientific writings was a Dominican friar, William of Moerbeke. He was a northerner—there are several villages of the name in Flanders—but, in the service of his order and of the papacy, he spent most of his adult life in Asia Minor, Italy and France. From 1278 until his death in 1286 he was archbishop of Corinth. Despite an extraordinarily active ecclesiastical career, he managed to immerse himself not only in the Greek language but in the philosophical context of the works he chose to translate. William became intimately acquainted with three highly important scientific writers and their work: Witelo, a Silesian writer on optics; Henry Bate of Mechelen, a Flemish astronomer; and Campanus of Novara, also an astronomer. He went far towards achieving his ambition to give Latin Christendom a complete collection of Aristotle's writings: some of them had never been translated before, while others had been quite inadequately handled. To his rich output of new Aristotelian translations he added others of works by the best Greek commentators on Aristotle. He translated works by Proclus—including the commentary on Plato's great scientific treatise, the *Timaeus*—and no fewer than seven by Archimedes, the greatest mathematician of Greek antiquity. As well as commentaries on Archimedes, William of Moerbeke translated works by Hero on optics, and a remarkable work by Ptolemy on an instrument known as the analemma. Some of his work is now the only evidence we have for lost Greek texts. From the point of view of the changes wrought in medieval awareness of the Greek scientific achievement, his work was unparallelled. Some hundreds of manuscript copies, and a dozen early printed editions, speak volumes for his influence.[127]

When Gerard of Cremona's and other early Latin versions of the cream of Greek and Arabic philosophical and scientific writings began to work their way into the world of western scholarship, they yielded a treasury of astonishing novelty and richness. The first great Aristotelian in the West was an Oxford man, Robert Grosseteste. By 1255, all of the philosophical works of Aristotle then known had become a required part of the curriculum in arts in Paris. The effect of the fifty or so brilliant productions of William of Moerbeke was to consolidate these early gains. The great Dominican friar Albertus Magnus set himself the task of making all of Aristotle's natural philosophy available to the Latins. His even more influential pupil, Thomas Aquinas, went much further along the same road. The names of the great masters are those now best remembered, but without the labours of the translators they would have had to pursue very different paths. The nature of the territory explored by the translators was strange, foreign, and exciting. To many it must have seemed as though they had turned the hard library bench into a magic carpet.

Jewish Contributions

One element in the transmission of scientific ideas and ideals to the West which is all too often overlooked is the part played by Jewish scholars. Political power in Spain was in the hands of either Muslims or Christians, which makes it easy to present the Jews as nothing more than intermediaries between those two alien cultures, able to act as such by virtue of the linguistic knowledge that circumstance had forced upon them. This does poor justice to the case. Members of the medieval Jewish community were also of use to Islamic and Christian scholarship for their original writings.ABRAHAM BAR HIYYA, for example, known to the Latins as Savasorda, was a Catalan scholar who illustrates both qualities. He wrote Hebrew works on astronomy and cosmography, and an abbreviated version of a Hebrew geometry he composed was translated into Latin in 1145. This much-copied work was prepared by PLATO OF TIVOLI, and was only one of several instances of close collaboration between the two men, Jew and Christian.

The Jews as a whole did not, of course, have the resources or political power of the Abbasids, so we must not expect their methods for acquiring scientific texts to follow the same pattern. Jewish scholars had access to the texts in many cases only because they were in service to those who owned them.[128] As early as the ninth and tenth centuries, Jews and Christians who had no intention of converting to Islam helped in the translation of scientific and philosophical works into Arabic for their rulers. Under Islam there were numerous later instances in which Jewish communities were tolerated and their academic pursuits allowed, in ways that put many Christian monarchs to shame. It is easy to forget that during the Moorish invasions of the Iberian peninsula in the early eighth century, the Jews welcomed the invaders simply because they had been so badly treated by the Christian Visigothic kings. At all events, it is to Spain that we must look for the truly spectacular, catalytic, effects of mingling the scholarship of three great religions—and some of these effects we have already seen. The dramatic cross-fertilisation of cultures continued in the Peninsula into the fifteenth century, which is of course not to say that relations between the different communities were never strained.[129]

The Jewish community is historically important because it sent out not only texts but people as carriers of ideas. We have already met PETRUS ALFONSI, but a Jewish astronomer more original than he, who visited English shores at much the same period in the twelfth century, was ABRAHAM IBN EZRA. Poet, mystic, biblical commentator, mathematician, and astrologer, Abraham too helped to quench a growing thirst for astronomical knowledge, and to make astrology respectable among Christians as well as

Jews. He wrote works of his own in Hebrew, on the calendar and on the astrolabe, and translated others, but more important was a new set of astronomical tables he composed. He adapted them to at least two of his places of residence, Pisa and Winchester, and perhaps to others. (Over a period of thirty years he lived in at least six Italian towns, and five in France or Normandy, as well as London and Winchester.) For some decades, Abraham's tables occupied a modest place in the Christian world at the side of al-Khwārizmī's and the *Toledan Tables*. He was not without influence on Christian theology, although this was after his death. He preached science, but was against the habit of interpreting the Bible as though it was a scientific encyclopaedia.[130]

Another Spanish Jew who greatly influenced the West was Maimonides, whose *Guide for the Perplexed*, written at the end of the twelfth century, was one of the high points of Jewish learning in the middle ages. To modern eyes, the work seems to be lacking in structure, but every chapter gives signs of a sharply critical mind, capable of turning authors—Aristotle, for instance—against themselves. Maimonides was a native of Córdoba who, with his father, had been forced to leave for Fez to avoid compulsory conversion to Islam. Forced to leave there, he settled in Cairo, where he practised medicine and eventually became court physician to Saladin. The *Guide* was written in Egypt around 1190—Maimonides died in 1204. It reached Christian Europe by a most unusual route, reminding us that the transmission of exotic books was not always a case of 'Sicily or Spain'. Maimonides wrote it in Arabic (in Hebrew characters), and when pressed for a translation into Hebrew arranged for this to be done by Rabbi Samuel Ibn Tibbon, in Provence. (Maimonides long kept up a correspondence with the Jews of Provence.) It was translated into Hebrew a second time within a decade, and into Latin by 1234 at the latest.

The *Guide* was by far the most important work of Jewish medieval philosophy, but it was also one of the most controversial. As with the writings of Averroes, his near contemporary, its readers fell into the categories of those who strongly supported it and those who just as strongly condemned it. It was not hostile to science, but offered a gentle qualification of science's claims: its whole purpose was to keep the eyes of those attracted by philosophy and the sciences fixed firmly on the Jewish faith and religious law. Maimonides had studied astronomy with a disciple of the Muslim Ibn Bājja (Avempace, to the Latins), and had a reasonable understanding of the Ptolemaic system and of various innovations made by Islamic astronomers. In criticising it all, he accepted certain Aristotelian principles, but he never showed the same exaggerated veneration of Aristotle as Averroes was then doing. Like several Muslim astronomers of the time, Maimonides believed that the Ptolemaic model fell foul of the physical principles behind Aristotle's vision of the cosmos, but he refused to be dogmatic. In a much-quoted passage, he said that we know nothing of the nature of what is in the heavens beyond mathematical calculations. 'God alone', he went on, 'has a perfect and true knowledge of the heavens, their nature, their substance, their form, their motions, and their causes.'[131] This is somewhat reminiscent of remarks made long afterwards by Bernard of Verdun at the end of the thirteenth century, and by Richard of Wallingford and John Buridan in the fourteenth, about the non-existence of the geometrical circles of astronomy in the heavens, but in their case Averroes is another likely source.

Parallel Worlds: Theology as Censor

The way in which Maimonides presented the study of science and philosophy as a religious obligation appealed to some Christian scholastics. Later in the thirteenth century, William of Auvergne, bishop of Paris, the Franciscan Alexander of Hales, the Dominicans Albertus Magnus and his pupil Thomas Aquinas, and the theologian-philosopher John Duns Scotus (who died a Franciscan), all made much use of it.[132] Maimonides claimed in his *Guide* to have read every Arabic work on astrology, and he accepted much of what was then virtually unquestioned astrological doctrine. He took a similarly ambiguous position in 'a letter to the men of Marseilles', again translated into Hebrew by Samuel Ibn Tibbon, who was then living in Arles. He never became a standard source in the West on these matters, but an idea he put forward, that there is a natural human faculty which allows us to predict the future, was repeated from time to time by western writers.

In the early fourteenth century there was a notorious controversy between two Jewish groups, led on the one side by the rabbi of Perpignan, a sympathiser with natural philosophy, and on the other by an obscurantist Montpellier scholar and the rabbi of Barcelona.[133] It was a controversy which illustrates some of the tensions in the intellectual atmosphere of the time, tensions which were strangely parallel to those resulting in the Tempier prohibitions of 1277. The outcome of the contest would have been very different had it not been for the Christian persecution of Jews in the Rhineland, which caused Asher ben Yehiel (1250-1327) to move from Germany to Toledo. There, as rabbi, he joined the party hostile to science, and in 1305 that anti-science party achieved great success in Barcelona, banning secular study, even the study of Maimonides. There were a few enlightened voices, but the outcome was hostile to science. The irony of it all is that at the very time when Maimonides was beginning to exert influence on Christian scholasticism, his work was being denounced in some Jewish circles as irreligious.[134]

Such accidents of history may clearly matter more for a time than general trends. The hostility to scientific study evidenced in the Barcelona edict was not effective in the longer term, even in that great centre of mystical Judaism, Perpignan. Another process parallel to one that had begun earlier in the Christian scholastic world was that in which the ambiguities and inconsistencies in Aristotle's writings developed into conflicting 'Aristotelian' traditions. This must have endeared the disputants to their Christian counterparts. Certainly one Jewish scholar who went down well in the Christian world for his doctrines was the Perpignan-educated Levi ben Gershom, a close contemporary of Richard of Wallingford.[135] Author of a famous philosophical work, *Wars of the Lord*, he directly influenced western scholastic thought many times over, for instance through his pronouncements on the nature of the spheres, the eternity of matter, and the idea of a world with a beginning but no end. Widespread anti-Semitism notwithstanding, Pope Clement VI ordered Levi's works to be translated into Latin, and in return Levi dedicated a Latin trigonometrical treatise to Clement.

Provence and Profatius

Two cultures will interact most readily when the people concerned mingle freely. Western Christian scholars outside Spain were far more likely to encounter Jews than Muslims. In Spain, where Jews lived in much larger numbers, Hebrew astronomical tables give evidence of strong intellectual interaction there. We know of such tables from twelfth-century Barcelona, and later related examples from Orange, Tarascon, Catalonia, Lisbon and Salamanca.[136] Christian scholarship could still not match this. Provence was a key region in the transmission of eastern knowledge, because of its strong links with Iberian Jewish communities and the peculiar way in which it was shuttled between political powers. It was defended against Muslim invasion by a local dynasty that ended in 1113, when the house of Barcelona took over, and it was ruled from Catalonia for more than a century. Science was received there in three phases. The first was when Spanish scholars put Arabic scientific texts into Hebrew, around the beginning of the twelfth century; the second was when Andalusian Jews, fleeing the Almohad persecutions of the late 1140s, fled to Provence and accelerated the process of translation; and the third, from the first decades of the thirteenth century up to the mid-fourteenth, was when the writings of

Maimonides added a new impetus.[137] Turning Arabic works into Hebrew greatly increased the chances of important works being put into Latin.

Twelfth-century Provence flourished through trade with the Levant, and its Latinate literature and architecture became an influence on neighbouring France. Its chief scientific influence on the Latin world came only at the end of the thirteenth century. First there were religious difficulties to be confronted on the home front: the new learning in poetry, literature, philosophy, and natural science, was often out of key with traditional Talmudic learning, on to which it was eventually grafted, but with some difficulty. Astronomy, by and large, escaped censure, except when it touched on cosmological matters which might be expected to clash with scripture. The political state of affairs in Provence, however, was less rosy than the intellectual situation. In the thirteenth century, the papacy and France were active in bringing the whole region to heel: the harsh methods of the Albigensian crusades are a well-known aspect of this movement. Not for the first time did Jewish scholars travel northwards and eastwards as a result, carrying much of their newly acquired knowledge with them. Here we must always keep in mind the relative rarity of scientific scholarship.[138] When we ask what science the slowly drifting Jewish communities had in their baggage, it would be unrealistic to expect very much more than they left behind. Jewish scientific works newly composed far from the Mediterranean were relatively few.

The year 1290 saw the expulsion of the Jews from England. Richard of Wallingford lived in a society riddled with anti-Semitism, influenced as it was by the manifest prejudices of a succession of popes and senior English clergy, as well as members of the royal house. He was aware, however, of debts he owed to one Jewish Provençal astronomer in particular, namely Jacob ben Makhir ibn Tibbon, known in Latin works as Profatius Judaeus. As it happens, Profatius was one of those who spoke out for his subject in the Barcelona dispute mentioned earlier, but he died in 1305 (at the age of about seventy) when it seemed that the battle was lost. His family had moved from Granada to Languedoc and Provence around 1150, and for at least four generations its members earned fame for their translations into Hebrew from Arabic. We have already come across one of their productions, namely Maimonides' *Guide*, first written in Arabic; and there were many others, in mathematics, medicine, physics and astronomy.[139] Profatius, who studied medicine at Montpellier and spent time in Spain improving his knowledge of Arabic, was the most original member of the family. Not only did he translate into Hebrew works by Euclid, Alhazen, Azarchel, Averroes, and half a dozen other scientific authors, he composed two very influential treatises of his own. One was an almanac, originally prepared in 1293 for the meridian of Montpellier, from which reasonably good planetary positions could be derived more easily than from a full set of more accurate planetary tables. He said it was inspired by the almanac of King Ptolemy, but in fact it owed much to a work by Azarchel. In a Latin version, the Profatius almanac was used by Dante. Often copied, the almanac was actually expanded in 1301, translated once more into Latin by Peter of St-Omer, and then later translated back again into Hebrew. If nothing else, this tells us much about the sheer difficulty of obtaining books in the middle ages.

The other renowned treatise by Profatius was one in which he explained the construction and use of his astrolabe quadrant, which he called his 'quadrant of Israel'. In the Christian world it was called the 'new quadrant', or the 'Profatius quadrant'. Apart from a thread, which can be drawn out along an arbitrary radius of the quarter-circle, and a small pearl marker sliding on the thread, it requires no moving parts, and so was relatively easy to make—even out of wood, covered with parchment or paper. It was actively used in Europe as an alternative to more expensive brass astrolabes for three centuries; and in Turkey almost to the present day. Its design and use are not to be easily explained, but one may imagine it as the result of folding a plane astrolabe along a diameter, and then folding it again—having first put the moving rete in a symmetrical position. When designing the albion, Richard of Wallingford made use of some of the ideas embodied in it, but he owed even more to Profatius, albeit indirectly, for his part in producing a translation of Azarchel's treatise on the universal astrolabe plate known as the saphea. A remark made at the end of that Latin translation is very revealing, for it tells us that the work was translated out of Arabic into Latin by John of Brescia in the year 1263, Profatius the Jew having first turned it into the vernacular.[140] There is plenty of evidence that this technique of double translation was often used. At the court of Alfonso X of León and Castile, at much the same time, Castilian was being used as an intermediary, and often as a final language.

Bertrand Russell once made the point that the spectacular achievements of the Greeks owed much to the fact that they did not need to trouble themselves with other languages than their own. To those who are less fortunate, when language is an intellectual barrier, translation is the only means of crossing it. Having a book and using it are not, of course, the same thing. Intellectual debt is virtually indefinable, and our long list of translated works is only the beginning of the balance sheet. We should also remember that a mastery of languages and the art of translation served other purposes than the transmission of knowledge. The Paris Dominicans taught languages for missionary purposes in the mid-thirteenth century. At about the same time, the Franciscan Roger Bacon pleaded for languages as a means to converting the infidel; and in 1315 Ramon Llull died while trying to implement a programme with a similar purpose. In 1311, the church council of Vienne, in order to promote missionary work, decreed that chairs of Greek, Hebrew, Arabic, and Chaldaic should be established at the universities of Paris, Oxford, Bologna and Salamanca. The convocation of Canterbury actually raised money from the southern clergy to pay a converted Jew to teach Greek and Hebrew at Oxford, but it came to nothing, for want of a competent candidate.[141]

There were of course those who wanted texts for the sake of the knowledge they contained. When such scholars lacked the advantage of living in a multilingual society, it was useful to have the means of paying for assistance. Robert Grosseteste, for example, when bishop of Lincoln, was able to employ men of Greek origin to aid him, and he also had the help of an Englishman who had studied in Greece. Princes could of course afford to pay for help, but only a few of them felt strongly enough about learning to do so. Science was fortunate in Frederick II in Sicily and Alfonso X in Spain; and in a few princes of the church, such as two archbishops of Toledo, Raymond and Rodrigo. As for the quality of the translations produced, this rarely depended on the source of funds. Translations were usually very literal, except where they simply left passages untranslated, and were often very poor. They improved as they began to lean on one another, but often an inferior version was copied far more often than its replacement. Whether a translation reads well or badly is something that was often decided by the translator's understanding of the subject matter, or lack of it. When manuscript diagrams are found to be poor, as is very often the case, the fact usually signals either a lack of understanding of the subject matter on the part of either the translator or the copyist.

In time, western science decided that it no longer seriously needed more texts. The great surge of the Renaissance ran counter to that general rule, and produced a few new Greek scientific texts as well as many new translations of old ones. Those chiefly responsible for translating Greek sources in the fifteenth and sixteenth centuries were Italian, and their motives were more often literary than scientific. With the close of the sixteenth century, the Jewish community gradually lost its importance as an intermediary between the civilisations of East and West. Several Christian scholars, having acquired a knowledge of Hebrew and Arabic, were now beginning to make their own translations. They acquired these languages for the sake of what they could reveal, and without needing the excuse that their work might help to convert Jews and Arabs to Christianity. People generally translate only what they value, although occasionally they speculate and are disappointed. In the sixteenth century, the Christian world began to value Jewish cabbalistic works. One cannot say that medieval taste excluded such things, but here chance seems to have played just as important a part in the transmission of ideas. Chance as well as taste decided which texts were put into scholars' hands. Western civilisation owes much to the chance which brought science, fully grown, into Europe. As texts were mastered, scholars learned to value their less obvious contents. They hungered for more; and the Greek and Arabic scientific corpus passed with extraordinary rapidity into Latin. That it did so changed the face of history.

17

A Primer in Aristotelian Natural Philosophy

> No one can object if Allah assembled the world in one individual.
>
> Abū Nāwās, referring to Aristotle, before AD 810[142]

GREEK SCIENTIFIC LEARNING and its later offshoots were nowhere more welcome in medieval western Christendom than in Oxford, and one man above all others bears responsibility for preparing the way. We shall later have more to say about Robert Grosseteste and the influential positions he held in the period during which the status of Oxford's university was made secure, but his importance to both his university and to western science depended heavily on one aspect of his taste: his predilection for Aristotle. His commentary on the *Posterior Analytics*, the work in which Aristotle set out his views on the nature of scientific thinking, was written between 1228 and 1230, shortly before the commentary on it by Averroes became available in Latin. As we shall see in the next chapter, Grosseteste had a good grasp of the overall pattern of Aristotle's philosophy of the sciences. The present chapter is meant for those without Grosseteste's advantage. It presents a bare summary of some of the most important of Aristotle's, to avoid having to interleave them with Grosseteste's own views.

Aristotle divided the sciences into the theoretical and the practical, the first giving knowledge for its own sake, the second giving knowledge of a sort that can guide our behaviour—such as how to make a tool or a beautiful statue. As academic man incarnate, he naturally gave more attention to the theoretical sciences, which were said to include metaphysics, physics, and mathematics. Mathematics, he said, deals with unchanging things lacking separate existence, such as numbers and lines; physics deals with changing things that do have separate existence; and metaphysics, also called theology, deals with unchanging and separately existing things. Physics was a subject Aristotle addressed in many of his works. In the book of that name, for instance, he tried to get at the first causes of all change in nature. In the book *On the Heavens*, he discussed the movement and arrangement of the planets and stars, but also with the bodily elements and how they could be transformed into one another. In his *On Generation and Corruption*, he gave a highly original analysis of what we mean when we say that a thing begins to be, or that it passes away. At a logical level, this subject taxed the minds of many an Oxford scholar of the later middle ages. In Aristotle, it was intimately related to his theories of what it is that bodies share, the common nature of bodies which have in themselves a source of movement and rest. Animals and plants plainly have a source of movement and rest; but Aristotle thought that the elements (earth, air, fire, and water) and their compounds have an inborn tendency to move. This is where he formulated an idea which obsessed most of his followers for the next two thousand years, in some cases well into the seventeenth century. He made a sharp distinction between the celestial regions, above the Moon, where the tendency of bodies is to move in circles around the centre of the universe—that is, the centre of the Earth—and the terrestrial regions below the Moon, where heavy bodies have a natural motion towards the centre of the Earth and where light bodies (such as smoke) move naturally upwards and away from that point. (This does not rule out other motions, but they are in some way forced.) He spoke of all natural movements, not only those of animals, as though they are in some way initiated from within the moving thing; and he spoke of nature as an inborn impulse to movement. To understand the natures of all types of body, working harmoniously together, is to understand nature. This, as we might express it, is to have a mastery of the sciences.

Aristotle wrote at great length, his argument is fine-grained, and much of what he wrote was by way of criticism of his predecessors. While he is not a writer to be summarised readily, he cannot be ignored, for some of his principles and conceptual distinctions—such as that which divides the universe into two regions, each with its own laws—were tacitly assumed by almost all who took part in academic debate in the later middle ages. One thesis on which he insisted was that the aim of physics is to be universal, not a study of this object or that object but of what they have in common, the types of matter which objects have in common. In the scholastic tradition, the contrast was expressed as one between individual matter (*materia individualis*) and a common sensible matter (*materia sensibilis communis*). When we start to study a complex thing, we may analyse it into its parts only up to a certain point. In the end we get down to the four elements, and the only analysis we can offer of them will be into a fundamental prime matter, which will be hot or cold, dry or moist. Prime matter is not something we can sense directly. We know it only through abstract thought. Students of physics must study nature in two different ways: they must study matter and form. They study forms which are embodied in matter but separable in thought.

This brings us to another distinction. Aristotle considered physics to be different from metaphysics, which studies separately existing (that is, non-embodied) forms. Of these, three in particular caught the medieval imagination. God is pure form; so are the intelligences which move the spheres carrying the stars and planets round the heavens; and so is the rational element in the human soul. (In the middle ages, the intelligences were often supplemented with ranks of angels above them but below God.) By comparison, physics may seem an inferior science, but it is a science in a way that a study concerned only with matter would not be.

Another of Aristotle's fundamental distinctions was that between physics and mathematics. It forced him to confront the problem of how to treat what we often call 'applied mathematics'. In the *Posterior Analytics*, he called such studies 'the more physical parts of mathematics'. Four types of study in particular fell into this troublesome category: astronomy, optics, harmonics (the mathematical side of musical theory), and mechanics of the mathematical type. They have an intermediate character, for while they proceed by mathematical methods, and had often been considered by his predecessors to belong to mathematics, he treated them as physical. Lines in optics he considered physical, those in geometry not. He is not always very clear on this question, but the general pattern he imposed on the sciences is plain enough: the lower sciences study facts, and the higher sciences study the reasons for the facts. The higher sciences have a structure that may be borrowed from logic or mathematics or other formal theories.

Aristotle was engaged in a search for knowledge, and physics he considered to be a search for the causes which operate in the natural world. He treated of four different kinds of cause, and his classification became a standard part of every university clerk's education. Where we usually speak of events as what are caused, he spoke of the causes of things. A *material cause* he defined as that out of which a thing comes to be—in this sense the bronze is the cause of a statue. The *formal cause* is the pattern or form of the thing—the shape of a statue, for instance, or the mathematical ratio of 2:1 in the case of a musical octave. The *efficient cause* is what gives rise to movement or brings it to an end. The last is closest to our commonest usage, although where we might say the cause of a ball's moving is the cricket bat's motion, he would have said it was the bat. In the same way, the father was the cause of the child. The fourth type was the *final cause*, the end or aim of something. (Teleology is the study of such final causes.) In this sense, health might be considered the cause of our walking.

Things may have causes of more than one kind—indeed, they were probably thought to have all four sorts. Aristotle, in insisting on physics as a search for knowledge of causes, was not oblivious to the fact that apart from things that always happen in the same way, there are things that in general do, but not always. He has much to say about chance, exceptions to rules, and luck, but most commentators agree that here he is not at his best. Medieval theological discussions of miracles could make use of him here, even so.

Regarding a final cause as a final state, as something that is going to be the case, it is not especially problematical, but in the middle ages it was often given religious overtones, making things yearn for a final condition—such as perfection or an utterly harmonious world—as men and women yearn for God. Aristotle himself engaged in a defence of the idea that nature operates teleologically, and he used anthropomorphic language in doing so. Why he did so is interesting, for there is a curious parallel with the religious side of the Darwininan debate of the nineteenth and twentieth centuries. Empedocles had previously argued that nature produced a great variety of species, and that only those which were fittest to survive did indeed survive. This seemed to Aristotle to deny final causes in nature, and here a discussion of chance (and therefore of necessity) was inevitable. What he wrote on the subject of necessity was profound, and highly relevant to a view of science which he helped to propagate. He addressed the question in different ways in different books. In the *De interpretatione*, a work with a logical character, he asked whether the law of the excluded middle (either A will become B or A will not become B) can be applied to particular events. He will not accept that it can, since he believes that some things happen by chance; but philosophers are still debating the problems he bequeathed in the course of this work. In the *Physics*, he distinguished between two different sorts of necessity, absolute and hypothetical. Suppose, again, that A becomes B. In simple mechanical cases, such as when light particles pass through the pores of a piece of horn, B comes about because A has been. By and large, however, physics is concerned with cases where A *must* be because B is to be. If B, then necessarily A. If there is to be a bronze statue then necessarily there must first be bronze.

That Aristotle was at heart a biologist shows up in his frequent use of examples from the animal world. His entire matter-form doctrine (hylomorphism) was designed as an answer to the atomists, however, who wanted to explain everything material in terms of arrangements of smaller things, often invoking chance—indeed, too often, for Aristotle's liking. Hylomorphism explains sensible things as composites of matter and form (Greek *hulê*, matter, and *morphê*, form). They were not meant as bodies or physical entities that could exist or act independently. They were considered to exist and act only within the composite, and they can be known only indirectly, by some sort of intellectual analysis. In broad outline, the doctrine was already familiar in the early middle ages. As Isidore of Seville had it, in his popular encyclopedia dating from the seventh century, the prime matter of things is not formed, but has a potential to take on all bodily forms. The visible elements are formed out of it, and sensible things out of them. This simpler story became popular in the middle ages, when Aristotle's wishes were often conveniently forgotten, because it was easy to grasp.

A third discussion of necessity is that in Aristotle's *Metaphysics*, where he considers chains of (supposedly) necessary causes, and insists that a point must always be reached at which the chain ends, so that there is a cause which has no cause. As innumerable philosophers have done since, he used a certain sleight of hand to get out of what, to many a student of nature, seemed like an insuperable difficulty. If there is no first cause, going backwards in time, are we to suppose that time is infinite? Doing so got him out of the difficulty of having events in his account which were considered absolutely necessary. He lived at a time when great progress was being made in finding explanations of the motions visible in the heavens. They were circular, and so endless. To find absolutely necessary causes we must turn our thoughts to the heavens. Other cyclical changes, as in the seasons, might be a second-best example; but it is easy to understand why Aristotle came to the idea that the region of the universe above the Moon was unchanging in general character, and somehow perfect, harmonious, and endless both forwards and backwards in time. All within the terrestrial regions, by contrast, he considered imperfect, subject to change, birth, growth, and decay. Aristotle's view of these contrasting worlds was easily incorporated into the medieval Christian vision of heaven and earth—indeed it had earlier helped to form that vision.

From here, Aristotle turned to a study of movement itself, in a highly abstract and brilliant, but complex, discussion. It antecedents were arguments proposed by the atomists that needed answering. There were, for instance, Zeno's paradoxical denials of the existence of any kind of movement and change. There were the atomists, with a more limited denial, saying that there can be no changes of quality, only a sort of reshuffling of atoms. Plato had asked whether motions take place discontinuously, happening in fits and starts, in one instant, then in another, and so forth. Aristotle insisted that motion is continuous, and that it takes place in a continuum that is infinitely divisible. He introduced two concepts which were bandied around in the middle ages, those of the potential and the actual. In speaking of motion, he was thinking not just of locomotion, movement from place to place, but of bringing about change, as in casting a statue. Motion, he said, is the actualisation of what exists potentially as such. After much discussion, he decided that there are only three sorts of movement, those in quality, quantity, and place.

There are other fundamental doctrines which distinguished Aristotle from his predecessors, and from many of his medieval opponents, and which sprang from his analysis of motion. One concerned the concepts of the infinitely large and the infinitely small, or better, the infinite in respect of addition (so that we cannot work our way through it by adding part to part) and in respect of division. He decided against actual bodies that are infinitely great, but in discussing this question, and time, magnitude, and number, he introduced an ingenious conceptual distinction between actual and potential infinities. Broadly speaking, he rejected the idea of simultaneously existing actually infinite wholes, while allowing the potentially infinite by addition: one can just keep going on and on, according to a general rule of adding. This, at least, is where his doctrine seems to be leading, but unfortunately he had already settled for a universe of finite size, so limiting the length of any straight line within it. Time offers another problem. It cannot be an infinite given whole, he said, since it is not of time's nature for its parts to coexist. He argued that it is (potentially) both infinitely divisible and infinite in regard to addition. Here, needless to say, he caused many a headache for Jewish, Muslim, and Christian scholars, almost all of whom believed in the world's creation at a particular moment, the beginning of history, when (as some argued) time itself began.

Aristotle's doctrine of place is too subtle to be dealt with in a summary fashion, but since it has a bearing on his cosmology we should at least be aware of his final definition of the place of a body. It is the inner boundary of the first unmoved body that contains it. Everything in the universe has a place, but the universe does not, if by 'universe' we mean everything that there is. This definition has consequences for his views on the impossibility of a vacuum or void, empty place. He might simply have argued that a void is a contradiction in terms, saying that without body there can be no place; but his several arguments for the impossibility of a void are very varied, and of varying quality. One of them was often repeated in the middle ages. He believed that the speed of movement of a falling body depended on the ratio of the weight to the resistance of the medium through which it fell. (This is of course quite different from Galileo's law that a body falling without resistance does so with a constant acceleration.) A body falling through a void, Aristotle thought, could encounter no resistance, in which case—he thought—the body's motion would be instantaneous. This he considered to be impossible, for nothing can move in zero time. The conclusion was plain: there can be no void. The maxim that 'nature abhors a vacuum' was interpreted by later writers in many different ways, but it is worth noting that Aristotle's view of matter as a continuum (in opposition to the atomists' view of matter as made of atoms moving through a void) allowed him to speak of a (potentially) infinite thinning out of matter.

Aristotle's closely integrated discussions of continuity, of time, and of movement ('of which time is the number'), represent some of the high points of ancient philosophy, and some of the most difficult. Their intrinsic merits were not fully appreciated until the nineteenth century—when it was also discovered that several fourteenth-century Oxford philosophers had made important additions to what they found in the *Physics*. Putting them aside for the time being, we may consider some of the cosmological consequences drawn by Aristotle from his doctrine of motion. From it he was led to the notion of a 'First Mover', a *Primum Mobile*, an especially important concept, since he gave to the First Mover ultimate responsibility for all movement in the world.

Aristotle's argument is finely textured, but it rests on such principles as that there has always been, and always will be, motion, and that what is moved is moved by something. It was to end the chain of responsibility for movement that he needed his First Mover. It was unmoved, but in the sense that it moved by itself and was moved by nothing else. It had to be eternal for him, since the world and movement are eternal; and since he thought only circular movements can be continuous and endless, he concluded that the First Mover is at the circumference of the world. (He thought it must either be there or at the centre, since they were end-points of the chain of motions; and he opted for the outer limit since there the motions were fastest and more powerful.) The First Mover was the outermost heaven in his cosmological picture, but it was unlike the rest in that it was incorporeal. How did it impart motion, if it could not provide a bodily push or pull? In the *Metaphysics*, he played with the idea that it caused motion as an object of desire or love. This made it even easier for medieval writers to identify this, the cause of all motions, with God. Aristotle would

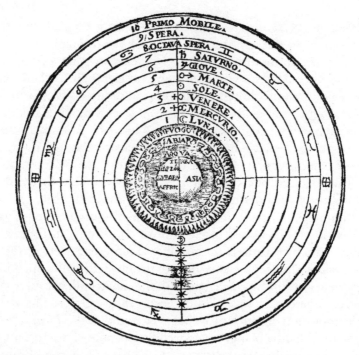

53. A typical illustration of the Aristotelian cosmos. From the first printed star atlas: Alessandro Piccolomini, *De la sfera del mondo. De le stelle fisse* (Venice, 1540). The language is Tuscan. The tenth (outermost) sphere is the First Mover (*Primum Mobile*), the ninth is not Aristotelian at all, but stems from the theory of the slow (precessional) movement of the stars of the eighth sphere, where the signs of the zodiac are marked. The planetary, solar, and lunar spheres come next (in the order Saturn, Jupiter, Mars, Sun, Venus, Mercury and the Moon), while within them come the sublunar spheres of the elements (fire, air, water and earth).

have said that, being incorporeal, it has no place, but for many a later writer it was the most important place of all, the place where God resides.

This theological dimension to Aristotle's writings, which accounts in some measure for his later influence, was perceived also in his astronomical work, *De caelo*, 'On the Heavens'. Some of the details will be touched upon in a later chapter, but here we may anticipate the overall Aristotelian picture, as it was generally understood in the middle ages (Fig. 53). The universe is taken to comprise a series of concentric spheres, with the Earth at rest at the centre and the First Mover on the outside. The outermost (corporeal) shell is that which carries the fixed stars with the daily rotation. The shells within it carry the planets, Saturn, Jupiter, Mars, the Sun, Venus, Mercury, and the Moon, working inwards. The motions of the planets are complex, and will be put aside for the time being, but what has already been said about the relations of the celestial region (where motions are circular) to the sublunary region (where natural motions is up or down) needs further comment, since there were certain phenomena which it was difficult to know how to classify.

Meteorology is by its Greek etymology 'the study of things above', but not only of those things that are now deemed meteorological. Earthly phenomena such as volcanic eruptions, tides, and even the formation of mineral deposits, all of which were supposedly influenced from above, came under the same heading, and in Aristotle's treatment of meteorology everything below the sphere of the Moon was included. It was in his *Meteorologica* rather than in the broader context of cosmology that he offered explanations of shooting stars, comets, and even the Milky Way, since these were thought to be in the upper atmosphere. The most influential later Greek writer on such questions was Posidonius of Apamea, but his work is known only through Seneca and Strabo. The fact that Aristotle's 'meteorological' ideas were (for him) unusually speculative encouraged others to embroider them. If there are two forms of exhalation rising upwards under solar influence from the earth (vaporous, moist, cold and windy, smoky, hot, dry), why not more, thick and thin versions of both? There were Arab paraphrases and commentaries on Aristotle's key work, and Averroes made a careful comparison of some of them. Avicenna was stimulated to measure Aristotle against the phenomena as he observed them himself, and as a result he found Aristotle wanting. Two notable instances, of great historical interest, relate to the rainbow and the Milky Way. Aristotle is surprisingly weak on the Milky Way, as later commentators did not hesitate to point out. Why, if it comes from burning exhalation, is it so stable, and why, if it is so near, does it include point-like stars that reveal no parallax? They were happy to provide their own answers.

When we come to phenomena in the lower atmosphere due to moisture, we reach meteorology as the word is now understood. It is surprising that the different experiences of the weather by the Greeks, Arabs, and Europeans did not lead to more disagreement. There was some, especially on questions of precipitation, hail and snow, winds, and the habitable regions of the Earth. With earthquakes, thunder, lightning, thunderbolts, hurricanes and whirlwinds, however, we are back in the foggy regions of exhalation theory, the vagaries of which, one might say, have even to this day not been thoroughly mastered. There was one group of phenomena, however, which offered practitioners of the exact sciences in Islam an opportunity to excercise their analytical skills. Haloes, rainbows, and mock suns begged to be investigated by optical theories to which Aristotle did not have access. Avicenna and Ibn al-Haytham were two of those who passed this subject on to later western science, to Grosseteste, Bacon, Pecham, and others; and while to us they appear to belong to an isolated branch of physics, it is as well to remember their context, namely Aristotle's *Meteorologica*. As far as the Aristotelian world view is concerned, there is no such thing as an isolated branch of physics.

18

Natural Philosophy in Oxford

> One who constructs a sundial or water-clock wrongly is refuted by the clear evidence of the facts; but the refutation of philosophical propositions is not so immediately clear. People may say whatever they wish, when they are shameless enough to abandon logical method.
>
> Galen, *The Affections and Errors of the Soul*, ch. 5[143]

NO HISTORY of the rise of western science can ignore Oxford's contribution to natural philosophy in the later middle ages, and no history of Oxford can ignore the writings of Aristotle, on which so much rested. While Robert Grosseteste was by far the most influential early Oxford Aristotelian, he was not the first to teach there from the newly translated Latin works of the philosopher. If we are to believe Roger Bacon, Edmund of Abingdon was the first to lecture on the *Sophistical Refutations*, and a certain Master Hugh was the first to teach from the *Posterior Analytics*. John Blund was very probably the first with the *De anima*. All of these courses would have been held in the first decade of the thirteenth century.[144] Aristotle's ideas were viewed with suspicion in many quarters from the beginning. In Paris there were condemnations of his books as early as 1210 and 1215, which suggests that the new learning was introduced to the two universities at much the same time. Most of the new translations prior to William of Moerbeke's had probably arrived in both places before the end of the 1230s, by which time many of the commentaries of Averroes were also widely known. By then, Grosseteste's influence was such that the Aristotle's place in the curriculum was assured. There has recently been much debate concerning the chronology of Grosseteste's writings, but here it will be assumed that most of his scientific writings are to be assigned to the thirty years before 1225, and that some of the more metaphysical scientific writings—especially those in which light is made the key to understanding the cosmos—belong to his 'theological period', from 1225 onwards.[145] What might at first sight appear to be a dry question of chronology is clearly much more, since it concerns the motivation of the most important person in the English scientific movement of the thirteenth century. The broad conclusion is that Grosseteste was a natural philosopher looking for ways of applying his knowledge to a higher study, rather than a theologian seeking scientific knowledge because he had theological questions to answer. The pattern was one that was often repeated in Oxford thereafter.

Grosseteste's lectures to the Oxford Franciscans in 1229-30, his own writings, and later his bequeathed library, influenced a number of excellent Oxford scholars directly or otherwise—Adam Marsh, Thomas of York, Bartholomeus Anglicus, for example, and later Adam of Buckfield, Roger Bacon and John Pecham. They were inspired by the unusual breadth of his learning, but they lighted on four theses in particular with a bearing on the sciences. Grosseteste taught that Christian ideals could be furthered in new ways by those who were prepared to study the sciences, and the languages—especially Greek—in which so many crucial scientific works had first been written. Second, he taught the importance of mathematics in the study of nature, and put this doctrine into effect in his own investigation of optics. This was not merely by way of a case study to illustrate his point, for he gave to light a special cosmological significance. Third, he taught the importance of experience—and even, in a very limited sense, experiment—in the advancement of the physical sciences. And fourth, he taught an invaluable Aristotelian lesson on the formal structure of the sciences. We shall consider all of these theses, for all were influential to a greater or lesser degree.

A Metaphysics of Light

We have seen that Aristotle found optics hard to classify, describing it simply as one of 'the more physical parts of mathematics'. Grosseteste was no expert mathematician, but he could see how powerful were mathematical methods in providing explanations of physical phenomena. One particular idea he found appealing, which had been ascribed to Plato, was that atomism could be reduced to mathematics. Here he knew that he was playing a dangerous game: atomism had been opposed by Aristotle, and was also associated with atheism. In a work he wrote *On the Heat of the Sun* (*De calore solis*), Grosseteste nevertheless touched on the idea that the world is made of particles, and he there explained solar heat in terms of the falling motion of small bodies. His ambition was to produce a cosmology which would somehow build up the world out of elements more akin to points than to particles.[146]

His was no simple problem: he had to reconcile the Bible with the arguments provided by his philosophical heroes—Plato, Aristotle, and Augustine. He advocated a Christianised Aristotelian scheme of the celestial spheres, but where was he to find the matter to fill the spheres? He found it in light. In the creation story of Genesis, the words 'Let there be light!' come only after mention of God's creation of formless matter, but with a certain sleight of hand Grosseteste argued that light is a necessary and sufficient condition for bodies to exist, having the 'function of multiplying itself and diffusing itself instantaneously in all directions'. Light, he said, was 'the first corporeal form': it requires three dimensions, and without three dimensions no body can exist. (He is here borrowing from Averroes and various Jewish or Arab authors whom he mistakenly took to be Aristotelian, such as Avicebron, Algazel and Avicenna.)

It would be easy to brush this essay of his aside, but for all its weaknesses it introduced the Oxford community to an important theme for discussion. Aristotle had shown—against Plato—that it is impossible to build up a line out of points, or a surface out of lines, or a volume out of surfaces, by simple addition. Grosseteste's answer was ingenious but flawed. He held that the addition would in each case produce the required result if done an infinite number of times. He thought that light, through the *infinite* multiplication of itself, can extend matter into some finite-dimensional thing. In his commentary on the *Physics* he again struggled with the same idea, but he was forced to resort to an unsatisfactory theological device: while we cannot do so, God can build up finite things out of the infinitely small. Not until the following century would such problems be tackled in an adequate fashion, and then it was by Thomas Bradwardine. In a discussion of these questions unequalled in logical power before the seventeenth century, Bradwardine argued against the idea of what he called 'non-extended indivisibles', the sort of points that his predecessors had considered capable of being built up into extended continua. It seems very likely that he had Grosseteste in his thoughts.[147]

Having persuaded himself that he had an adequate theory for building matter out of light, Grosseteste could begin to explain the process of the creation of the world. A point of light first gave rise to a material sphere. As the sphere spread outwards it was said to have become more and more rarefied until, at the outermost edge of the universe, there was nothing left to the matter in it but 'first matter and first form'. The language here is Aristotle's, as is that of Grosseteste's next step: when the firmament became 'completely actualised' it acted like the light at the centre, and radiated light back to the centre. At this point he began a long explanation of how, by a process of repeated retransmission and condensation at lesser and lesser distances from the centre, a series of spheres—the set of nine heavenly and four sublunar elemental spheres of Aristotelian cosmology—is formed out of the central light. The argument is impressionistic: it might have been inspired by the idea of standing waves in a circular pond, but at least he considered that he had done what he set out to do. He saw himself as having combined Aristotle and Genesis in a single unified scheme. Grosseteste's cosmos, arising through the infinite multiplication of a point atom of light, allowed of no void, no empty spaces. In this respect too it was true to Aristotle. Whatever his sources, and for all its shortcomings, Grosseteste's light-geometric cosmogony was the most original cosmogony of the middle ages, a theory of creation based on plausible natural principles. It even has a curious resemblance to the theory of the primaeval atom proposed by the twentieth-century cosmologist Georges Lemaître.

Grosseteste's theory was offered without the remotest hint that it should be tested against experience of the real world. (As it happens, this is something that was often said, mistakenly, about Lemaître's.) We might excuse him if we could show that it was a composite of theories, each of which was well supported by observation and experience. After all, Aristotle's cosmos was imagined to have a firm experiential base, and his theory of the elements likewise. In a tract, *On the Generation of Stars* (*De generatione stellarum*), Grosseteste tried to persuade his readers that the heavenly bodies are composed of the ordinary Aristotelian elements. He tacitly rejected the Aristotelian aether, the fifth element (quintessence) appropriate to the heavens. The astrologers often assigned elemental properties to the planets, and perhaps Grosseteste was fastening on that doctrine; but what of those largely empirical theories of optics in which he was himself expert? Did they not influence him? It is not hard to find hints of the doctrine known as the 'multiplication of species' in the 'light metaphysics' of his *De luce*.[148] The link is tenuous, and hardly worth pursuing, but Grosseteste's contributions to optics had important consequences for the later development of Oxford science.

Grosseteste and Thirteenth-Century Optics

At the very heart of philosophy lie the many problems which concern the nature and acquisition of knowledge. Many traditional theories had made use of the idea that some thing or things mediates or mediate between the knower and the thing known. In the language of the time, these intermediaries were sometimes called 'sensible species' (*species sensibiles*). In a series of brilliant lectures he delivered in the period 1317-19 on Peter Lombard's *Sentences*, William of Ockham would raise serious objections to the concept of a mediating species, but in Grosseteste's day the notion was widely accepted, and the theory of optics was used to underpin it. Throughout all of history, even to the present day, epistemology (the theory of knowledge) has given undue attention to the kinds of knowledge that are obtained through sight. Aristotle did so in his *De anima*, for instance, and Augustine even insisted that the other senses actually make use of vision. It is not hard, therefore, to understand why the science of optics, *perspectiva*, obtained such an important place in the medieval natural sciences. Once Aristotle's writings had entered the curriculum, two or three of his works in particular (*De anima*, *De sensu*, and *Meteorologica*, for instance) forced scholars to discuss his theory of vision, but even had that not happened, the writings of the perspectivists would no doubt have been valued for the theory's epistemological content.

There were in Greek antiquity four main schools of thought on the question of the nature of vision. They are best described in logical rather than chronological order, although that of the early atomists came first in point of time. The atomists had the idea that objects could slough off a succession of thin films of atoms, each like the skin of a snake, which passed to the observer and conveyed a series of images of what the object was like. As an alternative, which Euclid and Ptolemy both favoured, it was believed that radiation is sent out by the observer's eye in straight lines (except when they were reflected or refracted), and that somehow the radiation perceives (or feels) the object, and passes back the perception to the observer's senses. The other two types of theory correspond, in a rough sense, to these two, but postulate a medium which connects the object and the observer. For Aristotle (who lived fifty years before Euclid and five centuries before Ptolemy), the object transmits its properties through some sort of medium (perhaps air) which intervenes between it and the eye. The medium was conceived to actually change the properties of the eye on arrival. The fourth type of theory was put forward by the second-century medical writer Galen. He thought that there was some sort of spirit passing from the brain to the eye and out into the air. The idea was that the air so becomes an extension of the eye, and capable of perceiving the object.

Of these theories, the two intromission alternatives—where something passes from the object to the eye—come closest to our own, although none comes very close. Most of the obvious objections to all of them were raised in antiquity, and later by Arab writers, but at one time or another there were supporters to be found for all types of theory. Despite its extramission character, the theory proposed by Euclid and Ptolemy had the great merit of being combined with a geometrical analysis of the visual rays, allowing a number of exact laws to be formulated. By far the most important contributions to geometrical optics in the Islamic world were made by Ibn al-Haytham (965-1039), the distinguished mathematician, natural philosopher and astronomer who, we recall, was known in the West as Alhazen.[149] He took advantage of a principle put forward more than a century before his time by al-Kindī, who saw the futility of analysing emission from the surface of the perceived object as a whole, but who said rather that each point on the surface of that object sends out light in straight lines in all directions, independently of the rest. What seems a very obvious idea—at least to those who have been taught it—was not at all obvious in the past, and it presented a problem which al-Kindī found himself unable to solve: since every point in the eye seems bound to receive rays from every point of the visible part of the object, why is its appearance to us not utterly confused? It was Alhazen who offered the first plausible (although mistaken) explanation: of the many rays the eye receives from a given point on the object, he said, only that which falls at right angles to the eye is received at full strength, the oblique rays being too weak to stimulate vision.[150] Each point seen in the outside world may be assumed to correspond, therefore, to a single point on the eye; and vice-versa.

The intricacies of Alhazen's many contributions to optics, in particular those offered in his principal text *Optics*, do not lend themselves to a brief summary, but part of the reason for his great influence on later generations is easy to understand. It is that he was a good judge of what was best in his predecessors. He took geometrical methods from Euclid and Ptolemy, the doctrine of forms from the philosophers, and the physiology of the eye from Galen, and combined these elements into an intromission theory which quickly superseded everything that had gone before it. With Galen, he thought that sensation took place in the crystalline humor (the fluid behind the lens). He still spoke of light as a form, conveyed to the eye through a transparent medium, and yet his main aim was not to produce a theory of light, as Aristotle had done, but rather to produce an experimental and mathematical account of light and colour in vision. Among his many valuable contributions to the optical tradition, not least was a set of rules he proposed which may be regarded as embodying a law of refraction. (Our familiar sine law of refraction had to wait nearly six centuries, for Thomas Harriot.)

Alhazen's *Optics* was translated into Latin under the title *Perspectiva*—so providing the Latin and English name for the subject, 'perspective', until this was gradually changed to 'optics', under the influence of new Renaissance translations of ancient texts.[151] It is not known precisely when, or by whom, the first translation was made, but this was probably produced in the early thirteenth century, in Spain. There is at most one oblique reference to it in the whole of Grosseteste's writings. He knew Euclid's optical writings, and al-Kindī's, but he certainly does not give evidence of having known Alhazen's work well. Roger Bacon, however, did; and the resulting difference between Grosseteste's style and Bacon's is revealing. As in his cosmology of light, so in many other writings, Grosseteste writes about light in generalities. With biblical support, he could write of God as light, the uncreated light, and of Christ as the true light, the light of the world. When he provided an analysis of the optics of the rainbow, he analysed the cloud rather than the individual droplets, as Bacon was to do. Grosseteste still adhered to the concept of vision through the transmission of species. Bacon took over Alhazen's analysis into points. He had mastered enough of Alhazen's work by the time he began—around 1260—to produce a series of treatises of his own in which he gave the new ideas publicity. Not only did he draw on Augustine and pseudo-Dionysius, Avicenna and Averroes for his theory of visual perception, but he tried to keep the memory of his hero Grosseteste green, by blurring over the differences between his ideas and Alhazen's. This was not easy: Alhazen had denied the existence of visual rays, while Grosseteste had insisted that all natural bodies, eyes as well as perceived objects, emitted them. Both Roger Bacon and John Pecham, Franciscans both, put forward the Grosseteste view as an explanation of vision, and then ignored it as they developed their theories of sight further. There is in fact enough of Alhazen in the writings of Bacon and Pecham, and of the contemporary Silesian writer on optics Witelo, for it to be said that, in this respect too, western science owed a great debt to the Islamic rendering of the Greek tradition.[152]

In the course of the thirteenth century, Aristotelian writings transformed Oxford science, and their influence continued to grow in the fourteenth. There are perhaps two dozen scholars whose commentaries on the Aristotelian works of natural philosophy would be worthy of discussion in any detailed account. One especially assiduous Oxford scholar in this connection was Adam of Buckfield, who—perhaps too much influenced by Averroes—was lecturing on the entire corpus in the early 1240s. This was before his contemporary Albertus Magnus performed a similar service in Paris, there helping to ensure that Aristotle would dominate the Paris arts curriculum.[153] By the second half of the century, no Paris student could become master of arts without hearing lectures on all of the Philosopher's works then available in Latin. Commentaries on them were produced in abundance; and, while optics occupied only a small place in those, the fact that Aristotle had dismissed the extramission theory of vision also helped the Alhazen tendency along. His works, and the new western writings by Grosseteste, Bacon, Witelo, and Pecham, shared responsibility for promoting the subject of optics, and with it an awareness of how an empirical scientific theory might be set out. Their writings became standard university fare, taught with few additions in the later middle ages. The most significant fourteenth-century work on optics was by Theodoric of Freiburg. Writing around 1304, he confirmed his geometrical analysis of the rainbow through an experiment using a translucent sphere as proxy for a raindrop. Theodoric's work was seemingly still unknown in Oxford half a century later, however.[154] A renewed and deeper interest in these medieval texts came in the sixteenth century, stimulated by the writings of Francesco Maurolico. At the beginning of the seventeenth century, Johannes Kepler took up the challenge of explaining vision, neatly dividing the problem into physiological and perspectivist components, the latter based on an intromission theory. In very many respects, Kepler's work marks the beginning of modern optics, but it nevertheless rested heavily on medieval foundations with a strong Oxonian component.

Aristotle and Geometry

Aristotle's influence extended even into geometry, and there is no better example of this than a highly original work by Thomas Bradwardine, his *Geometria speculativa*. (This was written within Richard of Wallingford's lifetime, although possibly after he left Oxford.) Bradwardine began by following Aristotle in making geometry second to arithmetic—a subject on which he says he has written, although the work cannot now be identified. (Of surviving treatises in manuscript there are three candidates, one in verse aimed at remedying the lamentable modern neglect of theoretical arithmetic.) His view of geometry is not entirely true to the spirit of Euclid; he does not try to reduce his list of axioms, 'common notions', to a minimum, which is usually taken to be the mark of good mathematical taste. (As a specimen of Euclid's axioms, the first reads 'Things which are equal to the same thing are also equal to one another'.) In this connection, however, it is interesting to see him quoting Alhazen on the psychology of vision, when he says that the 'common notions' are propositions known to us by almost imperceptibly rapid reasoning. It is almost certain that he took this reference from Roger Bacon, an author from whom he seems to draw in several other places. When writing on circles, and their perfection, Bradwardine quoted Aristotle's assertion that mathematics was concerned with the good and the beautiful. Quite casually, after one of his demonstrations, he tells the reader that they will know this sort of thing from Aristotle's *De anima* and *Metaphysics*. There are times when he seems to be using Aristotle merely to give his work an acceptable cachet, although his modern editor has seen the work rather as 'an aid to lectures on Aristotle for explicating his frequent mathematical allusions'.[155]

For the geometry proper, of course, Bradwardine needed to base himself not on Aristotle but either on Euclid—which he did through the Campanus version—or on some derivative work, such as that by Theodosius on the geometry of the sphere. Bradwardine's text is largely descriptive, it has few lettered diagrams, and it looks deceptively easy until its technicalities are examined at close quarters. That 'the property of being rectangular does not allow of degrees' is as much a philosophical as a geometrical proposition. Like many of his modern philosophical successors, Bradwardine is often impatient of the technicalities of geometry, so that, for instance, he regularly omits proofs of constructibility, or indeed of any sort. The work, even so, gives a very good idea of new scholastic ways of looking at the subject, and it contains several exciting ideas not to be found in Euclid. There is a pragmatic side to it all, as when he comes to deal with problems of isoperimetric figures (figures having equal perimeters), and there is a smattering of spherical geometry of a qualitative sort, relating to another text Bradwardine seems to be suggesting that he has written. His love of geometry for its own sake is evident in what he writes on problems of star polygons, solid angles, space-filling by pyramids, and the number of spheres touching a given sphere. To see how fertile Bradwardine's mind was, one might turn to his consideration of the paradoxes of curvilinear angles, and incommensurability. Judged by modern or even ancient mathematics, he may appear ill-disciplined, but that is so often the case with fertile minds. Above all, he leaves us in no doubt that he was motivated by a philosophical desire to get beneath the surface truths of geometry, and that his philosophical model was Aristotle.

Aristotle and Scientific System

Aristotle would perhaps have been surprised at the place occupied by optics in the university science of the thirteenth century, although—as we saw in the last chapter—it was found a place in his division of the sciences. It was for him merely one of the more physical parts of mathematics. Those who, after him, classified and expounded the liberal arts by which the arts curriculum was structured—Varro, Augustine, Boethius, and the rest—had likewise made no special provision for it. In Hugh of St Victor's classification, philosophy was divided into theoretical, practical, mechanical, and logical; and from the middle of the twelfth century this list was often repeated, albeit much extended by subdivisions, some of them deriving from writers in Arabic. With a little ingenuity, optics could have been found a niche in these classifications. In a more refined scheme proposed by Robert Kilwardby, when he was still a Dominican friar, we find optics placed side by side with geometry, astronomy, music and arithmetic. As we have seen, Kilwardby was no lightweight. He was to become provincial prior of his order before his translation to the archbishopric of Canterbury in 1273, and it was in a guide to the sciences written around 1250 that he gave optics this special place, grouping it with the arts of the quadrivium under mathematics. With the natural sciences and metaphysics, mathematics was given an honourable place under 'divine things'. The practical arts—which included even ethics, grammar, logic, and rhetoric—were classified as merely 'human things'. If Richard of Wallingford had looked at this work, he might have wondered where Kilwardby would have placed clock-making. As a mechanical art, it would have been placed with farming, armament, commerce, and even medicine; as astronomy, it would have belonged to the philosophy of divine things.

Kilwardby's treatise *On the Origins of the Sciences* (*De ortu scientiarum*), in which this scheme was presented, seems to have been widely circulated in Paris and Oxford. No doubt optics was included because so much attention was being given to it by his contemporaries, but he cites no other work on the subject than Alhazen's. The sixty-seven chapters of the treatise provide us with many similar insights into academic attitudes in mid-century. To the student, these chapters offered knowledge without tears, in the form of superficial surveys of the definition, scope, sources, and divisions of the sciences and arts. The medieval scholar loved nothing so much as a classificatory scheme, especially one that had a hierarchical character.[156] The Boethian stages by which the mind ascended from changeable material nature, through immutable abstract mathematics, to immaterial metaphysical being, was an especial favourite. Kilwardby had Aristotle to hand, and cited him more than a hundred times, which may surprise us, in view of his prohibitions made in 1277 in the wake of Tempier's. The explanation is that he clearly saw no issues of faith at stake in his Aristotelianism in the *De ortu*.

The interlocking of science and theology throughout the middle ages is part of the reason why so many writers have considered the period doomed to be scientifically barren. Kilwardby no sooner states in his first chapter that he plans to exclude theology from his book than he declares a belief in the Augustinian thesis that all of human knowledge comes from the action of divine illumination. Whether or not the idea of divine illumination is essentially inimical to science, however, is not a simple question. It should certainly not be equated with mystical supernaturalism. Some have seen it as encouraging anti-scientific rationalism, and the notion that it is not necessary to go out into the world to discover how nature behaves. Many acknowledged scientists of the middle ages and even of the early modern period accepted it. In the thirteenth century some tried to combine it with Aristotelianism. It was certainly not a Dominican prerogative—many Franciscans promoted it strongly, and many Dominicans qualified it heavily.[157] It was not without supporters in the Jewish and Muslim worlds. Avicebron was one popular Neoplatonist writer who found favour with Christian theologians of many persuasions largely because it was so easy to read him as an Augustinian. In Oxford, the fact that Grosseteste had accepted the divine illumination of the soul with truth was for many scholars authority enough.

What was quite understandably—but regrettably—absent from Kilwardby's survey of the sciences as a whole was a sense of the importance of an overarching system within an empirical science. He perceived the interdependence and mutual need of theory and practice, but he seems to have only dimly appreciated that if a branch of learning is to count as science it must be bound together by general principles, and that they must themselves be systematically ordered. Taking this idea on board was a slow and painful business for the scholastic world. Even harder to appreciate was the idea that seemingly disparate scientific theories could be united, or at least should be united, to the best of human ability. Slowly, however, these messages were absorbed, and scholars became aware that, in the long run, a rationally ordered system was more important than a heap of facts.

The new awareness came from two sources—example and precept. It is easier to describe them than to say which, in the long term, was to prove to be the more important. Good examples of scientific systems were in short supply—and it is by example that most scientists down the ages have learned what they know of the logic of science. There is no doubt that the most important source of precept was Aristotle's *Posterior Analytics*—although his attitudes to this question were detectible at many other places in his writings.

When western clerics first embraced the knowledge to be had from Near Eastern cultures, it was not that they saw in them great architectonic virtues. It was rather that they could recognise parallel practical and religious needs. Jews and Muslims too needed the hour and the day and the time of full moon; they too needed to mend broken bones, to lay out arches, to colour glass, and so forth. New methods of working were required, but in the first instance for practical reasons, not for the sake of the theory from which those rules had originally been derived, which was often barely discernible. Much the same could be said about the other empirical sciences, and about the rules of working in mathematics. Roman and medieval western geometry, for example, took over many of the results of Greek geometry, but without the supporting theoretical (that is, axiomatic) system by which they were justified. Even when Euclid's great work of geometry, *The Elements* , was discussed by western scholars in the twelfth century, its remarkable logical structure often completely escaped them. In the physical sciences of the thirteenth century—excluding astronomy, which is a special case, since it allows of no straightforward controlled experiment—optics was the only branch of activity where experimental confirmation or test was combined with a theoretical underpinning of any depth. Looked at in this way, while the early medieval sciences are historically interesting, they were far from being the chief glory of the age.

An awareness by western scholars of the need for a rational system in an empirical science was provided by the writings of Aristotle and his commentators, although it was partly masked by Aristotle's emphasis on a search for causes. Aristotle set forth in his *Posterior Analytics* what he considered to be the mark of a demonstative science. Mathematics, in particular geometry, gave him the basis of his theory, and those who had been taught geometry in the Euclidean tradition, with its first principles (definitions, axioms, and postulates), and logically derived propositions, would have known instinctively what he had in mind. The question was then whether it is possible to have a demonstrative science of the natural world, and if so, how it should be acquired and secured.

According to Aristotle, full scientific knowledge of something requires an understanding of its necessitating causes. In offering proof that we have this, we must give a proof of it (a demonstrative syllogism) that shows it to be a 'reasoned fact'. We must show *that* it is so, and *why* it is so. (The first demonstration was called a *demonstratio quia,* the second a *demonstratio propter quid,* 'on account of which'.) Those who accepted the broad outline of the nature of an Aristotelian-style demonstrative science were more or less agreed on its overall pattern. They were also in general agreement on the logic by which the propositions (theorems) were to be deduced—it was to be the logic, cast in syllogistic form, which owed so much to Aristotle himself. Where they were least often unanimous was on the status of the first principles of any particular science. What are their sources, how are they related to truths in other sciences, are they were necessarily true, and if so, what does necessity mean in this case? As always, when there was talk of necessity, talk of God was never far to seek. When elaborating upon Aristotle, therefore, there was plenty of scope for disagreement.

The many different sorts of debate which clustered around Aristotle's conception of a demonstrative science continued, and indeed still continue in other dress. While they did not lead to anything approaching a consensus, it was through them that the crucial scientific ideal of a formal, logical system, underlying all worthwhile scientific systems, entered into the mentality of western European scholarship. Its adoption should be counted as one of the most important of all episodes in medieval history, for it marks the stirrings of a western scientific consciousness. To quote Einstein, science is 'an attempt to make the chaotic diversity of our sense experience correspond to a logically uniform system of thought'.[158] Innumerable modern writers have made the same point, and very many more would have said that he was uttering a platitude, but that shows how far we have come round to an idea that would not have seemed at all obvious in the earlier middle ages—the simple idea that science on the small scale seeks to create a network of acceptable generalisations. That the unification of different fields of scientific activity is no more than a natural extension of the universalising tendency of science on the smaller scale would have then seemed even less obvious.

Rationalists, Empiricists, and God

The earliest known Oxford statutes for the arts faculty, dated February 1268, list the books of Aristotle's natural philosophy which every bachelor was expected to read. The arts curriculum was a propaedeutic to theology. Even those whose main university allegiance was to the theology faculty, indeed, even the most conservative of them, such as Bonaventure, recognised the value of Aristotle's natural philosophy. There was irony in the post-1277 situation in Paris, when masters in arts were forbidden to teach these books while the theologians continued to make much use of them in their theology. There, as they had long been doing in Oxford too, they imported Aristotelian natural philosophy into their commentaries on the *Sentences*. Even when they could not agree with it, they delighted in using it as a foil for debate. Such theological debate, before and after 1277, introduced numerous tensions into natural philosophy itself.

Greek geometry had given to the sciences—through Aristotle—a lesson in logical structure, but the lesson stopped short at experience of the world. As the thirteenth century wore on, more and more western scholars realised that the kind of necessity applying to geometrical propositions was different from that be found in the overtly empirical sciences. Responses to this situation varied greatly. The common scientist's way has always been to treat firm conclusions as necessary, and to paper over the cracks in the argument until (if ever) experience forces a retreat. The rationalist looks for advance guarantees of truth, but rationalists come in many different sorts. Aquinas, for instance, with his love of final cause, argued quite illogically that if the end result occurred (the tree, say), then the efficient cause must have occurred (in this case the seed). Throughout history, this fallacy (technically known as 'affirmation of the consequent') has been at the divide between rationalists and empiricists. Duns Scotus was no empiricist, but he saw the fallacy, and avoided it by looking for guarantees of necessity in the premiss, not only those guarantees provided by divine illumination but others of his own making. Divine illumination was a common route to certainty, in principle if not in practice. Aristotle's was a very different account from that of Augustine, who supposed that we may have access to knowledge by divine illumination. Grosseteste's dilemma was that he wanted to follow both authorities. In his wish to believe that something is true if it conforms with a reason in God's mind, he was a sort of rationalist, believing that there are necessary truths about the world, but that God decides them.[159] The belief was not as dangerous as certain other forms of rationalism, for the simple reason that it is flexible. When the observed world does not agree with our theory, we simply confess that we are imperfect receivers of the light from God.

Those who are seeking a medieval pioneer of the experimental method frequently end up with Roger Bacon, but how true to his present-day reputation was he? The omens are mixed. He owed much to Aristotle, who had no agenda for experiment. As a young lecturer in Paris, Bacon began by interpreting Aristotle along theologically conformist lines, giving due deference to Avicenna and the doctrine of our illumination by God, the 'active intellect'. Having resigned his position in Paris in 1247, he spent ten years in scientific study, much of the time probably in Oxford, where he fell under the spell of the Grosseteste tradition. It is with much bombast that he tells us of his investment of £2,000 on books, experiments, scientific instruments and astronomical tables. He was, even so, a product of his age, a believer in doctrines bordering on the occult. He was greatly influenced by reading a pseudo-Aristotelian work which had been recently translated from the Arabic, the *Secretum secretorum*, or 'Secret of Secrets', for instance.

Bacon entered the Franciscan order in Paris, probably around 1257, but for reasons unknown he quarrelled with his superiors. Those who wish to cast him in the role of martyr of science have here an opportunity to supply the theologian's native hatred of science as the reason, but of course that is nonsense. His fortunes improved greatly as a result of proposals he made to a clerk in the employ of Cardinal Guy de Foulkes for the writing of a work in philosophy. The cardinal was elected pope, as Clement IV, in 1265; and before long he asked for a copy of Bacon's philosophical writings. Bacon eventually replied with his three famous works, the *Opus maius*, *Opus minus*, and *Opus tertium*. These monumental writings contained not only his views on the character of natural philosophy but proposals for educational reform and missionary planning.

There were theological tensions in plenty there. Bacon still insisted on the primacy of wisdom given to us from God, through the Bible, but insisted that it should be subjected to tests of reason and experience, whether inner and mystical, or sensory. In the second case we should, he said, supplement out experience with measurement made with the help of instruments. He made out a strong case for the necessary use of mathematics—he had geometry in mind—for the discovery of efficient causes in the material world. How revolutionary was all this? It has proved all too tempting to some of his biographers to air-brush God out of the story, but Bacon's proposed programme of research clearly owed much to his predecessors, especially Grosseteste, and their theology had much in common. With the methods of the experimental sciences in view, his arguments against the excessive use of authority in science have been found refreshing, although as Hastings Rashdall pointed out long ago, it is ironical that his argument rested on a series of citations. While it is true that Bacon used the word *experimentum* to mean experience rather than controlled experiment, he did insist that it was necessary for confirming the conclusions of deductive reasoning from general principles. In that sense he was an empiricist, although like all Christian scholastics he made exceptions for the knowledge to be had from holy scripture. He seems to have picked up the phrase 'experimental science' (*scientia experimentalis*) from an introduction to Ptolemy's astronomical work, *Almagest*. He did not use it in our modern sense, but it was undoubtedly a happy phrase that would eventually come into its own. We recall his references to a renowned Parisian 'master of experiments' (or 'experiences') who understood the requirements of a sound empirical science, a man whom we conjectured to be Pierre de Maricourt. Qualified empiricist though Bacon was, the works he sent to the pope were not in circulation in his own lifetime—he died around 1292—and he was then known mainly for his optical writings. When his fame escalated in the sixteenth and seventeenth centuries, it was helped along by an undeserved reputation for occult practices in alchemy and magic.[160]

The theological constraints on a practising scientist like Bacon were no doubt more strongly felt than those on epistemologists who theorised on our means of acquiring knowledge, but there is a detectible change of style in their writings after the Tempier condemnations. The renewed emphasis on the reality of God's absolute power and its relevance to scientific explanation affected the writings of some of the best Parisian natural philosophers—of whom Richard de Mediavilla (Middleton), Duns Scotus, William of Ockham, and Walter Burley all had an Oxford connection. To take just two specimens of the resulting arguments from the mid-fourteenth century: Nicholas of Autrecourt considered sense experience and logic to be the only certainties. Most true propositions, he said, were only probable: they could be falsified by God's absolute power. In reply, John Buridan acknowledged that God could falsify them if he wished—he could make exceptions to the law that all fire is hot, for instance—but that there is 'firmness of truth' in the basic proposition, which relates to the 'common course of nature'.[161] There speaks the voice of one who has no wish to use God's absolute power as an excuse for making no effort to reach the truth. On the contrary, he could use the notion of God's absolute power in a far more constructive way—in a way that many others had been doing in Oxford and Paris for decades. (He was writing around the year 1340.) Since, subject to the law of non-contradiction, God could do as he wished, why not investigate the consequences of as many hypotheses as possible, even those which conflicted with Aristotle's opinions? This tendency to investigate what we might see as imaginary worlds, a tendency much derided in the Renaissance, would prove in the long run to have many scientific virtues. It was a search for potentially acceptable hypotheses, between which experience could decide at a later date. Of course without experience it was not enough, but it was a necessary part of the scientific process.

There are many examples of this important type of activity from the late middle ages, examples which are all too easily dismissed as exercises in rationalism. The most striking example will be considered in the following section, but there is another important case that is perhaps easier to grasp. It concerned Aristotle's utterances on the impossibility of a vacuum or void, and one of the strangest consequences of the 1277 affair was the enormous interest this question aroused. We recall that Aristotle had argued that the notion of a void was absurd. He presented a whole array of arguments. Some people had said that a void was a place without matter, but his definition of place made this a contradiction in terms. Some of his predecessors had introduced movement into the argument. Without following them, he did the same in an extremely influential passage of the *Physics*. He said that speed of movement varies in the ratio of the weight of the moved body to the resistance of the medium through which it moves. A void, being without resistance, would yield a motion that took no time, an impossibility. Added to this, light and heavy bodies should move at the same speed through a void, another supposed nonsense. Whether or not it is right to detach such principles as these from their context, and put them into a system of 'Aristotelian mechanics', is a moot point.[162] The fact remains that the question of the void and the law of motion both gave posterity much food for thought.

This is where the theologians came into the debate. With God's absolute power in mind, could he not create a void if he wished? Did he require an empty space in which to create the world? If he had none, does that mean that the world is eternal, uncreated? Even having answered these questions, what are we to say of the limits of our existing world? Is there no empty space outside it? If not, is it improper to speak of the rectilinear motion of our own world? And otherwise, may there be other worlds besides ours in the void? Leaving Aristotle far behind, such questions could be discussed in entirely hypothetical terms, without the need to make any heretical assertion, and many a theologian found the activity irresistible. For such reasons as this, the 1277 condemnations included several articles directed against discussion of the vacuum. The articles had little effect on the theologians, but they gave rise to much valuable discussion of a kind we may perhaps best describe as conceptual exploration. The medieval contribution to this type of activity was a significant factor in the development of science, and its influence continued to be felt, well into the seventeenth century.[163]

A New Dynamics

The most striking modifications of Aristotelian science in the later middle ages were those made to the Philosopher's account of motion, natural and forced (sometimes called violent). Here was a prime example of conceptual exploration. A word of warning is called for here. When today we think of the concept of motion, we are likely to think of it as only a small part of the physical sciences as a whole. To the informed scholar of the late middle ages, however, there was a sense in which the science of motion was synonymous with the whole of natural philosophy. Aristotle had defined nature as a principle of motion, its cause or source, and his medieval commentators followed suit. As they not implausibly pointed out, those of his works which related to nature (the *libri naturales*) were organised around different kinds of motion. The *Physics* considered mobile bodies in general; *On the Heavens* dealt with the circular motions of the heavens and the straight-line motions below them; the work *On Generation and Corruption* treated of those motions, in the sense of changes, which take place when bodies are generated or decay. And so the list would typically go on, through the works on meteorology, plants, the soul, and animals. One could say that there was no one science of motion, but many; or that there was only one, and that it was identical with natural philosophy as a whole. In what follows we must dispense with such niceties, and consider only theories of local motions (motions from place to place) and their causes.

The law of motion Aristotle had formulated in the context of his discussion of the void, which with an obvious notation we may perhaps write as $V = F/R$, was severely criticised by a succession of scholastic writers. Some criticised his wording of the law: we must bear in mind here the difficulty many had with interpreting proportionalities, especially in the absence of an appropriate mathematical notation. Almost all had an intuitive dislike of the law's implications. Did Aristotle believe that any finite force, however small, could move a body against any finite resistance, however large? Something that we have learned to take in our stride, helped as we are by the simple mathematical relationship, seemed to them impossible. To avoid the impasse—a psychological rather than a mathematical one—Thomas Bradwardine in 1328 put forward a new relationship between velocity, force, and resistance. His law was extraordinarily complicated, in view of the limitations of the conventional mathematics available to him, and it is this fact, rather than the law's applicability to the real world, that has most effectively dazzled modern commentators. It is not necessary to go into any great detail here, although the law may be roughly summarised using our own notation. If we denote by f the ratio of force to resistance (F/R), and by v the velocity, and use a zero subscript to denote initial values, then the law may be generously expressed as follows: f is equal to f_o raised to the power v/v_o.[164]

Long before this time, Aristotle's ideas on the dynamics of motion had come in for much criticism. Palpable hits were scored by John Philoponus, a Greek writer of the sixth century, Ibn Bājja (Avempace) in twelfth-century Spain, and Averroes himself, who died at the very end of the twelfth century. In the West, Aquinas tried his hand at the problem, but all suffered from an inability to cast aside common misconceptions or to define their terms carefully. How was motion to be measured? What counts as resistance? As post-Newtonians we can conceive of inertia as a resistance, even in a vacuum, but they had no such advantage, and it would be foolish to pretend that arriving at that idea was a simple matter. In fact a far more complex explanation was found, one which we cannot accept, but which was extremely ingenious. Newton's dynamics, as we know, made no distinction between inertial masses on the basis of their chemical constitution. A good Aristotelian, however, when considering a (sublunar) compound of different elements, could argue that the light should rise and the heavy fall, so that the resulting motion of the compound body should ultimately depend on the proportions of the elements in it. Its inertia was therefore an internal affair, even in a void. Thomas Bradwardine was even able to draw the conclusion that two homogeneous bodies of different size and weight could fall with equal speed in a void, every piece of such bodies having the same proportion of heavy and light elements. There are tenuous historical lines to be traced from this result, and Albert of Saxony's elaboration of it, to Galileo's familiar theory of free fall two centuries later. At the very least we can say that Galileo lost his dread of motion in a void by reading medieval authors.

On the subject of forced motion, progress was less impressive. Islamic writers had tried to formulate general principles relating their ideas of impressed force, the weight of the moving thing, and the gradual dissipation of the force. They had met with little success. Bacon and Aquinas refused to accept the idea that impressed force could explain forced motion, but in the fourteenth century the idea was taken up again in earnest. In the 1320s, for example, Francis de Marchia suggested that the air surrounding a projectile had a part to play after the projecting force had been left behind. The air, somehow received an impressed force that kept the body moving. A superior explanation was presented by John Buridan in his *Questions* on the *Physics* and *Metaphysics*, both composed within a few years either way of 1330. Buridan, born in Béthune (in Artois, a French fief) around 1295, was undoubtedly the most talented Parisian natural philosopher of his century. He explained how a projectile may be considered to have been given a certain impetus by the initial mover. He conceived this impetus to be proportional to velocity and quantity of matter, so that—in Newtonian language—it is roughly the same as our momentum. Like Newtonian impetus (a product of a force and a short interval of time) it was seen as a cause. Buridan even hinted at a law of its conservation. He sketched an application of his theory to the celestial spheres. As for the sublunar realm, he did not restrict his attention to forced motion, but also tried to explain the acceleration of freely falling bodies. His analysis was presented in terms of successive increments of impetus—reminiscent of some of Newton's work—but he remained loyal to the Aristotelian principle that force is proportional directly to velocity and inversely to resistance. Once more, we can trace the influence of his ideas on impetus to the young Galileo, through sixteenth-century intermediaries.

A New Kinematics: The Mertonians

Any good disciple of Aristotle who sought an acceptable science of local motions would have hoped to give an account of how they are caused. The examples we have just given here are among the most important. So much attention was paid to this type of causal analysis—the dynamics of motion—that the mathematical analysis of spatial movements regardless of their causes, namely kinematics, was for long overlooked. (The exception, ironically, was the analysis of motions in the heavens, where in general they were much more complex. I will not pursue the question of how it came about that the Mertonians made such a clean break between kinematics and dynamics.) Asked to imagine a historical situation which might have spawned a theory of kinematics for describing motions—saying, for example, what we mean by constant acceleration—we might guess that some such thing originated with Zeno's paradoxes, or perhaps that it stemmed from the Archimedean tradition of generating geometrical figures in time. In fact there was a more important source, one that was far less obvious; and yet again it was a work by Aristotle.

It is easy enough to appreciate how, with sensible qualities, we may assign degrees to them. In his *Categories*, Aristotle gives the example of 'whiteness'. He insists that we may assign different degrees to abstract qualities too: his example is justice. He notes that there are those who are not prepared to describe this as a case of a variable quality. 'They maintain that justice or health cannot very well admit of variation of degree themselves, but that people vary in the degree in which they possess these qualities.'[165] Here are two different ways of looking at the case, and the difference may not seem particularly significant, but when Peter Lombard touched on the question, in his first book of the *Sentences*, university theologians were obliged to take note. He asked whether the Holy Spirit could be increased in degree in a person, and answered that the things the Holy Spirit gives (such as charity and grace) could not vary in themselves without supposing that the giver was not constant. The idea that the Holy Spirit is inconstant was considered unacceptable, and therefore it became a theological commonplace that human beings simply share in these things in varying degrees. In time, however, some philosophers switched their allegiance, and took it to be the quality itself which varied.

This change of heart had very fortunate consequences for the course of physics in general, and kinematics in particular, in the later middle ages. Early in the fourteenth century we find the Oxford Benedictine Walter of Odington—a man Richard of Wallingford very probably knew, although Walter was a monk from nearby Evesham Abbey—discussing degrees of heat and cold.[166] In the fourth and fifth decades of the century, an influential group of scholars at Merton College, most notably William Heytesbury, John Dumbleton, and Richard Swineshead, found it helpful to treat local motions as though they were qualities, which could vary in intensity. Speed was a sort of intensity to which a number might be assigned, and acceleration (our word, not theirs) was an intensity of an intensity, which again could be assigned a number. The resulting theory of 'the intension and remission of forms (or qualities)' made use of a vocabulary that we might find rebarbitive, but it was surprisingly complex, and was operated without the help of any compact mathematical notation. Its importance to the growth of the scientific consciousness should not be underestimated. It was carefully studied and modified by a select few for another three centuries. Galileo began by using it, and some of the rules for accelerated motion that we ascribe to him were derived by the Mertonians. Explaining the concept of instantaneous velocity in terms of distances attained hypothetically was something Galileo found useful in his *Two New Sciences*.[167]

One memorable finding of the Mertonians, now often called the 'mean speed theorem', was that the distance traversed in a uniformly accelerated (or decelerated) motion of a point is equal to the distance which would be traversed in unit time in a uniform motion of half the initial and final velocities. Heytesbury did not prove the theorem, but stated simply that it could be proved. Richard Swineshead, known to posterity as 'the Calculator', later offered a proof, and many others followed in the fourteenth and fifteenth centuries, some with arithmetical proofs in the Mertonian style, and some with geometrical proofs. The latter had Mertonian roots, but were more directly dependent on graphical methods later developed in

Paris in the 1350s.[168] The mean speed theorem came into its own, as we know, with Galileo, but less familiar is a commentary on Aristotle written around 1545 by Domingo Soto, who speculated that a body falling freely through a homogeneous medium will fall with constant acceleration, not with constant speed. Unfortunately he failed to pursue the idea further, and it was left to Galileo to show that all bodies in nature (frictional resistances being ignored) fall with the same constant acceleration.

How were Heytesbury and the other Mertonians first led to their result? One conjecture is that it originated by analogy with Gerard of Brussels's identification of the velocity of a rotating disc with the speed of the mid-point of a radius. A possibility which cannot be ruled out is that the Oxford men had in their thoughts the techniques of the astronomers, to whom it was second nature to analyse irregular motions. The context of Heytesbury's work was the logic of sophisms, but he discussed maxima and minima in motions, a subject which was also familiar to the few who were expert in using astronomical tables.[169]

The Rise and Fall of Aristotelian Science

The scale and rate of changes in the prevailing scholastic mentality brought about by the discovery of Aristotle should not be exaggerated, nor should we suppose that the principles we are discussing were very conspicuous. To focus our attention only on high-profile scholars—such men as Robert Grosseteste, Albertus Magnus, Roger Bacon, John Pecham, and Thomas Bradwardine—is to risk overlooking the fact that the number of scholars writing on scientific subjects was very small, and that many who did so were very conservative. Judging by those Oxford writings that have survived, we get the impression that interest in natural philosophy did not rise at all steadily, but peaked in the middle of the thirteenth century, fell back in the last quarter, and then rose fairly steadily until the mid-fourteenth century. After that, it collapsed dramatically, in a way suggesting a connection with the Black Death.[170] Writings on logic follow a roughly similar pattern. Those on astronomy and the life sciences were fewer, and a trend is harder to specify there, but if we may use these crude measures of scientific interest, we can at least say that the period in which Richard of Wallingford was writing his now lost commentary on the *Sentences*, in which he would have engaged in natural philosophy, was one of burgeoning interest. If we define a scientific writer of note to be one who probes the meaning of the text, rather than one who merely quotes it as a recognised authority, then we might estimate the number in Oxford to have been at best of the order of half a dozen per decade—an insignificant group by comparison with the phalanx writing on more straightforwardly theological and philosophical subjects, such as the soul and the nature of universals.

The traditional curriculum in the Oxford arts faculty continued to apply, at least nominally, long after the death of Richard of Wallingford, but university statutes do not give the whole story. The importance of Averroist thought is one imponderable. No account of medieval Aristotelian cosmology can ignore it entirely, but the importance it was perceived to have to what we think of as the sciences should not be exaggerated. Averroes helped scholars to understand Aristotle, and provided numerous minor alternatives, but the truly controversial issues were those with a bearing on Christian theology. The theses condemned by Tempier in 1277 contained several which had been professed by Thomas Aquinas, who like his master Albertus Magnus had relied heavily on Averroes, and yet they themselves had found much in Averroes that was theologically unpalatable. The eternity of the created world remained a thorny question. Aquinas risked opprobrium by saying that there was no proof either way, and that it must be left as a matter for faith. The Averroist thesis of the numerical identity of the intellect in all human beings was discussed tirelessly in the university world: those who rejected it did so on the grounds that it was incompatible with the notion of a person, and with personal immortality. A third bone of endless contention was the two-truths doctrine—one which owed more to the Latins than to Averroes himself—that a proposition may be philosophically true but theologically false.[171] This might seem to have been potentially advantageous to natural philosophy, guaranteeing freedom of passage, intellectually speaking. If it was heeded, however, it was as a blanket permission, and scholars very rarely gave any sign that they were availing themselves of it.

Tempier and Kilwardby notwithstanding, the thirteenth century did not see a high-water mark of Averroism. The tide rose relentlessly, and it was still rising during Richard of Wallingford's lifetime. Whether he knew it or not, his young Carmelite Oxford neighbour John Baconthorpe would become a renowned apologist for the seeming heterodoxies of Averroes across a broad spectrum. Oxford and Paris was paying increasing attention to the views expressed by Averroes on such questions as the existence of celestial intelligences, and the means by which the heavens may influence sublunar affairs. 'Averroist' is a description that needs qualification in almost every case, but it can be applied to an increasing number of notable philosophers, including Walter Burley and several followers of William of Ockham. If we must name a high-water mark for this movement, Padua in the sixteenth century would be a reasonable answer. It is better by far, however, to focus on the bedrock of Aristotle's own works.

Both in Oxford and in Paris, in the early fourteenth century especially, the cult of Aristotle brought about a marked improvement in the standard of teaching in logic and in the quadrivium. There was a steady drift of interest on the part of regent masters, and as a result, the old curriculum was slowly encroached upon by the new logic and the three philosophies—natural, metaphysical, and moral. Standards in astronomy improved, but it is hard to avoid the feeling that mathematical astronomy would have progressed more rapidly had Aristotelian cosmology not enjoyed such a high profile. The rich geometrical character of Aristotle's system of planetary astronomy had been almost entirely forgotten, and Aristotelian science, as taught, was almost entirely unmathematical in character until a new mathematics was created for it. Through logic, epistemology, and even theology, Aristotelian science in Oxford and Paris was mathematised; and in its new forms, it eventually passed into the scientific tradition of early modern Europe, with very great effect.

Needless to say, this is not the usual picture of the blinkered, medieval, hair-splitting Aristotle. His reputation was indeed victim to his great success. The adoption of so many of his ideas only served, in the end, to obscure his achievement. To take a simple example: while it was modified in endless ways, his general picture of the universe—or rather a debased form of it—was almost universally accepted throughout the middle ages. It survived in many educational courses well into the seventeenth century, if only as a topic for debate. There is no doubt that, with a number of related ideas, it stood in the way of much of the empirical science of the sixteenth and early seventeenth centuries: several important scientists had to fight hard against forms of Aristotelianism. When their downfall was eventually secured by Copernicus, Tycho Brahe, Galileo, and others, Aristotle's doctrines were two thousand years old, a fact which did nothing to moderate the enthusiasm with which his last followers defended his central arguments. It was largely a consequence of their enthusiasm that it became—and long remained—a commonplace to speak of Aristotle as though he was in some way anti-scientific. Even presenting the rise of medieval science as heavily dependent on the recovery of Aristotle may therefore be taken to reflect badly on medieval science itself. This is a misconception. Even for our own attitudes to the deductive structure of what we call science, we owe more to Aristotle than to any other person, and nowhere were his views more intensively debated than in the universities of the middle ages.

19

The Astronomers

IN THE MIDDLE AGES, much as today, those who studied, wrote, and lectured on the astronomical sciences fell into two intersecting classes. On the one hand there were those whose aim was to provide a mathematically exact account of the visible heavens, often, but not necessarily, with a view to applying their exact knowledge. The cosmologists, by contrast, had much grander ambitions: they wished to account for the totality of things in the material world, including the Sun, Moon, planets, stars, and possible entities beyond. It is convenient to distinguish these groups by the simple titles 'astronomer' and 'cosmologist', but there was no equivalent of the latter word, either in Latin or the vernacular languages of medieval Europe. Belief about the world as a whole was a medley of biblical knowledge and pagan ideas, some of them stemming from astronomy in the narrower sense, especially from ancient Greece and Rome.

Scriptural cosmology was more problematical than the simple astronomy needed to regulate the church. The rigid structure of the Julian calendar on which the liturgical calendar had been based allowed for ritual cycles to be superimposed on it without too much difficulty. The cosmological content of the Bible, however, allowed of—even encouraged—far greater flexibility in its interpretation. There were helpful texts, such as the first chapter of the Book of Genesis, with its emphasis on an orderly creation. There was the repeated emphasis placed on God's wisdom as the source of regularity in the heavens; and the view expressed in the Book of Wisdom that the ordering is by weight and number and measure. Against all this, there is the occasional anti-scientific hint that the behaviour of the heavens is beyond human understanding. Throughout scripture, the reader is made aware that God may, arbitrarily and miraculously, alter the regular order of things. Joshua ordered the Sun to stand still; God answered the request made by Isaiah that the shadow of the Sun move back ten degrees, as a sign to Hezekiah; the Sun was mysteriously darkened at the Crucifixion, in a way that could not have signalled a solar eclipse. Here are some notorious instances of passages that begged for, and that were often provided with, astronomical exegeses. Occasionally those were astronomically well informed, as when Bede refused to dispense with the miraculous element, but tempered it with allegory and a good understanding of basic astronomy. He was not typical of the early middle ages.

Biblical knowledge was the concern of every Christian clerk. In this sense, all were cosmologists, and no special term was needed to describe their cosmology. The arrival of Aristotle scarcely changed the situation, since his account of the principal divisions of the universe—the spheres of elements in the sublunar realm, and the planetary spheres in the celestial region—went virtually unchallenged. Advanced cosmological discussion certainly took place, but it came to fall squarely within the province of natural philosophy and metaphysics. Are the heavens made of a solid substance? Is there a void between them? Is it possible for there to be more worlds than one? Is the world finite or infinite? Is it eternal or otherwise? What are the causes of the celestial motions? The chief starting point for those attempting to answer such questions was Aristotle's *De caelo* (*On the Heavens*). The very formulation of those problems owed most to him. His doctrine became tacitly accepted by almost everyone, partly because some of its concepts are implicit in the phrasing of the questions, but chiefly because the elements of it were taught to all university students alongside the elements of geometrical astronomy—what is still called 'spherical astronomy'. The chief text here was that by Sacrobosco.

Early Western Astronomy

Long before the arrival of the new Aristotelian texts, there was a far from trivial European astronomical tradition. The central principles of the most advanced Greek astronomy had not yet been mastered, but very many of its operational procedures had already been in use with astronomical tables from the Islamic world, and the general character of the Greek geometrical models had been dimly recognised for about three centuries before the Aristotelian material began to arrive in quantity. What little knowledge there was at the beginning of this period came from Roman authors. It is convenient to take the death of Boethius in AD 524 as separating the classical Roman past from a period during which scriptural wisdom replaced almost every last vestige of classical astronomy. The idea that the liberals arts should be a training for the service of the state faded quickly. Computus was kept alive in Ireland, and in the eighth century in England too. The writings of the Venerable Bede 'On the Nature of Things' (*c.* 703), drawing largely on Pliny, could be described as both astronomical and cosmological, while his 'On the Reckoning of Time' (725) provided the best guide to calendar compilation before the thirteenth century; but planetary theory, even in Bede, was purveyed only in qualitative terms.

The first clear signs of a recovery date from the Carolingian period, when four Roman authors in particular provided inspiration for the very thinly spread community of scholars. Those authors were Pliny the Elder, that extraordinarily prolific first-century writer, most famous for his *Natural History*; Calcidius, a fourth-century Christian commentator and translator of much of Plato's *Timaeus*; Macrobius, a fifth-century writer who interests us chiefly on account of his commentary on Cicero's *Dream of Scipio*; and Martianus Capella, a late fifth-century writer of a Latin encyclopedia of the seven liberal arts, a work generally known as *The Marriage of Philology and Mercury*. Each of them had included an assortment of elements of planetary astronomy in his writings, and while none had a deep technical understanding of ancient Greek methods, their works were seized upon by new generations of ecclesiastics who had been persuaded that the calendar mattered to the church, and that current Christian practice was largely inept. The single most important cause of this newfound awareness was a meeting of churchmen in Aix-la-Chapelle, called together by Charlemagne in 809. Leading members of the conference immediately set up a programme for the compilation of relevant texts, and from these texts later generations were able to acquire a bare outline of Greek theories, drawn from Roman sources.

The new texts were supplemented with many diagrams, presumably most of them newly drawn, although in some cases conceivably under the guidance of Byzantine visitors to the Frankish court in 810-11. Some of these diagrams contain hidden surprises, such as the use of stereographic projection—of the kind we encountered in connection with the astrolabe and anaphoric clocks—and what at first sight appear to be graphs of planetary movements, some drawn in rectangular coordinates, some in polar coordinates.[172] Many of the diagrams share a peculiar property, however, which is not at first apparent. When we see a planetary figure with, for example, an epicycle on a deferent circle, we naturally think that those responsible for it were framing a model of what happens in the heavens. In fact it would be truer to say that they were trying to illustrate the worded descriptions of such writers as Calcidius. There are other surprises, even so. There is an illustrated report of the discussion by Martianus Capella of three different interpretations of a heliocentric world, attributed in the ninth century to Pliny, Martianus, and Bede. Another classical source available at this time was the poem by Aratus (*c.* 315-pre 240 BC) entitled *Phaenomena*. The first and main part of this was a versified catalogue of the constellations due to Eudoxus, the great contemporary of Plato and Aristotle in the previous century. A new Latin version of the astronomical material by Aratus was prepared in Corbie in the eighth century, and in due course luxurious manuscripts of the new version were produced, in some cases not only illustrated with mythological figures but also annotated in places with technical planetary information, such as apogee positions.

The practical needs of the monastic communities where the necessary materials were preserved were naturally uppermost in the thoughts of those who composed the new works. Aristocratic patronage was one sort of practical need, but the need to keep time and the calendar were more immediate. Besides texts on sundials, there are others dealing with the instrument known as a nocturnal—by which the time is found at night by observing the disposition of the stars around the north celestial pole. (Those stars behave almost as the hour-hand of a 24-hour clock.)

It would be wrong to think that the Roman texts which assumed such great importance in the Carolingian period were put aside when translations from Greek and Arabic began to arrive on the scene. On the contrary, they survived and were widely read, chiefly, but not entirely, for their literary and supposed didactic virtues. One antiquated (Plinian) method for calculating planetary positions, known from an English text of around 975 AD, was still being used in an English monastery more than five centuries later, an astonishing example of learned ignorance.[173] *On the Marriage of Philology and Mercury*, by Martianus Capella, was still a popular Latin text-book in the later middle ages. At the end of the fourteenth century, for example, the poet—and expert astronomer—Geoffrey Chaucer was still happy to cite 'Marcian'. In it, each of seven bridesmaids is presented one of the seven liberal arts. The astronomy was very basic, and after the arrival of Islamic texts might have been judged insufferably vague by those who truly wished to get to grips with them, but the image of Astronomy's bridesmaid arriving at the wedding party in a hollow sphere of heavenly light, filled with a transparent fire and gently rotating, was one that most readers surely found appealing. She carried a pair of dividers and a globe, traditional symbols of astronomy. Macrobius too, with his commentary on Cicero's *The Dream of Scipio*, remained popular. Here the dreamer was portrayed as undertaking a journey through the spheres, allowing him at length to look down on the entire universe and allowing Macrobius to expound astronomy in a literary vein. Chaucer quoted him, as did Dante had done in his *Divine Comedy*—a work written in Richard of Wallingford's lifetime. Those who want evidence for the central place occupied by astronomy in the medieval consciousness need to look no further than Dante's *Divine Comedy*, the greatest of all medieval allegories, in which a supremely important moral subject is set squarely within a framework of Aristotelian cosmology. Paradise itself is there given an astronomical structure, extending that of the Aristotelian world: the souls there are placed in ranks corresponding directly to the planetary spheres; and the Inferno is given a structure mirroring that of Paradise.

The Renaissance of the Greek Tradition

The dramatic changes brought about by the acquisition of new texts between the tenth century and the thirteenth were such that, in concentrating on them, we are inclined to overlook the slow pace of change, as measured in terms of the human life span. Scholars who concerned themselves with astronomy were few, and even those who made use of everything they could copy were still likely to be in possession of an ill-assorted mixture of materials. This was true of the best tenth-century scholars, men like Notker of St Gall, Abbo of Fleury, and Gerbert of Aurillac. When Gerbert brought Iberian materials across the Pyrenees, and introduced it to Rheims, Ravenna, and Rome, he may be said to have greatly enhanced European knowledge of the astrolabe and quadrant, but only in the way that introducing a speck of dust into a cloud of vapour enhances the chances of rain. At first, the lives of only a handful of people were affected. Even Gerbert's own later correspondence on astronomical matters shows how deeply immersed in the Romano-Carolingian tradition he remained.

We have already seen that the earliest clear northern European witnesses to the great power of Greek and Muslim astronomy were Petrus Alfonsi at the court of Henry I of England, the Lorraine scholar Walcher, prior of Malvern, and the erudite Jewish scholar, Abraham Ibn Ezra. The astronomical tables prepared by Abraham had some currency in England, in versions based on London and Winchester, while through Petrus came an awareness of the tables of al-Khwārizmī, which Adelard of Bath translated and partially adapted. It is probably true to say that in the entire history of Ptolemaic astronomy, covering more than fourteen centuries, very few astronomers, Islamic or European, realised that issuing handbooks containing tables, with canons for their use, but independently of a major text, was a tradition which Ptolemy himself had begun. He split off his so-called *Handy Tables* from his monumental *Almagest*, modifying them somewhat in the process. Adelard knew that the techniques used in al-Khwārizmī's work were mainly Ptolemaic, but he is unlikely to have realised that they contained Hindu materials as well.

Another curious medley which bid to enter the Latin world, but with much less success, was by the twelfth-century Barcelona scholar Abraham bar Ḥiyya (Savasorda). There is at least one surviving Latin manuscript containing what he called his *Tables of the Prince*. It was not long, however, before all of these began to be gradually replaced in the affections of northern scholars by the basic *Toledan Tables*, with the canons Azarchel had written for them late in the eleventh century. In slightly different versions, they came to hold the field alone. Well over a hundred manuscripts of them survive to this day, an unusually large number for a genre of text that was of virtually useless once it was superseded. Scholars in many towns adapted the *Toledan Tables* to their local meridians. Abraham had made use of them. Tables for Marseilles were based on them in 1140. Roger of Hereford used them for his tables for Hereford, dated 1178. Of other derivative sets, one more widely used than these was for Toulouse. A fourteenth-century Latin version of the Toledan work was even translated into Greek. The cultural circle was complete.[174]

The *Toledan Tables* and their canons do not survive in the original Arabic form, but in their early history they influenced western Islam in various compilations, the last distinctive set being produced in al-Andalus around the year 1205. In the rest of Europe this left no mark, although more than a century passed before it was overshadowed there by another set, the most famous of all from Spain, the *Alfonsine Tables*, first drafted in the 1270s.

No matter which set of the earlier tables was adopted, it was bound to draw attention to the fact that the old Roman texts were quite inadequate for purposes of accurate calculation. No matter which tables were used, to use them was to be made aware that they were based on theories which the canons—which were merely rules for their use—did not explain. The greatest astronomical compilation from antiquity, Ptolemy's *Almagest*, available in Latin only by 1160, was eventually recognised as the essential key. It was recognised as a monument to past greatness, but honoured more by name than by consultation, for it was quite beyond most scholars' comprehension. By its mere existence it underscored the poverty of existing learning, but its length and difficulty ensured that it was little copied. It was eventually taught—when it was taught at all—through digests, with such titles as *Abbreviated Almagest*.

Ptolemy's Almagest

Ptolemy's work might have been unfamiliar, but—even more effectively than Aristotle's writings had done for philosophy—it laid down the general theoretical structure for its subject between the second century, when it was written, and the sixteenth, well beyond the time of Copernicus. We shall need an outline of some of its most basic principles, if we are to appreciate Richard of Wallingford's contribution to astronomy. What follows is our own *Almagestum abbreviatissimum*.

The astronomer, mathematician, astrologer and geographer Claudius Ptolemaeus was born around AD 100, and died about seventy years later. He was of Greek, or at least Hellenic Egyptian, descent, and held Roman citizenship. In view of his name—a common one, in his time—Arabic and medieval European writers often mistook him for a king, and he is often graced with a crown in manuscript and early printed portraits of him (Fig. 54). His extensive writings suggest that his ambition was to produce a monumental encyclopedia of applied mathematics. What he wrote on mechanics is now lost, and his *Optics* and *Planetary Hypotheses* can only be pieced together from Greek or Arabic versions, although they leave no room for doubt as to his high calibre. Some minor works on projection (his *Analemma* and *Planisphere*), as well as a monumental *Geography*, survive in Greek, as does his greatest achievement, the *Almagest*.

54. Ptolemy as king, observing the heavens with a quadrant, under the guidance of Astronomy. From Gregorius Reisch, *Margarita philosophica* (Strasbourg, 1504).

Our title for it tells us something of the work's migration through history. It began in Greek simply as *Mathematical Compilation*, and then became *The Great Compilation* (or *Greatest Compilation*). When the Arabs translated it in the ninth century, only the adjective was kept, but more or less in the form of the Greek word (*megiste*), so that it now became *al-majisti*. From there it entered Latin as *Almagesti* or *Almagestum* in the twelfth century, eventually providing us with our *Almagest*.

Ptolemy began this long and complex work with his reasons for adhering to a largely Aristotelian philosophy. This simple fact is testimony to the power of Aristotle's thought: five centuries separated the two men, five centuries that had seen almost the whole of what we now consider the greatest achievements in Greek astronomy.[175] Ptolemy's style also shows the influence of Stoic thought, but he is not out to preach to us. Our daily affairs may give us moral insight, he tells us, but to attain a knowledge of the universe we must study theoretical astronomy. He therefore begins his task. He accepts Aristotle's dual-region universe, and his general picture of the workings of the First Cause of celestial motions, which Ptolemy is happy to accept as the divine First Mover. From such relatively brief philosophical beginnings he turns to some rather general cosmological arguments of a qualitative sort, concerning the heavenly sphere and the various motions observed in it. Again he follows Aristotle, more or less, in his physical arguments for the spherical shape, central position, and fixity of the Earth. Both writers considered it to be of insignificant size in relation to the heavens. Few readers are likely to have had difficulty with these first few pages of the text, but all too soon the situation changes.

A mathematical introduction follows, with a table of chords to three sexagesimal places, and other items that we should now classify as 'trigonometry'.[176] Astronomers, then as now, make constant reference to geometrical figures on the surface of the celestial sphere, when calculating, since it is easier to visualise arcs on a spherical surface than the corresponding angles at the centre of the sphere. (The distances of most celestial bodies is so great in comparison with the Earth that for most, but not all, astronomical purposes the observer may be taken as measuring angles at the sphere's centre.) Fortunately for Ptolemy, a beautiful and extremely powerful theorem which greatly simplified this type of calculation had been developed in the previous century by Menelaus of Alexandria.[177] It relates to what was known in the middle ages as the 'sector figure', a configuration of arcs of great circles on the surface of a sphere, such as is illustrated in Fig. 55. Richard of Wallingford's *Quadripartitum* gave not only a very thorough account of the theorem of Menelaus, but used it to prove a series of new trigonometrical results.

Ptolemy wasted no time in his first two books of the *Almagest* in applying his various mathematical techniques to fundamental astronomical problems. Broadly speaking, knowing five of the angles in the sector figure, the sixth is deduced. The case is frequently simplified when one angle is a right angle. Ptolemy also showed how it is occasionally possible to find a solution when only four angles are known, together with the sum or difference of the other two. One problem to which the theorem was applied, and which played an important part in all of the books that followed, was to calculate the obliquity of the ecliptic, the angle between the Sun's path and the celestial equator (or equinoctial). Too heavily influenced by his predecessor Hipparchus, he took the answer to be 21°51'20". A better figure would have been 23°40'42", but the figure is slowly changing, and later generations found themselves needing to revise it often.

In Book III of the *Almagest*, Ptolemy accepted the solar theory drafted by Hipparchus in the second century BC. Again, Ptolemy made a small error when calculating the length of the tropical year—a mere day in six centuries, but a mistake that persuaded him yet again to accept Hipparchus' figure for the year. This was about 11 minutes of time too large—only one part in fifty thousand. The theory accounted for most solar phenomena so well that Ptolemy can have had little incentive to change it. He added ta-

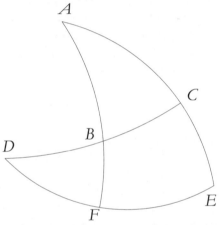

55. A spherical triangle *ABC* (one created out of arcs of great circles on the surface of a sphere) cut by a transversal *DEF* (another such arc), illustrating the theorem of Menelaus. By crd(*AB*) we denote the chord of the angle subtended by the arc *AB* at the centre of the sphere. Today we should express the theorem as {crd(2*AE*) : crd(2*EC*)}.{crd(2*CD*) : crd(2*DB*)}.{crd(2*BF*) : crd(2*FA*)}=-1, using signed distances and leaving the question of internal and external intersection to look after itself. Here the colons are used to express the proportionality: there was much discussion of the nature of proportion and proportionality, following the writings of Thābit Ibn Qurra and Campanus of Novara, and of what modes were acceptable. In medieval texts, separate proofs were thought necessary for each of the eighteen acceptable modes distinguished, each being written out in the pattern {p:q}={r:s}.{t:u}. Transforming this into equivalent modes, to us a trivial matter, was not so when every mode had to be written out in a worded paragraph, and when the rules for cross-multiplying in an equation were not appreciated.

bles to allow the rapid calculation of two especially important angles that are needed to settle the Sun's position. The techniques Ptolemy used for the Sun will be explained shortly by way of an introduction to his theories of planetary motion generally, since he extended them to account for the more complicated motions of the planets. The general pattern of his planetary models would turn out to be of crucial importance to the course of astronomical history, remaining so long after the death of Copernicus. Before giving the barest of outlines of what they involved, however, we shall take a fleeting glance at the structure of the remainder of *Almagest*, and indicate how much and how little of it was familiar to scholars in the universities of the middle ages.

Book IV contains a careful discussion of the lunar theory of Hipparchus, with new parameters obtained from observation. In Book V, when he came to compare the lunar theory with his own observations, Ptolemy found that it fitted well only when the Sun, Earth and Moon were in line (at conjunctions and oppositions, or syzygies, as they are collectively called). He therefore supplied a remarkable lunar theory of his own. Book V ends with a discussion of the distances of the Sun and Moon, and includes the first known theoretical discussion of parallax, that is, of the correction it is necessary to apply to the Moon's apparent position to obtain its position relative to the Earth's centre. This was of use in Book VI, where he set out the procedures to be followed for calculating solar and lunar eclipses.

Books VII and VIII of the *Almagest* were given over to the fixed stars, conceived as situated on a sphere, to their simple daily motions around the poles, and to their more complex long-term motions, subtle motions first identified by Hipparchus. Here Ptolemy gave the methods used to record their positions, and long and detailed lists of the coordinates and magnitudes (relative brightnesses) of over a thousand stars. Many of the star coordinates too were drawn from the work of Hipparchus.

The planetary theories follow. In Books IX, X, and XI of the *Almagest* he accounted for the longitudes of the inferior planets (Mercury and Venus) and the superior (Mars, Jupiter, and Saturn). Thus far, the treatment was two-dimensional: each planetary model was on its own sheet of papyrus, so to speak. In Book XIII, however, Ptolemy introduced planetary latitudes (angular distances from the ecliptic) into his scheme. He had earlier performed a similar exercise for the Moon, which showed him the way.

A Painful Climb: Student Texts

The number of western medieval scholars who made themselves conversant with Ptolemy's great work, or with revised but broadly equivalent material from Arabic sources, was never very large. On the other hand, the number was not large in Ptolemy's day: we know of scarcely any comparable work in the two centuries and more separating him from Hipparchus, for example. Even by the fourteenth century, at any one time, the number equal to the challenge of reading Ptolemy with full understanding in Paris and Oxford together would probably have been fewer than a dozen, and most of them would have been a disappointment to the author. This was the situation in two universities where almost every student had been obliged to study astronomy as a part of the quadrivium. From the text-books then in general use, it is easy to understand why.

Grosseteste's book 'On the Sphere' is one such guide, but by far the most popular, even in Grosseteste's Oxford, was that of the same title, *De sphera*, by John of Sacrobosco (Hollybush or Hollywood). It was a work written expressly for students in the faculty of arts, one of a series he wrote—others being devoted to computus and computational arithmetic, including the use of Hindu-Arabic numerals and root extraction.[178] Very little is known of its author, an Englishman who had a certain reputation as teacher at the university of Paris in the first half of the thirteenth century. Judged as a scientific text, to modern eyes it is a strange work, interspersed as it is with passages from classical authors and poets. This quality helped to sugar what for many students must have been a bitter pill. The text was not particularly long—in its abbreviated script it occupied only two dozen pages, say three sheets of parchment, folded quarto—so that many students would have been able to buy or make their own copies. Ptolemy's *Almagest* might have filled thirty times as much parchment.

Sacrobosco told of the sphere of the heavens, considered to carry the stars round the sky with its daily rotation around earth, the sky being pivoted at the poles. He set out the barest essentials of Aristotle's view of the cosmos, with the four elements and the fifth 'quintessential' substance of the heavenly regions beyond them. He described in qualitative terms the properties of the chief circles which the astronomer imagined drawn on the celestial sphere, especially important being the equinoctial (the equator through the sphere's middle) and the ecliptic (the path of the Sun as it traverses the sphere of stars in the course of a year). He told of the zodiac, the band of sky through the middle of which the ecliptic passes, which was divided into twelve equal segments. each of thirty degrees, the signs of the zodiac. They had been named in antiquity after the constellations that occupied them, more or less, and their very names—Aries (the Ram), Taurus (the Bull), Gemini (the Twins), and so forth—allowed scope for fantasy, then as now; but Sacrobosco had a more serious purpose. He wanted to convey, as painlessly as possible, a feeling for the intricacies of the daily risings and settings of stars, of the Sun, Moon, planets and signs of the zodiac. He wanted to explain how these are related to the seasons—something that was probably better understood by educated people then than now. There is something in Sacrobosco on the divisions of the globe of the Earth that depend on astronomy—the tropics of Cancer and Capricorn, and the bands of the Earth's surface known from antiquity as the climates. Finally, near the end of the book, there is a very superficial account of the pattern of movements of the planets, and an explanation of the reasons for eclipses of the Sun and Moon. The text closes with an explanation of why the darkened Sun at the time of Christ's crucifixion cannot have been a natural eclipse.

It should by now be clear that the student who had to study astronomy in order to incept in arts needed to have barely passed through the portals of Ptolemaic astronomy. From Sacrobosco he would have learned less than the equivalent of half of the first book of the *Almagest*, and nothing of serious mathematical astronomy. What of the alternatives? There were several unsuccessful attempts to rival its success in the next three centuries. One such work was written by John Pecham, whom we encountered as writer of an important work of optics. His tract on astronomy is especially interesting, for what it might tell us about interchanges between Oxford and Paris.[179] The number of surviving manuscript copies, a dozen or so, may be compared with some hundreds of copies of Sacrobosco's work, not to mention at least 180 printings between the fifteenth century and the seventeenth. Part of the reason is no doubt the relative difficulty of the two works, and Pecham's relatively dry and unadorned style. He was writing for young students, and yet he touched on a theory that some of his senior colleagues found difficult, concerning the slow movement of the stars supposedly on the eighth sphere. This was a movement which Hipparchus had discovered and explained as a movement in ecliptic longitude only. (It is the movement we now ascribe to the precession of the equinoxes. It changes the longitudes of the stars by a degree or so every seventy years. It is actually not constant, but for most purposes can be treated as such.) In due course, for quite erroneous reasons, Muslim astronomers decided that this subtle motion was variable. They designed different models to account for its variability, the best of them known from a text later ascribed (wrongly) to Thābit Ibn Qurra. Pecham, with a surfeit of academic optimism, introduced this theory into his work, not particularly well, but the fact that he tried to do so at all must surely have been enough to discourage the novice astronomer. For the rest, Pecham entered into the geometry of the sphere, with diagrams rather than the worded descriptions of Sacrobosco.

Strangely enough, Pecham twice apologised to the knowledgeable reader, pleading the need to write for young students, and insisting that his own book was easier than some alternatives and more accurate than others.

For which university did he write it? He had several opportunities to teach on such a subject. By the mid-1250s, before he had turned thirty, Pecham joined the Oxford Franciscans, having studied there and perhaps earlier in Paris.[180] As an Oxford Franciscan he would have had the advantages of Grosseteste's library and scientifically well-informed associates, including Bacon, but by the late 1250s we find him in Paris. There he became a controversial figure, noted for his trenchant debating style—which he used both for and against Aquinas, for example. He was back in Oxford by 1271, however, and by 1275 was made provincial of the English Franciscans. His teaching career was not finished: two years later he was made lector in theology at the papal university of Viterbo; he was for a time in Rome; and then, in 1278—against the wishes of King Edward I—the pope nominated him archbishop of Canterbury. Pecham was consecrated the following year, and wrote no further works of scholarship, unless we are to count a little commentary in Norman-French written for the queen on the *Celestial Hierarchy* of the pseudo-Dionysius.[181] It is not that he chose to rest on his laurels: on the contrary, henceforth he devoted himself to a rigorous programme of ecclesiastical reform, as Grosseteste had done before him, albeit more successfully.

Was Pecham's astronomy written for students in Oxford or Paris, or even for readers in Viterbo or Rome? The matter is not without interest, for it has been suggested that the papal court in Viterbo—partly by chance and partly through the activity of William of Moerbeke—served as a centre for the transmission of optical literature in the 1260s and 1270s, decades to which the optical writings of Bacon, Witelo, and Pecham all belong.[182] One of the curious aspects of Pecham's astronomy text is that in it he pays much attention to optical matters, especially the conceptually difficult question of pinhole images, which again might favour a Viterbo origin. One astronomical example Pecham gives, however, may be used to show that in it he was almost certainly using a Paris latitude for his calculation. The book therefore most probably stems from his period of teaching in Paris, after 1257 and probably in the early 1260s.[183] He and Bacon were moved from the Oxford convent to that in Paris at about the same time, and since they must have known each other well, it is inconceivable that they did not occasionally discuss optical questions in both places.

Sacrobosco had devoted only a few lines to the geometrical models for the motions of the planets, and Pecham included little more, although he gave the theories of the Sun and Moon in some detail, albeit not without errors. Most students at this period obtained their knowledge of planetary theory from a type of treatise with the generic name *Theorica planetarum* (*Theory of the Planets*), and it has been suggested—but probably mistakenly—that Pecham wrote a work of this type. More than two hundred manuscript copies of the most popular work with this title are still extant, far surpassing the numbers of its rivals combined, and yet that version is of unknown authorship. It was often associated with a writer named as 'Gerard', who has usually been assumed to be the famous twelfth-century translator Gerard of Cremona, but an unknown thirteenth-century writer seems more likely. There were other works of *Theorica planetarum* type, including two available in Latin in the twelfth century, and much circulated: one the prototype manual by al-Farghānī (Alfraganus), as translated by John of Seville, and another by the English scholar Roger of Hereford. A fourteenth-century text by Andalò di Negro gained some currency; and in Oxford at the end of that century another version of the basic *Theorica* was produced, with much numerical material. There were several others of the same sort produced after Richard of Wallingford's death, one of them probably by the Merton scholar Simon Bredon. The chances are that, as a student in arts, Richard was first introduced to planetary astronomy through the famous anonymous treatise. For want of an author's name, this is usually known by its opening words: '*Circulus eccentricus vel egresse cuspidis ...*', and university statutes often name it in this way.[184]

Neither this treatise nor those on the sphere, such as Pecham's, were entirely accurate, judged by the *Almagest* from which they ultimately derived, but together they brought students up to a level at which Ptolemy's work, and others of an advanced nature, could be attempted. They provided something more: their authors tried to reconcile the physical approach of Aristotelian cosmology and the geometrical approach of Ptolemy's *Almagest*. What was not appreciated at the time was that Ptolemy had attempted to do precisely this. When Pecham wrote *On the Sphere* for his students, the author at the front of his thoughts on the subject of the physics of the heavens was al-Biṭrūjī (Alpetragius). We shall return later to that problem, to see why it was felt to be so pressing throughout the middle ages and into the Renaissance. First, however, we must take a bird's eye view of Ptolemy's planetary astronomy.

Ptolemaic Planetary Theory

The problem of calculating the positions of the planets in the heavens, whether pursued for astrological reasons or for its own sake, was one which

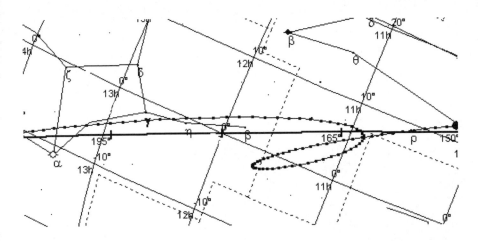

56. A characteristic loop in the motion of a planet as seen against the background of fixed stars. In this example the positions of the planet Mercury are marked at daily intervals as it moved through the constellations of Leo and Virgo during August 2003. Except for the retrograde section, the planet's movement was from right to left of the figure.

had engrossed mathematically minded astronomers for two thousand years by Richard of Wallingford's time. The Aristotelian spheres were considered physically important, but they had been stripped of the astronomical detail Aristotle had carefully worked out for them on the basis of the earlier work of Eudoxus, and as far as most medieval scholars were concerned, the details of planetary motion were to be had from the Ptolemaic tradition, and not from Aristotle. That Ptolemy had drawn heavily on his predecessors, such as Apollonius, Hipparchus, and Menelaus, was almost entirely unknown in the middle ages.

For at least two millennia, there was a basic problem which had intrigued many of those who had no great desire to make precise calculations: it was that of explaining in general terms why the planets behave in the odd way they do. The 'fixed' stars seem to move rapidly as a whole, circling the sky once every day. (They actually take about four minutes less than twenty-four hours of solar time, since the Sun moves slowly relative to the stars, with its annual motion, in the opposite sense.) The planets share that general daily movement, but they have other movements of their own relative to the stellar background. These movements are generally so slow that they will only be detected over a period of several nights. By and large, they are in the same general direction as the slow yearly motion of the Sun, from

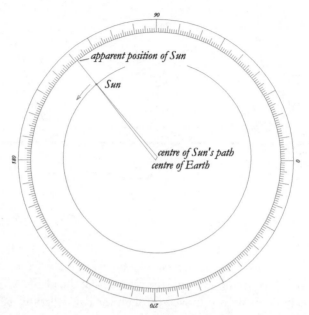

57. The Greek model for the motion of the Sun. It was assumed to move at constant speed around a circle eccentric to the Earth.

west to east. From time to time, however, a planet will be found to slow down against the stellar background, stop, and then begin to move in the opposite sense, relative to the stars. After a further interval of time, a few weeks or a few months, it will change direction once more, and again move in its normal direction from west to east (Fig. 56). Occasional retrograde movement against the stellar background is a property of all the familiar planets, although the only planets known before the eighteenth century were Mercury, Venus, Mars, Jupiter, and Saturn. The motions of the Sun and Moon vary in subtle ways, but they are unlike the planets, in that they always move in the same west-east direction relative to the fixed stars.

Why do the planets behave in this strange way? The modern answer, one that has been generally available since Copernicus, is that our observations are made from the Earth, which is itself moving around the Sun, just as the planets are. Even in antiquity, Greek astronomers had another, highly successful explanation, fitting our instinctive belief that the Earth is fixed at the centre of the visible universe. The Sun, which seems to move round the Earth annually, was at first taken to be moving at uniform speed through the stars, around a circle centred on the Earth. Ancient astronomers who wanted to account more precisely for the Sun's movement, and who knew that it varied slightly over the course of the year, decided that the Sun's

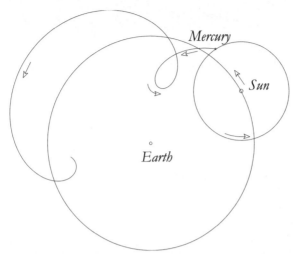

58. A simplified version of the Greek model used to account for the apparent motions of Mercury. (Ptolemy's theory was appreciably more complex than this.) The planet is considered to move around a small circle (the epicycle), the centre of which moves round a larger carrying circle (the deferent circle). As seen from the Earth (whether at or near the centre of the deferent) the planet will occasionally appear to move backwards, relative to its general movement. As seen from outer space, if the view were always to be centred on the Earth, the planet would seem to spiral.

path was circular but not exactly centred on the Earth (Fig. 57). Making the same kind of adjustment, that is, making the motion eccentric, was a useful device for matching planetary theory to the observed motions, but still it could not explain the occasional reversals in the direction of planetary motions. To explain those retrograde motions another device was needed. It was supposed that each planet moves around a small circle (an epicycle), the centre of which moves around a larger circle (the deferent circle) centred in the neighbourhood of the Earth. It should be intuitively obvious that on this hypothesis the planet will have a spiralling motion through space, such as that drawn in Fig. 58, which gives a fair representation of the movement of Mercury as it could have been observed in August 2003, and as illustrated in Fig. 56.

It is not hard for a modern reader to see why this scheme works so well. The one thing that was not generally recognised before Copernicus is that the Sun is at the centre of the Mercury epicycle. In each of the separate models for the motions of the planets, as presented by Ptolemy, there is always a point that can be identified with the Sun. Fig. 59 is included here to

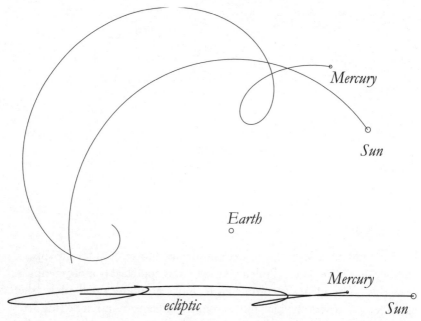

59. The upper figure is a view from outer space of the movements of the Sun and Mercury. Our view is fixed in the direction of the Earth, and at right angles to the plane in which the other two move. Looked at from the Earth, the two would seem to move against the stellar background as in the lower figure. A successful theory will reproduce this second situation, as it unfolds in time.

demonstrate this point with reference to Mercury: the upper part of the figure is an accurate representation of the movements of the Sun and Mercury as they would be observed from outer space by someone whose field of view was fixed in relation to the Earth. Here, then, we have the beginnings of a two-dimensional theory of planetary motion, a theory for finding the apparent position of the planet on the ecliptic.

Mercury's obit is not quite in the same plane as the Sun's, but the fact will not be appreciated from this (upper) two-dimensional image. Looked at from the Earth, however, the Sun and Mercury move against the stellar background as shown in the lower part of the figure, which is simply a version of Fig. 56, extended to cover a longer time interval. The ecliptic is the apparent path of the Sun, and ecliptic latitude is measured from it. It should be clear, even from these simple illustrative diagrams, that a theory of planetary latitude will not be easy to formulate. Most elementary texts avoided it entirely, except in discussion of the movement of the Moon, or when explaining the reason for giving breadth to the zodiac. The Moon's ecliptic latitude is an important factor in eclipse calculation. As for the zodiac band, through the middle of which the ecliptic passes, it is not astronomically important, but in principle it marks accommodates the planets in their wanderings from the ecliptic in latitude.

We chose Mercury here, the innermost planet, only because it conveys the epicycle idea most readily, since Mercury presents us with more frequent retrogradations than the other planets beyond it. The theoretical model Ptolemy formulated for Mercury was far more complicated than a simple epicyclic model would suggest. Every planet was assumed to have a deferent circle and epicycle, but the deferent was itself taken to be eccentric to the Earth. The centre of the deferent was placed at a carefully calculated distance from the Earth; every epicycle was assigned a carefully chosen radius; and the speed of the planet's rotation around its epicycle was another important parameter that needed to be established. More complicated still was Ptolemy's idea that the centre of every epicycle should move around its deferent circle at a variable rate, at least when measured around its centre. It was said to move so that it would appear to move round at constant speed from another centre, known as the equant centre. The Moon and Mercury were assigned still more complicated models than the other four known planets. It is not necessary to explain the fine detail of Ptolemy's models for planetary motion here, as long as it is appreciated that they served two purposes. First, they offered a broad measure of scientific satisfaction, for those who want to know the qualitative workings of the world. More importantly, for the expert astronomer, they permitted the accurate calculation of solar, lunar, and planetary positions for times past, present, and future.[185]

The geometrical models of the separate planetary motions were only scientifically valuable—that is to say, able to give a good fit with the observed positions of the planets over long periods of time—once the relative sizes of the various circles, the relative distances between their centres and the angular motions around them had been accurately specified. The search not only for broad explanations but for satisfactory parameters had been pursued with extraordinary success in the early Greek world and Mesopotamia, and culminated in the work of Ptolemy, in Alexandria in the second century. Throughout the middle ages, the broad picture remained that set out by Ptolemy, despite various attempts to modify it in small respects. The parameters of the planetary models—circle sizes, motions, and so forth—were repeatedly modified, however, usually for the better. The most significant medieval improvements to the Ptolemaic inheritance, however, were in the methods used in calculation.

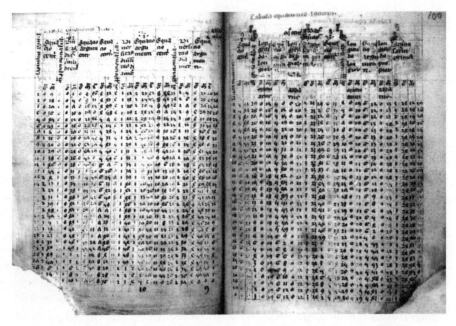

60. Specimen pages from the Alfonsine tables, as copied into the manuscript most closely connected with Richard of Wallingford during his period as abbot (now Bodleian Library, MS Ashmole 1796). The pages shown here contain tables of equations (tables of anomaly).

Astronomical Tables and Techniques

Performing from first principles the geometrical calculations needed to find the positions of the planets for a given time was difficult and time-consuming. Already in the ancient world astronomers devised elaborate sets of rules to simplify their task, rules which were applied in conjunction with a large number of pre-calculated tables. Astronomers of lesser talent who could follow the rules did not need a particularly good understanding of what they were doing. The task was still time-consuming and tedious, but the number of university scholars of the late middle ages who were capable of ploughing through the necessary processes grew steadily, especially in the fourteenth century and after. The predictive accuracy of this kind of astronomy, often good for a small fraction of a degree in ecliptic position, was steadily improved. We have already encountered, at least by name, some of the more famous sets of astronomical tables in the western middle ages, mostly coming from Spain, but all heavily dependent on Islamic precursors, in a chains of transmission mostly leading back to Ptolemy. We recall that the two most widespread were the *Toledan Tables*, dating from the eleventh century, and the Alfonsine, also strongly connected with Toledo but later modified in Paris and Oxford, dating from around 1270. Richard of Wallingford worked with both sets, but seems not to have had the Alfonsine until after he became abbot in 1327. Fig. 60 illustrates facing pages from a St Albans copy of them. To a casual observer, the two, separated in history by nearly two centuries, would have been scarcely distinguishable, apart from the fact that the older was more likely to be written in Roman numerals than not. A full set of tables of either sort would normally have occupied a space of the order of eighty pages of a manuscript, or even more.

It is a mystery why the *Alfonsine Tables* took so long to reach the rest of Europe, but they did reach Paris around 1320, and in Parisian versions they arrived in Oxford soon after.[186] They had been put together between 1263 and 1272, those chiefly responsible being the Jewish scholars Isaac ben Sid and Judah ben Moses ha-Cohen, the latter a physician. The work was done under the patronage of the Christian King Alfonso X of León and Castile, who encouraged the translation from Arabic into Castilian of many philosophical and scientific writings.[187] His father San Fernando had taught him the virtues of patronage, and the two kings no doubt hoped to be classed with the great patrons of Islamic astronomy, east and west. The very introduction to the Alfonsine canons uses phrases that present Alfonso in this light, and the tables themselves make use of an Alfonsine epoch of noon, 31 May 1252, the eve of his coronation. Perhaps some readers here gave a thought to 'King Ptolemy'. The epoch of the Christian king was related in the tables to the old Spanish era, to the Islamic Hijra, and to the (Persian) era of Yazdijird, all for reasons of chronological continuity. This chronological aid, whether intentionally or otherwise, served as a reminder of cultural continuity and dependence, and must have been perceived as such when the tables eventually crossed the Pyrenees—leaving their somewhat inconvenient epoch to be readjusted for a wider Christian world.

One of the principal achievements of medieval western astronomers was to devise new mathematical techniques for processing planetary calculation. Richard of Wallingford's albion was another kind of answer to this problem, but that instrument was meant for speed, rather than the highest accuracy. There were circumstances under which Richard would have used tables; and indeed, had he not been intimately familiar with their use he could not have designed the albion in the way he did. It is impossible to say how familiar he was with those Parisian developments in his time,

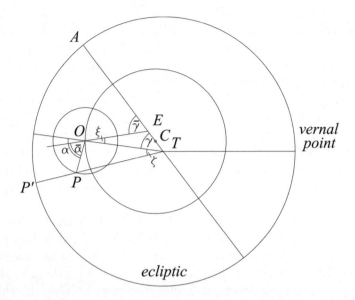

which were resulting in a transfer of affection to the Alfonsine material. One important document in this movement was written by John of Murs, an *Exposition of the Meaning of King Alfonso in Regard to his Tables* (1321). A year later, John of Lignères—whose pupils John of Murs and John of Saxony both were—wrote a work still heavily dependent on the canons to the old *Toledan Tables*, but showing certain characteristics of the Parisian versions of the *Alfonsine Tables* to come. At some time between then and 1327, he and his pupils assembled the essential ingredients of what eventually became the most popular version of all, an edition composed by John of Saxony in 1327. This came into its own in the age of printing, but still it followed a pattern which users of the old tables would have recognised instantly. John of Lignères wrote his own canons for the tables between 1322 and 1327, but during these years—and perhaps as early as 1320—he and his circle were working on a way of streamlining them for use with tables of combined planetary equations. This was an important step, and is in need of a brief explanation; but first it will be necessary to explain the technical medieval term 'planetary equation', in the context of standard Ptolemaic planetary theory.

Consider the moderately straightforward Ptolemaic model for the planet Venus (Fig. 61), using the terminology and partial explanation provided in the caption to the figure.[188] Ecliptic longitudes are measured counter-clockwise from V, the vernal point. To place the planet P' correctly, that is, to find its longitude, we must effectively add the two angles marked at T (γ and ζ) to the longitude of A (aux, apogee). Those two angles are

61 (opposite). The standard Ptolemaic model for the movements of the planet Venus. The plane of this two-dimensional figure is taken to be that of the ecliptic, a second model being needed to give the latitude of the planet, which will not be considered here. The planet is on its epicycle at the point P, and the centre O of the epicycle moves round the deferent circle while the radius OP rotates. Both motions are counter-clockwise on the present figure. T represents the Earth, which is not at the centre C of the deferent circle. The eccentricity (TE) and the radius of the epicycle are important parameters of the model, varying from planet to planet. (As it happens, Venus has an unusually large epicycle.) They are usually expressed as fractions of the radius CO, this being typically taken to be 60 units. The epicycle moves around the deferent at a variable speed, but it is the line EO which rotates at constant speed, and not TO, as in earlier and simpler models. E is known as the equant point, and is placed so that C is the mid-point of ET. This ingenious way of introducing a variable velocity was devised by Ptolemy, but was disliked by many, even Copernicus, as supposedly violating the uniform circular motions called for by Aristotle. The two angles increasing at constant rate are those marked with a bar, $\bar{\alpha}$ and $\bar{\gamma}$, and the rates in question were important parameters of the model. The line TCE is clearly that along which the planet is capable of reaching its maximum and minimum distances from the Earth. This is today usually called the 'line of apsides' or 'apse line'. In medieval parlance it was the 'line of aux'. It was usually assumed to be fixed in the sphere of (unfortunately named) 'fixed stars', and therefore it too was moving, but very slowly, and the position of A (on the ecliptic) needed to be recalculated only occasionally. The angle $\bar{\gamma}$ at E was called the 'mean centre', and $\bar{\alpha}$ at O the 'mean argument'. The mean centre, strictly speaking, is not obtained directly from the tables, but as the difference between the mean motus, VTO, and the longitude of aux, VTA. (Historians prefer to use the Latin word *motus* in English in the present context, rather than its usual equivalent, 'motion', since it refers to an angle, a longitude, or even to a (changing) place in the ecliptic.) The astronomer's usual problem was that of finding the position of P' on the ecliptic (the longitude of the point), that is, the apparent place of the planet P when viewed from the Earth. Longitudes are measured counter-clockwise from the vernal point V. Given A, $\bar{\alpha}$ and $\bar{\gamma}$, a good geometer could have calculated this longitude from first principles, but with difficulty. Ptolemy and other astronomers established standard routines, as explained in our text, using tables. In principle, such tables required no reference to a diagram.

problematical, however, for they do not increase at a constant rate. Those which do, the mean centre ($\bar{\gamma}$) and the mean argument ($\bar{\alpha}$), we can easily find from tables of longitude against time, with entries increasing by equal steps. Such angles are simply proportional to the time interval from some basic reference time, when the angles had some specified values (each called the 'root' or 'radix' of the motion in question). We shall suppose that the mean values, $\bar{\gamma}$ and $\bar{\alpha}$ have been found.[189]

The next step is to find ξ, known as the *equation of the centre*. It can obviously be calculated from the known proportions of the figure by simple trigonometry, if we know $\bar{\gamma}$. Using such methods, we could indeed draw up a table for *all* values of $\bar{\gamma}$, or at least for values at, say, one-degree intervals. Subtracting* ξ from $\bar{\gamma}$ gives us our γ. The nature of this kind of 'equation' should now be clear. It can be treated as a correction term to be applied to a steadily increasing quantity.

There is one small complication in the present simple account: it relates strictly to the geometrical figure as drawn here. Asterisks are added to our explanation whenever a figure disposed differently might have required an addition rather than a subtraction, or vice-versa. Canons to tables usually included instructions as to how to decide between the alternatives.

The next stage is to calculate α. It is simply the sum* of $\bar{\alpha}$ and ξ. Now, however, the problem becomes more complicated. It is easily appreciated that the other equation, ζ, the so-called *equation of the argument*, depends on α (which is now known) and $\bar{\gamma}$. The mathematical relationship, unfortunately, is very complicated, but Ptolemy drew up an extremely ingenious series of tables by which the equation of the argument could be evaluated. His method depended on certain approximations, but the end result was quite accurate enough for ordinary purposes. Our final answer to the longitude question then comes almost as an anticlimax. The longitude is the mean motus minus* ξ plus* ζ.

These were the basic steps which every astronomer worthy of the name had to take, to calculate a single planet's position for a single moment of time. Calculating a full set—for the Sun, Moon, and planets—might have taken a true expert one or two hours, or in the case of a novice much longer. Since there are two main equations to be calculated, and since one of them is dependent on the other, might it not be possible to improve on Ptolemy and have, for each planet, just one table, giving one combined equation? The Parisians took this step, probably unaware that a similar step had already been taken by Ibn Yūnus, three centuries earlier. The user's task was now easier, but not that of the person who drew up the combined tables, which were now much longer and more difficult to draft than before.

Some of the more traditional Parisian tables were in use in various towns in England very soon after they were written. Even quite mild changes evidently made scholars uncomfortable. When, around 1340, William Rede of Merton College took a Parisian version and adapted it to the Oxford meridian, he preferred the old Toledan signs of 30° to the more consistently sexagesimal signs of 60°.[190] There are surviving versions of the tables for other English towns, some from perhaps as early as 1323 for Leicester and Northampton, and others for Colchester, Cambridge, York, and London. There was clearly astronomy being studied and practised in English religious institutions outside the universities—and, for that matter, outside St Albans.

Rede's were all relatively simple adaptations, and it is doubtful whether more radical revisions were made in England during Richard of Wallingford's lifetime. In Oxford around 1348, however, an unknown man, perhaps William Batecombe, produced tables going much further than the labour-saving work of John of Lignères. The 1348 tables allowed the planetary longitudes to be extracted more or less directly, apart from a small adjustment for precession. They were extensive, and could also be used to carry information about the direct motions, stations, retrogradations of the planets, and other matters that were of astrological interest. Their value was appreciated in many centres of learning. There are early manuscripts from Silesia and Prague. Henry Arnaut of Zwolle, in the northern Netherlands, used them, referring to them as 'the English tables'. A fifteenth-century Hebrew translation was done by Mordechai Finzi, assisted by an anonymous Christian from Mantua. Giovanni Bianchini, the most notable Italian astronomer of the mid-fifteenth century, was influenced by them when he produced a similar set, much used by such leading contemporaries as Peurbach and Regiomontanus. Through this intermediary, and a fourteenth-century eastern European version known as the *Tabulae resolutae*—all of which had been studied by Copernicus when a student at Cracow—they provided the core of John Schöner's tables of the same name (printed in 1536 and 1542). These in turn were widely used for many decades, although their Oxford origins had long been forgotten.

After the 1348 tables, one further set of Oxford tables of very great merit in the Alfonsine tradition was produced by the Merton College astronomer John Killingworth (*c.* 1410-45). His were meant to be used in calculating a full planetary almanac, an ephemeris listing all planetary positions day by day. That, of course, was what everybody wanted, but it came at a great cost in labour, not to mention parchment. A magnificent copy of

Killingworth's work made for Humfrey, duke of Gloucester, is still extant. It is heavily interlined with gold leaf, but hidden in the work itself is a greater treasure. It contains evidence of a type of theorising that Killingworth does not explain to us, but that few could reproduce today without recourse to the differential calculus. Did scholars like Killingworth work alone, in glorious isolation? It is hard to imagine what form lectures on the *Alfonsine Tables* might have taken, but we do have an isolated university record dated 1448 which shows that there were such things in Oxford at that time.[191] In Richard of Wallingford's day, it is unlikely that there were public lectures of the same sort.

Natural Philosophy and the Astronomers

There was a world of difference between those university scholars, like Richard of Wallingford and John Killingworth, who set a high value on the explanatory power and accuracy of Ptolemaic astronomy, and those who restricted themselves to Aristotelian cosmology, in some cases believing that it offered deeper truths than the mere geometry of the astronomers. There were various points on which the two approaches to understanding the world were incompatible. Aristotle's planet-carrying spheres were centred on the Earth. The Ptolemaic circles were supposedly carried by spheres that were eccentric to the Earth, or epicyclic. Did this mean that they were not a true representation of the physical world? The fact that the two sets of ideas were introduced in different parts of the curriculum meant that disagreement could usually be ignored with impunity.

Few early scientific explanations strayed far from common experience. It had always been tempting to treat the epicycles and deferents of astronomy as though they had a real existence, especially to those who thought of the universe as a working machine. Those who most often manipulated them in their dealings with astronomy were more inclined to treat them as a purely hypothetical means of yielding the right results, and since they were so successful, those who wished to cast doubt on them needed powerful arguments. When, in the middle of the fourteenth century, Henry of Langenstein, of Paris and Vienna, who inclined to the cosmological side of the fence, began to attack the astronomers, his argument ran: 'If epicycles existed, the same simple body would be moving at one and the same time with different motions', and this cannot be, 'since one and the same cause cannot produce different effects on the same body simultaneously'.[192] In other words, an inability to think of a motion as anything but a real, unique, and ultimate thing, something that could not be resolved into two parts, led him to a savage embargo on an entire way of thinking, the way of

Ptolemy and the Greek geometer-astronomers before him. Even in the seventeenth century we find excellent writers on the theory of motion tying themselves in knots on the question of a body sharing two motions simultaneously. Added to this, there were those who were deeply suspicious of a theory with numerous real but interpenetrating spheres. spheres. The Ptolemaic circles were rarely rejected, but they were kept under sufferance, as poor relations capable only of doing a hard day's work.

It is now known that Ptolemy himself was worried by the thought that his models did not conform to the broad Aristotelian viewpoint, but that he found a way of reconciling his models with the demands of Aristotelian cosmology, and included it in the middle of his *Planetary Hypotheses*. This work survives today, but in Greek only in part, the missing sections—including the crucial on—being now known only from medieval Arabic and Hebrew translations. These were not known to the Latin middle ages except indirectly, and even in modern times were not fully appreciated until the 1960s.[193]

True to Aristotelian ideas, Ptolemy made the assumption that there are no empty spaces in the universe; and he accepted that there can be no overlapping of matter with matter. Consider two planets, such as Mercury and Venus, adjacent in the overall hierarchy. Each has its model, but—complicated though each model may be—each planet has its maximum and minimum distance from the Earth. Ptolemy's idea was simply to make the outermost point reached by a planet in its epicycle equal to the minimum distance reached by the planet next above it. One can therefore imagine that Mercury is always within a certain spherical shell, a sphere with an appreciable thickness, and that Venus is in the shell next outside it. It became natural to assign real existence to these thick shells—which were different in character from those geometrical entities, the Ptolemaic circles.

There was much discussion in the middle ages of the nature of the planetary spheres, in the more material sense. There is no doubt that people were less happy with the abstract geometrical explanations of planetary motions—as found in the *Almagest*—than with the idea of material spheres in the heavens, spheres that could interact mechanically, transmitting their motions down from the First Mover (*Primum Mobile*) to the terrestrial core of the universe. We might criticise them today for their failure to distinguish clearly between scientific theories, models, hypotheses, and reality; but many a medieval scholar was conscious of the problem, and not all adopted naive materialist answers. In the course of laying the geometrical groundwork for the astronomy to be embedded in his albion, Richard of Wallingford makes an illuminating aside about his attitude to theoretical entities. It should not be thought, he says,

> that there are things among the heavenly bodies literally corresponding to an eccentric or epicycle, which the mathematical fancy contrives for its own purposes. No educated man could ever suppose that idea to have any semblance of truth. It is rather that, without mathematical concepts of the state of things, or at least without unfamiliar intellectual ideas, there can be no certain way of dealing with the regulated motions of the stars which will assure us of their places at any moment, places differing in nowise from what we actually see. (*Albion*, I.10.)

It is clear that he was here rejecting the crude idea that epicycles, equants, deferent circles, and all the rest, have to be real. For him, they are the boundaries of real spheres made out of a fifth element or other substance. On the other hand, he was not trivialising them as a mere means to producing the right numerical answers. They are intellectual ideas, theoretical terms, we might say, not necessary to producing a knowledge of planetary longitudes, and so forth, but the kind of concepts needed to this end. It is unfortunate that he did not say more on this question in *Albion*, but that was not really the appropriate place. At all events, this debate over the nature of theoretical terms and hypotheses was already stirring in the middle ages; it became especially heated in an astronomical context in the sixteenth century, and has gone on in science at large ever since.

At the risk of oversimplifying the situation, one might see the realists as people enamoured of the Aristotelian doctrine of material causation, and of the idea that such causes have to be things, even in the heavens. Those of Richard of Wallingford's type are likely to be content with formal causes. There were even some who bid to give the debate an astrological flavour, arguing, for example, that there could be no such thing as a ninth sphere—the sphere which had been introduced to explain the slow precessional motion of the stars in longitude—since a sphere without stars would be a sphere without influence, and so would have been made by nature in vain. Aristotle's division of the universe into sublunar and celestial realms also played a part. It suggested that the celestial world was outside the true province of physics, giving the mathematical model-makers free rein. Those who adopted the physical version of the Ptolemaic system can be seen as trying to bridge the gap between the two parts of the universe. They were insisting, in effect, that there were shared principles operating in the two worlds. This point of view prevailed only in the late sixteenth century and after.

62. The physical version of the Ptolemaic model for the planet Venus, from Georg Peurbach's *Theoricae*. Venus, on its epicycle, must always lie between the innermost and outermost circles. The epicycle will always lie inside the white area between those limits. There is a similar figure possible for Mercury, which fits exactly into the central (white) area, and a simpler figure for the Sun above Venus, into which the present figure fits. With the terrestrial and stellar regions, such nested figures were considered to fill the entire universe, leaving no empty space.

Ptolemy's physical model, and others deriving from it in the Islamic world, had a surprising and unexpected consequence within astronomy itself. Ptolemy's planetary models were conceived, and usually taught, as though they were independent of one another. The physical model united them within a single system, and in doing so it provided almost a full set of planetary distances without any need to make further measurements. A moment's thought will show how this came about. Since the relative sizes of the various circles in any planet's geometrical model had all been laid down by the sort of analysis presented in the *Almagest*, and since the furthest reach of one planet was now taken to be the nearest reach of the next, the scale of the circles of the one was seen to fix the scale of the circles of the one above it. Step by step, starting with the Moon, each theoretical model fixed the scale of the one above it, up to that of Saturn, the highest planet known. All distance were known, relative to the nearest distance of Moon; but Ptolemy had found a supposedly precise distance for the Moon, and it was therefore thought possible to quote equally precise distances for each and every planet. The answers do not now bear close examination, for the entire scheme is illusory, but the distances are at first sight plausibly large—they are in millions of miles—and no one in the middle ages was in a position to gainsay them.

Ptolemy's scheme, which gave beautiful effect to Aristotle's rejection of the concept of a vacuum, was seized upon by several Islamic writers. Through a text-book written by the ninth-century astronomer al-Farghānī (Alfraganus), it eventually became a standard item in the curriculum of western universities. In the intervening centuries, however, Ibn al-Haytham had studied it closely in his influential work *On the Configuration of the World*, and many other writers, east and west, made use of it. Al-Biṭrūjī (Alpetragius) wrote about it in a work which Michael Scot put into Latin in 1217 as *De motibus celorum* (*On the Motions of the Heavens*). Roger Bacon was another who was aware of it: he went into some detail when giving an account of the 'total orb' (*orbis totalis*) of the Moon—the thick shell, in our brief description—in his *Opus tertium*. He was fully aware of one unfortunate aspect of the model, namely, that some of the component orbs within the shell were eccentric to the Earth in their movements: in this they failed to conform to the principles of uniform, circular, Earth-centred motions, as taught by the Aristotelians. That unpleasant fact was easily ignored, but in the age of the printed book, the very striking diagrams in Georg Peurbach's *Theoricae novae planetarum* forced the book on the attention of a large readership (Fig. 62).[194] Written around 1460, and printed in 1472 or 1473, Peurbach's book quickly replaced the old *Theorica planetarum* texts of the late middle ages, which were less well illustrated and technically inferior. Indeed, judging western astronomy only on the basis of such elementary material, some writers are occasionally tempted to say that only by the end of the fifteenth century had western astronomy begun to approach its Islamic precursors. The conclusion is entirely spurious, however, if it is taken to apply to the best astronomers of the high middle ages.

Heaven and the Heavens

The question of the nature of the celestial spheres was considered relevant to much more than physics and astronomy. The theologians had made it their own at a very early date, when they wrote commentaries on the scriptural account of the creation of the world. As Genesis has it, on the second day, 'God said, Let there be a firmament in the midst of the waters ... And God called the firmament Heaven' (1:6-8). While the original text makes no mention of planetary spheres, the reference to a firmament was understood in two different ways, one meteorological—which here we shall ignore—and one relating to the sphere of fixed stars, in connection with which Greek philosophy was usually invoked. Plato's fiery element was an early candidate for the firmament, but inevitably Aristotle's fifth element was also strongly favoured. Many opinions were expressed, and in the thirteenth century, for example, Augustine and Bede were often quoted. Heavens spiritual or corporeal, sidereal or empty, corruptible or incorruptible, luminous or transparent—all gave scope for endless theological discussion. Wiser heads—Grosseteste and Aquinas, for example—hesitated to make firm pronouncements on such questions, but it is significant that with Grosseteste, and many who followed his lead, the firmament was taken to include the entire planetary and stellar region, from the lunar spheres upwards.[195] This image fitted comfortably with that of Ptolemy's *Planetary Hypotheses*, where there was continuity in the material of the spheres, whatever that was taken to be. Sacrobosco, Vincent of Beauvais, Aquinas, and many others, on the other hand, simply took the firmament to be the sphere of fixed stars, within the First Mover, the *primum mobile*. What is striking about the entire discussion is that almost all of those concerned made a valiant attempt to do justice to what astronomy they knew, but that none could honestly escape the fact that astronomical orthodoxy sat very uneasily with biblical exegesis.

It was to Aristotle that most late medieval theologians and astronomers turned when they wished to decide on the nature of the planetary spheres, but Aristotelian cosmology was not immune from difficulty. In his book *On the Heavens*, Aristotle had assumed the spheres to be corporeal, albeit of that fifth kind of substance which the medieval scholar called the *quinta essentia*. What were good Aristotelians supposed to say about the properties of this substance? They were not supposed to allow it any property which admitted of an opposite—such as the property of hardness. According to Aristotle, qualitative opposites can only be found in terrestrial matter. Judged in this way, there seem to have been no good Aristotelians, since both opinions—that the spheres are hard, and its opposite—were held at one time or another. There are admittedly problems of translation here: the word 'solid', for instance, could mean hard, or continuous, or three-dimensional. As a result, not only is it often difficult to decide on medieval writers' intentions, but they themselves often found it difficult, and slid from one meaning to another. It was possible for a person to hold that the heavens are fluid in one sense and solid in another, and only in the sixteenth century were the two descriptions generally considered to be mutually exclusive.

A decision on this question made no real difference to the practice of astronomy or astrology in the thirteenth and fourteenth centuries, except when it came to literary embellishment, but there remained that difficult question of how to interpret the opening verses of Genesis. Like the Church Fathers before them, most Christian clerics wanted to know the nature of the waters above the firmament. Were they like a crystalline stone or icy solid, as Jerome and Bede had thought? Or were they fluid, as Ambrose, John Damascene, Robert Grosseteste, and probably the majority of later scholastics thought? Bartholomew the Englishman argued that the word 'crystalline' meant only that they were luminous and transparent. Some reinterpreted even common words like 'water'. For want of science, philology was made to fill the breech, and almost any sophistry was possible, since the only measure of truth was a series of ambiguous texts. It was no doubt the thought of this that made Augustine, Grosseteste, Aquinas, and many others move away from the discussion as soon as they decently could. In the end, it would be left to astronomy to upset the arguments of most of those who opted for hard spheres. The new star of 1572, and the comet of 1577, both carefully observed and measured by the Danish astronomer Tycho Brahe, were both shown to be well beyond the Moon. His discoveries are now chiefly remembered because they put an end to the Aristotelian doctrine of incorruptibility in the celestial regions, but they also had serious implications for the debate about the nature of the planetary spheres themselves. The star and comet seemed to moved across the heavens in ways that made a hard sphere scarcely conceivable. Those who had argued for fluid spheres at least had half of their argument intact.[196]

Theological speculation which made use of astronomical theory did not stop at the planetary spheres. In commentaries on Genesis, the heaven created by God on the first day was often called the empyrean heaven (Latin *coelum empyræum*, from the Greek adjective for 'fiery'). This highest heaven was not an astronomical construct, except by analogy: it was a heaven beyond the outermost heaven of the astronomers. For those who spoke of it, it was the dwelling of God, the angels, and the elect—or at least some of them. In time it was provided with its own internal structure, making for numerous compartments in the heavens, by analogy with those of astronomy, heavens arranged in a hierarchy that could accommodate the various ranks of angels. With or without the name, some sort of empyrean had been discussed by the pseudo-Dionysius, Martianus Capella, Bede, and such influential early medieval theologians as Anselm and Peter

Lombard. By the time of the great scholastic movement of the later middle ages its existence was rarely considered open to debate.[197] This, however, did not rule out a discussion of where exactly the empyrean was.

Here astronomical theory came into play. We recall that it was supposed that the eighth sphere holds the fixed stars, and that the First Mover (*Primum Mobile*) lies beyond it. Once it was recognised that they have a slow movement of their own, it became usual to postulate a 'ninth sphere'. When a theory was devised to explain the supposed variability of that stellar movement, through a complex system of toing and froing, or 'access and recess', a tenth sphere was sometimes added—although many astronomers simply kept the old terminology, and supposed that the ninth sphere carried the machinery of access and recess within itself. Always a First Mover was added to provide the daily motion to the whole system, outside however many spheres had been postulated for the stars. Numbering spheres is a rather pointless exercise for our purposes, as long as we recognise the theological dilemma. Is the empyrean to be identified with one of the astronomers' spheres, or not? Early in the thirteenth century, it was often taken to be the ninth, or 'crystalline', sphere. In 1241, and again in 1244, the bishop of Paris condemned the idea that the Blessed Virgin could be there. Where then? Michael Scot tells us that the empyrean was introduced to act as a divine cause of motion, lying beyond the First Mover.[198] Astronomy did not provide this insight, but it provided the necessary analogue, and henceforth it set the boundary: the empyrean was to be one sphere beyond what they, the astronomers, knew about. The theologians had captured the high ground by ecclesiastical decree.

It was left to the theologians to speculate about the properties of the empyrean. The most dramatic distillation of what some of them decided was of course Dante's, in the *Divine Comedy*, written at the beginning of the fourteenth century. With Beatrice, the poet Dante leaves the ninth, crystalline heaven for the empyrean, where the two are surrounded by a great brightness. Paradise first appears as a river of light, then as a great white rose in which are the seats of the blest, beginning with the Virgin and including St Benedict and Augustine. The vision ends with the unbearable brightness of the deity, then the Trinity and Christ.[199] Almost every important theologian had something to say on these matters, What is it like to be there? Bartholomew the Englishman argued that since it is at the opposite end of the universe from our own region, it is therefore plainly the brightest, most aetherial, and least heavy body. Dante of course, had been even lower, inside the Earth, and down to Purgatory and the fires of Hell, but the contrast he drew was much the same. Speculation did not stop with portrayal of the glories of the place. It may be taken as read that it is the abode of the blest in a future life, but does it exert any influence on affairs in the planetary orbs, or on affairs on the Earth below? Here was one of several ways in which the astrologically-minded could account for celestial influences in what ssemed to be a wonderfully rational way.[200]

There were a few scholars, notably in Paris in the mid-fourteenth century, who were not prepared to accept such glorious speculation as all this, or to accept that even the existence of the empyrean can be demonstrated rationally. Some accepted it only as a matter of faith. John Buridan and Albert of Saxony produced arguments against it which were an extension of those they had learned in philosophy, arguments concerning mobility and nobility. These two Parisians were playing the theological game according to alien rules, but they were also, in this small way, marking out their intellectual territory, and their stance is just one of many signs of the growing independence of science and religion.

A far more subtle approach to the theological question of God's whereabouts, and one which also called upon the philosophy of the schools, was Thomas Bradwardine's. Bradwardine's *De causa Dei*—which we cited earlier in connection with the clock—is a long and complex work, heavily laden with quotations from Aristotle, Augustine, and other classical authors, and what he has to say in it about God and infinite space is easily lost to view. To its select readership, however, its subject gave it importance, as did the fact that the writer was taking up cudgels against William of Ockham, a thorn in the flesh of so many Oxford theologians. On the question of God's existence, Bradwardine had little new to say: it was something to be proved by physical fact, by the need for a first cause in the long sequence of cause and effect, at the end of which is the world we see. When he came to write of God's nature and powers, however, he supplied over forty corollaries, all answers to heretics and unbelievers. (He was a rather extreme example of the hypercritical academic, and few of his predecessors escaped censure entirely.) God, for him, was the highest good, every quality in God being of infinite perfection. Such statements seem almost to be tantamount to his definition of God. They have tripped readily off the tongue of generations of theologians, but they were not platitudes in Bradwardine's time. He accepted that God had infinite simplicity and unity, but that we need to ascribe different qualities to God in order to be able to talk about his infinite virtues. Ockham had taken a very different position, making God's essence a grouping of all perfections, with no real distinction or precedence among them. To distinguish wisdom from goodness, said Ockham, was merely a question of verbal convenience.

This doctrine had drastic consequences for the Ockhamist sceptics, who allowed God only one attribute, namely his will, which was of little help to the theologians who wanted to discuss how God acts in the world, working out the answers from a knowledge of his character. Bradwardine, on the contrary, thought it possible to make use of traditional categories—immutability, eternity (of a sort), omniscience, self-sufficiency, and the rest.[201] With what result? God, he said, must exist outside space and time, for these are categories that belong to the created world. (Like Augustine, he considered time to have been created with the world.) But where exactly does God exist?

In the context of a discussion of God's immutability, he set out five corollaries according to which God is (in essence and presence) everywhere in the world and also beyond the real world in an 'imaginary infinite void'. A void can therefore exist without a body, contrary to what Aristotle had said, as long as God is present there. This neatly sidestepped an article (201) of the 1277 condemnations, which forbade the teaching that there had been a void before the creation of the world. In a series of arguments aimed at proving God's ubiquity, Bradwardine developed the concept of an infinite place, although it was not considered to be a place depending on any created thing. He certainly did not wish to overturn the traditional astronomical picture of a finite created universe with a spherical boundary, nor did he want us to suppose that God is an actually extended magnitude: God is immense in the sense that he is unmeasured and unmeasurable. Those of us who need a mental image to help us through his argument are given one, in a reference to a passage from Augustine, who pictured God as an infinite sea, and the world as a sponge in the sea, but penetrated by it.[202] He could have added that the sponge was spherical. No one in the middle ages seems ever to have questioned that fundamental astronomical belief. Every age has its tacit presuppositions, which go almost unnoticed by virtue of the fact that they go unquestioned. In the later middle ages, the astronomers had a goodly share of them.

20

The Astrologers

> They say miracles are past; and we have our philosophical persons, to make modern and familiar, things supernatural and causeless. Hence is it that we make trifles of terrors, ensconcing ourselves into seeming knowledge, when we should submit ourselves to an unknown fear.
>
> Lafeu, in Shakespeare's *All's Well That Ends Well,* II.iii.

LAFEU'S REMARKS remind us of the many people who are offended at the idea that the sciences have a monopoly of knowledge, and in particular, of those who want to leave room for miracles, for the occult, and for the right to be afraid of the unknown. To a few modern readers, astrology has its place there. To many more, it is either a harmless concoction written by a secretary in the newspaper office, or something that is authentic only in the sense that it belongs only to history, an occult practice to be classed with magic, satanism, witchcraft, alchemy, and the like. Grouping such subjects together, however, obscures an important distinction among them. Occultism involves a belief in supernatural beings and powers, and usually presupposes that knowledge of them may be used to affect the natural world. While, during the past two millennia, various different opinions have been held as to the way in which the heavens affect the natural world, there has been a broad consensus that they did so in some sort of natural way, that could be studied as one studies nature. The stars and their properties at one extreme of the causal chain, like their effects at the other, are open to view. There are many problems of how to correlate the two, but astrology differs from those 'secret philosophies' which have no family likeness to the natural sciences—doctrines many of which were derived from Babylonian and Hellenistic magic and alchemy, Hermetic philosophy, Jewish mysticism, and the Cabbala.

Medieval astrology was not occultist, in the sense usually supposed, but it was all too often entangled with doctrines that were, doctrines designed to predict—or even bring about—future events. It began in this way, in the ancient Mesopotamian practice of finding omens in the heavens, where there was a clear belief that the stars reveal the will of the gods. Gods being gods, they were open to persuasion. There were medieval theologians in plenty who were prepared to argue that the influence exerted by the stars is an agent through which the one unique God chooses to affect the world. Such supernatural considerations were no doubt always present in the thoughts of some people, but there was naturalistic core of astrology largely untouched by such beliefs during a period stretching from Hellenistic times to the seventeenth century—when many of the best scientific intellects still took the subject seriously. When it entered the university curriculum in the middle ages, it was not in some hole-in-the-corner way, but as a subject to which the names of many of the great intellects of the past were attached. It is often treated—not without reason—as a classic case of medieval reverence for the received text, but there were just a few scholars who were prepared to try to improve on what they had inherited. Putting the matter crudely on their behalf: Why not? If the Moon can affect the tides, why should the stars and planets not affect the natural world in their own ways?

While many original thinkers took astrology to be akin to the natural sciences, there was one respect in which this idea frightened the moralists and theologians. If the Sun, Moon, stars, and planets, all influence human behaviour in a thoroughly deterministic way, then they remove our freedom of action, and thus our responsibility for our actions. In India and parts of Mesopotamia, rigid determinism seems not to have been a serious problem, since the planets were there widely considered deities, whose behaviour might be controlled through prayer and supplication. Determinism remained a serious objection to the subject in Christianity and Islam, however, with the result that various escape clauses were devised around the idea mentioned earlier, that the heavenly bodies 'incline but do not compel'. The moral strength of true believers was considered enough to allow the influence of the stars to be resisted. To a believer in the universality of laws of nature, needless to say, this doctrine was as troublesome as the belief in miracles.

Early History

Ancient Mesopotamian omen literature and derivative works spread by degrees across the Mediterranean world, through Persia and India, and as far as China in the east. The more mathematical sorts of astrology travelled with them. These were more in line with Greek thought, and in the long term were of much greater influence on the western medieval intellectual world than divination from omens. Simple forms of mathematical astrology had reached Egypt long before the founding of an Alexandrian empire there, and they were being practised in other Greek centres by the second century BC at the very latest. It was in Alexandria that the astronomer Ptolemy, in the second century AD, wrote what became a standard work of reference. This, his *Tetrabiblos* ('Four Books', or in Latin *Quadripartitum*), provided astrology with much-needed scientific coherence.

In its earliest stages, astrology was not directed at any and every individual. The ancient Mesopotamian reader of omens informed the royal court of impending good or bad fortune—in regard to the weather, epidemics, or military campaigns, for example—which might affect the state as a whole, or the ruler and his family as its representatives. Only at a much later period, and then chiefly in the Hellenistic world, was the science considered to be applicable to private individuals. Of the main branches of the science, what is now the most familiar dealt with the casting of nativities, *genethlialogy*. Here it was considered that every human life is governed by the heavens in a way that depends on the positions of the planets, thesigns of the zodiac (or the constellations of stars from which the signs derive), and the relation they all bear to the local horizon at the moment of birth or conception. To cast a nativity from first principles required much knowledge of mathematical astronomy, and as the centuries passed, new and increasingly complex mathematical principles for dividing up the heavens at that crucial moment were devised.[203]

Out of this fundamental form, others were developed. According to one, the science of *elections*, the heavens decide whether a course of action begun at a particular moment will be successful or not. What is known as *iatromathematics* is a special case of this: here the timing of medical events, disease and cure alike, were supposed to affect them according to the state of the heavens, and so in calculable ways. According to another doctrine, that of *interrogations*, a client's queries could be answered on the basis of the state of the heavens at the moment the question was posed. This bizarre idea was a very far cry from the view of astrology as a deterministic natural science, and seems always to have been near the fringe of intellectual respectability, although it would in time provide seventeenth-century astrologers with a rich source of income.

From Greece and Alexandria in particular, in the second and third centuries, astrology passed in a second wave to India, and there some new ideas were grafted on to it which in time worked their way westwards and into Europe. One such concept was that of enormously long periods of time into which history is divided, calculated in terms of the period over which some or all of the planets return to a particular configuration (such as a mutual conjunction). Another Indian notion was that of lunar mansions, each a subdivision of the zodiac occupied by the Moon during one day of the month. By the time these ideas were absorbed in the western astrological tradition, they had travelled along a path taking in Sasanian Iran, in the third century and later, and then in due course those regions conquered in the name of Islam. In Iran, additions of a peculiar historical character were made. All of human history was there analysed using techniques of nativity-casting: for every year, a 'Lord of the Year' was selected, that is, a planet that was supposed to have especial potency during the complete year. With the steady accretion of such material as this, the astrological corpus grew increasingly formless. A new synthesis was needed, of the sort Ptolemy had provided with his *Tetrabiblos*. Insofar as it ever materialised, it came with the work of Abū Maʿshar, the ninth-century Muslim astrologer.

Abū Maʿshar (Albumasar, to the Latins) became a great favourite of western astrologers, partly for his encyclopedic range, enhanced by some additions he made on his own account, and partly because he used Aristotelian physics to explain the influences of the heavens. Seen chronologically, some parts of Aristotelian physics were actually first encountered in the West in his writings. Islamic astrology was making its way into Europe, curiously enough, just as it was beginning to be criticised ever more effectively by Islamic theologians. In Byzantium this sort of attack had taken place much earlier, but both there and in Islam there was a recovery of sorts. A fourteenth-century Byzantine renaissance in astrology had important European repercussions, for it resulted in the production of new editions of classical Greek astrological treatises, which passed into European languages in the fifteenth and sixteenth centuries.

There had been very few astrological texts of any weight written in Latin during the period of the Roman Empire. The two most famous were the poem *Astronomica* by Manilius—written a century before Ptolemy—and the *Mathesis* (or *Astrologia*) of Firmicus Maternus, written in the early fourth century, summarising the wisdom of the Babylonians and Egyptians. The latter reveals an extraordinary degree of technical incompetence, considering its author's enthusiasm; but this does not seem to have diminished the work's popularity. What was available in the early western middle ages, however, was not good enough to give access to the principles needed to cast horoscopes by mathematical means. Divination was left to other devices. It is impossible to read English chronicles, even of the twelfth century, and escape the conclusion that the demand for prophecy was always strong, whether it was based on a decoding of the mysteries of the Bible, numerology, reports of the predictions of Merlin, prophecy from dreams, geomancy, or other nefarious means. Rationally warranted prophecy from the stars was perceived to be of a different intellectual order, and gave rise to an ambition long before scholars knew how to go about satisfying it. It was helped along by numerous intriguing asides in early Christian writings—such as when Rabanus Maurus insisted that the magi at the birth of Jesus were worthy astrologers. A few simple astrological principles had already crept in under the cover of medicine. Aratus and Pliny had written of the influences of stars and planets on plants and animals. An amalgam of such undisciplined ideas, comforting at the time, suddenly paled into insignificance with the arrival of new and exciting texts from Islam.

Henceforth, the academic astrology of Christendom would to all intents and purposes be filtered through the veil of eastern astrology, with its own extraordinary farrago of accretions. Some of these were recognised to be spiritually dubious, but there was a great temptation to copy whatever was available, and give it some sort of credence. We find many manuscript copies, for instance, of al-Kindī's *De radiis stellarum* (*On the Rays of the Stars*), which instructs the reader on how to use magic to concentrate astrologically calculated influences. At first, faced with such spiritually dubious material, there was a need to be apologetic about astrology as a whole. When Daniel of Morley, al-Kindī's translator, returned to England after studying astronomy and astrology in Toledo, and was questioned by an old friend who had become bishop of Norwich, he put a case for using astrology to avoid foreseeable disaster, and sidestepped more difficult religious problems.

Those who feared the charge of determinism on a personal plane could find numerous ways of calculating impersonal events. There are numerous instances, for example, of calculations being made of the times of especially significant conjunctions of Saturn, Jupiter, and Mars, which were considered important, among other things, because they were thought to chart the rise and fall of religions and sects. What is believed to be the oldest piece of writing in Grosseteste's own hand contains the results of such a calculation for the year 1216. Such things became commonplace. Nearly two centuries later we find Geoffrey Chaucer structuring some of his writings around a great conjunction, so called.[204] The discussion of great conjunctions was often placed alongside that of the portents of comets, although the latter tended to be political rather than religious. A York writer on alchemy whose works were often copied—Robert, known as

Perscrutator—discussed at one point a comet seen over York in 1313, and explained its consequences for a battle between the English and the Scots in which the Scots prevailed. The reference was plainly to Bannockburn.

Oxford Astrology

The Oxford astrological climate is not easy to judge. At first, most of those who were closely concerned with the subject came to it after they had mastered its astronomical underpinning. There is no evidence that it was ever formally required in the arts course in the fourteenth century, although records are fragmentary. The frequent references to astrology in required astronomical texts must have whetted the appetite of not a few readers. There is enough surviving manuscript material for us to be certain that astrology was being carefully studied. The Merton College library list of 1325 includes a dozen of the best astronomical writings then available, but no astrology. The fine collection of sixty books amassed by Simon Bredon, however, when he was fellow of Merton in the 1340s and afterwards, and listed in his will of 1368, included not only serious astronomy—*Almagest* and heavily annotated copies of Richard of Wallingford's *Albion* and Gebir (Jābir ibn Aflaḥ)—but astrology by Ptolemy and Albumasar. Bredon needed astrological knowledge to practise medicine, which he did—to his great financial benefit in court circles—after leaving Merton. Some of the astrological writings produced in Oxford were voluminous: the best example is John Ashenden's *Summa iudicialis de accidentibus mundi* (*Astrological Summary of Opinions on the Properties of the World*), an enormous work of reference, composed in 1346-48 while its author was a fellow of Merton College. Ashenden's chief concern was with large-scale events, such as tempest, flood, drought, earthquake, war, and famine. In later sections of the *Summa* he dealt with the plague, then the most topical and frightening of large-scale human events with a supposed celestial cause. Like almost all astrological texts, his was primarily a compilation, but it was one in which its author tried to weigh the evidence, often disagreeing with his eastern writers on the basis of his knowledge of weather patterns unknown to them. In their search for the roots of empirical science, modern writers often seek out, and lay emphasis on, the empirical element in such works as Ashenden's. His use of classical writers such as Pliny, Virgil, and even Servius—a fourth-century commentator on Virgil—should not pass unnoticed, for such ancient authors were often the sources of what he presented as common weather lore. There is something from nearer home, when he notes that hearing a bell tolling at a distance greater than normal is a sign of impending rain, but by and large he is unashamedly eclectic.

Good scholastic that he was, he often wove Aristotle into his account. His eclecticism was no doubt seen as a recommendation. Ashenden's great *Summa* was later translated into Middle Dutch, and was thought important enough to be printed (Venice, 1489)—this despite, or perhaps because of, its enormous length.[205]

Ptolemy's astrological *Tetrabiblos* (*Quadripartitum*) was being taught in Paris in mid-century, and the fact that at the same period Simon Bredon took the trouble to improve the coherence of the Latin of the standard translation suggests that it was very probably being taught as well as studied in Oxford too. Ashenden borrowed wholesale from his elder colleague's text. *Quadripartitum* was certainly taught in fifteenth-century Oxford: university records that survive from a fifteen-year period show that permission to give lectures on astrology was sought then by Oxford regent masters and scholars, and granted, on at least four occasions.[206] It is also quite clear from the great quantities of manuscript material originating in Oxford and Paris, and dating from the end of the thirteenth century onwards, that university records are a very poor indicator of the true levels of activity, which cannot all have been conducted at a private level.

It is often said that medieval scholars did not draw any distinction between astrology and astronomy, but this is entirely mistaken. Isidore of Seville was a widely-read author who drew a very clear distinction: astronomy, he said, deals with the motions and risings and settings of celestial bodies, while astrology, which could be natural or superstitious, interpreted such things. Over and over again, this simple distinction was repeated verbatim from Isidore. After the arrival of eastern astrology in the West, it became fashionable to lay stress on the boundary between licit and illicit. Kilwardby's classification of the sciences embroidered the Isidorean division a good deal, giving it an Aristotelian tinge, but added nothing really new. When Roger Bacon rephrased the old distinctions, he gave greater prominence to the claims of mathematical astronomy viewed as an abstract mathematical science—that is, a science not applied to judgements of earthly affairs.[207]

A man who rewrote the basic Isidorean division in a very different way, with the loss of something in the process, was the Carmelite John Baconthorpe, writing in the 1330s. He distinguished between four sorts of 'astronomy'. First there was 'vulgar astronomy', shared by laity and clerics, to know the time of year, when to plant, and so forth. This included forecasting future natural events, events—such as droughts and rains—which are naturally caused by the heavenly bodies, placed there by God for that purpose. Second, he took 'scientific astronomy', under which heading he placed the calculation of conjunctions in order to divine future events. Third, Baconthorpe named 'fatal astronomy', which was to divine from the constellations of stars (the planets) the manners and fortunes of men. In a word, genethlialogy. Finally, there was 'superstitious astronomy', for deciding the days and times on which it is best to perform certain acts. The last three, he said, are dangerous, and to lead to the sect of Antichrist, through the diabolical arts, the consultation of oracles, and encouraging magic. The first is licit in itself, but through man's natural curiosity it may lead on to the others.[208] Why Baconthorpe's distinctions are especially interesting is that they are drawn up by a very erudite scholar, familiar with the intellectual life of Oxford and Paris, and yet apparently quite incapable of appreciating the idea that astronomy could be anything other than an applied science.

His sometime Oxford neighbour, Richard of Wallingford, would have been appalled, but it is impossible to determine how many Oxford scholars of the time felt the same about astronomy as a self-sufficient subject. We do not have to suppose that it was necessary to have high expertise to share this feeling. A good example of a contemporary of Richard of Wallingford who was deeply interested in astronomy without being especially adept at it was the Dominican friar Robert Holcot. A highly respected theologian, Holcot left behind him a large quantity of theological, philosophical and logical writings, among them a *Quaestio de stellis*. It is not at all technical, but in it he seems to show slight familiarity with Ptolemy's *Almagest*, and with the difficult astronomical work by Jabīr ibn Aflaḥ used by Richard of Wallingford. He was familiar with the casting of nativities. He believed that he would die peacefully, since Jupiter had been in the house of Mars at his birth.[209] Did he remember this prediction, when he lay dying of the Black Death in 1349?

As time passed, university men became increasingly familiar with the mathematical techniques of the astrologer, and the methods of simplifying them with the help of simple instruments, and such astrological tables as those drawn up by John Walters, of New College, in the late fourteenth century. Astrologers courted more than spiritual danger: by claiming to cover so many aspects of human life with their powerful tools for reading the future, they were straying outside the lecture hall, and academic discussions of free will. Their subject might have clear political implications, with dire consequences, even for those within the university world. The idea was not new in England. There are political prognostications known from the royal court in the twelfth century. Ashenden himself surveyed

Muslim authorities on the celestial indicators of royal power, although he did not pursue the subject very far. Divination by interrogation was always a dangerous game, but it was often practised—there is a good example with a horoscope for 1376, cast to decide whether Richard of Bordeaux will become king. Three generations later, we meet with two Oxford masters, Thomas Southwell and Roger Bolingbroke, who suffered the ultimate penalty of mixing theory with practice. In 1441 they were found guilty of conspiring with Eleanor, duchess of Gloucester, to enquire 'by the black arts' into how King Henry VI would die. Southwell died in prison; Bolingbroke was hanged, drawn, and quartered for his crime. Another rather similar case of treason occurred in 1477, now involving two Merton astronomers, Thomas Blake and John Stacy. They had answered a query from a friend of the king's brother. Would the king die? By the time the trial finished they had been found guilty of trying by magic, necromancy, and astronomy to cause the death and destruction of the king and the Prince of Wales. Blake was pardoned at the request of the Bishop of Norwich, but Stacy was executed at Tyburn. These were dangerous times. In 1485, Lewys of Caerleon almost met the same end. He was an astronomer of much talent who greatly admired Richard of Wallingford, and built upon the *Albion* treatise. In his case, he escaped with a period of imprisonment in the Tower of London—and in the peace and tranquillity of that place used his time to great advantage to produce numerous astronomical tables.

The crime, in all of these cases, was not that of practising astrology, or even of mixing it with magic, but of allegedly doing so for treasonable purposes. No one blames an axe for a beheading. Astrology continued to be pursued in the universities. It was still being actively studied by the first president of the Royal Society; and since times past are so easily compressed by the mind, it is perhaps worth pointing out that the man in question, Lord Brouncker, was somewhat nearer in time to us than Richard of Wallingford was to him.[210] Richard was certainly convinced of the worth of astrology, but he lived at the more sober end of a historical development of which he would very probably not have approved.

Exafrenon

When Ptolemy wrote his *Tetrabiblos*, he had a broadly scientific programme in view. By the time the new astrology arrived in the West, the qualities he had tried to give it were largely debased. It made use of numerous procedural rules, many of them mathematical, but was formally very superficial: the fact that the mathematics was consistent and difficult gave the subject much of its renown, but effectively hid the fact that it had very little by way of bonding, unifying, principles. Its practitioners could rarely see why its rules were what they were, which perhaps explains why there was a general reluctance to question it. This gave it an almost religious aura. The theologians must have secretly feared it for this reason. Their safest mode of attack needed no expertise: it was on the venerable grounds that it led to moral determinism. When Richard of Wallingford wrote his *Exafrenon*, he chose the branch of astrology least likely to incur this charge. To write 'on the prognostication of times', that is, on the weather, was broadly innocuous. His essay would not have been judged entirely above had the few theologians who opposed astrology taken the trouble to read it, but in fact they almost invariably stuck to generalities.

Shortly after Richard of Wallingford had completed his *Exafrenon*, Robert of York, Perscrutator, his near-contemporary, showed how it was possible to give astrology a measure of respectability in the eyes of another academic group, the Aristotelian natural philosophers. Writing in 1325 on meteorology, he said he was proud of the fact that he based himself on reason and experiment rather than on mere authority. He was more speculative than experimental, but his discussion of the forces which operate on the elements to form compounds (*mixtiones*) out of them is interesting because it reflects on the language in use in Oxford at that time. Equally significant is the fact that its speculative element is an amalgam of Aristotle, astrology, and alchemy, which no doubt helps to explain why he was still being read three centuries later.[211] Here again, Grosseteste's influence was very real. He was the writer of the most influential work on astrological weather prediction in the thirteenth century, and his *De impressionibus aeris* (*On the Forces Affecting the Atmosphere*) was the source of parts of *Exafrenon*. Perscrutator's treatment of the subject is all the more interesting because it points to an opportunity missed, not only by Richard of Wallingford but by Robert Grosseteste, the chance to conjure with Aristotelian doctrines.

Oxford interest in the use of astrology for weather forecasting continued after *Exafrenon*. It is nowhere better illustrated than in a mid-century volume assembled by William Rede of Merton, for his own use. The volume was eventually bequeathed by him to those of his kin who studied at Oxford, and is now MS Digby 176 in the Bodleian Library. Written in various hands, some sections of the manuscript were probably obtained from colleagues. It includes a long series of weather records by William Merle, as well as weather prognostications and calculations by John Ashenden and Reginald Lambourne. Weather apart, the volume shows that Oxford academic astrology in the early fourteenth century was not above giving actual personal judgements—as opposed to the general principles of them as set out by standard writers. It includes a nativity for 1317, an astrological analysis of the causes of the murder of Walter Stapledon, bishop of Exeter, in 1326, and a rich set of horoscopes for the 1340s. It was in 1386 that tables for casting horoscopes were drawn up by John Walter, fellow of New College.[212] While a skilled astronomer does not need such tables, they were easy to use, and that they were valued is clear enough from the fact that they were often mentioned by those who later cast personal horoscopes.

The empirical element in such material as this, written after *Exafrenon*, contrasts with the more traditional character of that work, which was probably Richard of Wallingford's first, and which depended heavily on Grosseteste's. Ashenden was quick to note that last fact, and clearly found Perscrutator's approach much more exciting. Traditional though it may have been, *Exafrenon* was coherent and well structured, as astrological writings go. It was much more so, for example, than the more popular text by Leopold of Austria of the previous century.[213] *Exafrenon* drew heavily on Albumasar's *Flores* (*Flowers of Astrology*), and shared much of its structure, but Richard of Wallingford paid more attention to the astronomical problems posed by astrological doctrine. He concerned himself, for instance, with the slow drift of the solstices and equinoxes through the calendar, which affects decisions over which planet is to count as 'Lord of the Year'. In the second chapter, he gave a rule for casting the houses of a horoscope—dividing the zodiac into twelve parts—which is identical to a rule discussed at length in his *Quadripartitum*, a rule which he knew his contemporaries found it hard to master.

Exafrenon was not entirely without an Aristotelian component. The planets and the signs of the zodiac were related to the qualities of the elements, but this all came from Albumasar's astrology, and not from Latin translations of Aristotle. When in *Exafrenon* Richard likened the signs to a body and the planets to souls, a Neoplatonic idea, this too came indirectly, through Muslim astrology. Such philosophising, however, was little more than window-dressing. The hard core of the work, at this point, was the account given of domiciles (signs of the zodiac where planets were most at home, sometimes also called houses), exaltations and depressions (specific degrees where the planets rejoice or are dejected), triplicities (groupings of three signs with various shared properties), terms (five of them in every sign, being unequal divisions known also as limits), and faces (three to a sign, all of ten degrees, also called decans). The aim here was to establish the strength of a planet on the basis of its position in the zodiac relative to those divisions (see Fig. 63). Here Richard draws on another work by Albumasar, the extremely influential *Introductorium*. A more difficult doctrine follows, according to which the virtue of a planet is related to its apparent motion. In brief, this was taken to depend on its position on the

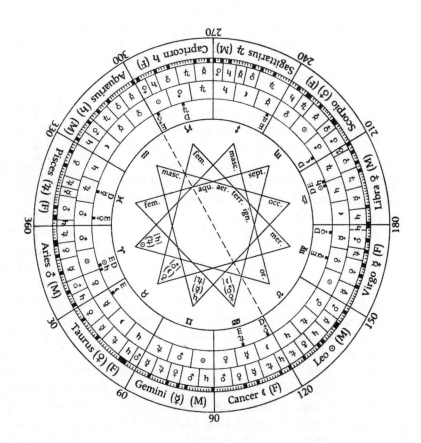

63. A summary diagram for the doctrine of planetary dignities. A planet was given five dignities when in its own domicile, four if in its exaltation, three if in its triplicity, two if in its term, and one if in its own face. Here domiciles are on the outer ring, next come the terms, then the faces, then the exaltations and dejections (marked *E* and *D*), and finally the triplicities, each indicated by a triangle. The planets are here denoted by their conventional symbols. Thus Aquarius belongs to Saturn, Pisces to Jupiter, Aries to Mars, Taurus to Venus, Gemini to Mercury, Cancer to the Moon, and Leo to the Sun. The familiar symbols in the innermost circle are for the signs.

epicycle, being greatest at the highest point and least at the lowest, and especially great if at the aux of the deferent.[214] There was supposedly a connection here with the tides, which we recall were shown on the clock at St Albans. (Albumasar had much more on this very subject, but Richard did not repeat it.) Weather prognostication was considered to depend on similar factors, being determined, as *Exafrenon* tells us, by planetary potencies, which were in turn decided by position and velocity, and planetary character. The Sun is hot and dry, Venus is warm and moist, and so forth down the list. This all comes from Ptolemy, and was widely known. From Ptolemy too comes the theory of planetary visibility dealt with in tables which Richard of Wallingford borrowed from some as yet unidentified astronomer, glossed as John of London, who was seemingly working in the 1290s.

No astrology was complete without an account of planetary aspects—angular separations in the zodiac determining inter-planetary friendships and hostilities. Thus a separation of 60° (two signs, a sextile aspect) was deemed friendly, and 90° unfriendly. For a planet to be in unfriendly aspect with the malevolent planet Saturn, for example, was highly unfortunate, although of course in all such relationships there were two planets involved, and the characters of both counted towards the result. Many an astrological treatise of this period gave long lists of all possible combinations of planets in aspect—chiefly conjunction, opposition, sextile, quartile and trine. Needless to say, this sort of thing gave great scope to the imagination, but Richard of Wallingford pressed on without it, to pick up again his account of lord of the year, and then lord of the month. From there he passed to his final chapter, first with a calculation for 15 April 1249—lifted out of Grosseteste, but perhaps with Bacon as intermediary. This was an interesting of date, the last day of the Muslim year (AH 646). It was therefore very probably taken from Muslim tables which had not been converted to the Christian calendar, and chosen because it made calculation easier. Another example for 1255 was probably chosen for the same reason: Saturn was in Capricorn, where it stayed, conveniently, for most of the year. It is quite obvious that at this period there was no great fund of expertise in planetary calculation, just as there would be little today, if it were not for the electronic computer. As for Saturn's position in 1255, it is used to account for the cold weather of that year, almost certainly retrospectively.

Exafrenon had the merit of conciseness, and was appreciated for that quality. Within a century of its writing, it was translated into English—the only work by Richard of Wallingford to be translated at an early date. This tells us something about the audience for its subject. Those who needed a translation were literate, but not in the academic ranks of those able to understand his mathematical writings. It is chastening to read the following solitary reference to Richard as an astrologer, in one of two dozen texts in English translation collected together by an unknown but callow Englishman of the early sixteenth century:

> Luna the Mone ... hath more myght in creaturs here beneathen then any other planet, as saith saint Thomas [Aquinas] and Richard of saint Albanes, Protholome [namely Ptolemy], Alphragani, and Algassel [al-Gazālī], both for it is most nere unto us of al planetis, and for he [the Moon] is most fast in his cours of al planetis ...[215]

With whatever justification, Richard was being placed in illustrious company. If he had written nothing beyond *Exafrenon*, however, he would not have merited the honour.

21

Instruments of Thought

LIKE HIS MODERN COUNTERPART, the medieval university scholar did not always have the privilege of teaching the subjects dearest to him, or even of being able to discuss their finer points with others in the same university. We suggested earlier that the person with whom Richard of Wallingford was most likely to have exchanged ideas was John Maudith of Merton College. He compiled tables for which Richard wrote canons—indeed, some of his work in the Ashmole manuscript is intimately linked with Richard.[216] Both men concerned themselves with instruments—Maudith with the Profatius quadrant—but the word 'instrument', like the word 'astronomer', means many things. The three main functions of the astronomer's instrumentarium were to assist with observation (in the middle ages usually associated with measurement), to assist with calculation, and to demonstrate a principle or a theory, through a model. These functions often overlapped. Even observations could serve many purposes: they might be made to disprove a theory, or to provide a theory with improved parameters, as in the case of two crucial observations made by Simon Bredon in 1347, in the hope of deriving the 'motion of the eighth sphere'.[217] Observations might have a less critical theoretical motive, such as when Maudith measured star positions so as to be able to put them on the quadrant; or they might have no theoretical purpose, as when, for example, they were used to find the time of day. We usually associate the word 'instrument' with a material artefact, but as Samuel Johnson said, language is an instrument of science. We tend to think of language as an invariant property of the sciences, but nothing could be further from the truth. The language of mathematics illustrates the point well.

Mathematics as Instrument

Richard of Wallingford's place in the history of medieval mathematics is made secure by his trigonometrical treatise, *Quadripartitum*, which like his albion was designed to ease the task of astronomical calculation. Its roots were in a series of astronomical treatises stretching back through

Arabic intermediaries to antiquity, to Ptolemy, to Menelaus, and beyond them no doubt to Hipparchus. We may describe it as trigonometry, as long as this does not conjure up a vision of the modern notations which go under that name. It is a work on the geometry of the sphere, but—like the now more familiar trigonometry of sines, cosines, tangents, and the rest, applied to plane figures—it makes use of techniques which simplify geometrical procedures to a point where the practitioner can almost lose sight of the geometrical scheme entirely.

There have been numerous histories of early trigonometry, many of them aimed at identifying the birth of the subject. As in Topsy's case, no one will accept the idea that it just growed. Ratios between the sides of similar triangles take some to the pyramids and others to Stonehenge; the use of sines in place of chords may take us to India; advanced geometrical methods using specially designed theorems for astronomical purposes will certainly not allow us to begin the history later than Hipparchus; but if we insist on finding a purely mathematical treatment which is at all advanced, and yet independent of any astronomical application, where then are we to look? Many a writer has named Regiomontanus in this connection. His masterly *De triangulis*—written in 1464 but not published until 1533—is sprinkled with numerical examples, and yet in it he does manage to present the subject systematically, in a way not unlike that of Euclid's *Elements*. He was plainly writing with a view to its astronomical application—one only has to see his correspondence with Bianchini to confirm this suspicion—but astronomy is not given as the justification for the text, which is more or less mathematically self-sufficient.[218] Richard's *Quadripartitum* could easily have its astronomy removed, and it too would then have system and self-sufficiency, although to a lesser degree. The fact is that the historical transformation in attitudes, from one which saw no clear distinction between pure and applied mathematics to one which considered them to be separate subjects, was very slow and uncertain, and took the best part of two millennia to complete.

The chief stimulus in the development of a subject most of us would regard as trigonometry was almost certainly the need to produce numerical answers to astronomical questions—to calculate angles, for example. To get an answer—let us call it A—the mathematician-astronomer found that it was necessary to perform a sequence of geometrical operations, combining certain terms in certain ways—let us call this combination C. The instruction was then 'If you form C, then you will get A'. In the fulness of time, this became viewed as an equation, $A = C$, and in due course it was appreciated that it was permissible to operate on both sides of the equation in the same way, while preserving the truth of the equation. $A = C$ was no longer an operational instruction; in customary parlance, it was a trigonometrical identity. Richard of Wallingford's *Quadripartitum* stands at the watershed between the two ways of looking at the case.

This subject, in the form it assumed at the boundary between mathematics and astronomy, was taken up and developed by a number of eastern Islamic scholars, whose writings remained largely unknown in medieval Europe. (Among them were Abū'l Wafā', Abū Naṣr Manṣūr, al-Bīrūnī, and Nāṣir al-Dīn al-Ṭūsī.) The canons to the tables of al-Khwārizmī, which were of course known, carried some trigonometrical knowledge from the East, as did Azarchel's canons from Spain, but in both instances it was of slight importance. The best western writer on this subject in Arabic was Jābir ibn Aflaḥ, Gebir.[219] He worked in Seville in the first half of the twelfth century, and compiled an important and influential reworking of Ptolemy's *Almagest*. It was translated into Latin by Gerard of Cremona, and became generally known in the West as the *Flores Almagesti* (*Flowers of the Almagest*). This was an interesting example of a work developed along roughly similar lines to those being followed elsewhere, in this case in eastern Islam. Richard of Wallingford wrote the first version of his *Quadripartitum* in ignorance of it. The fact that all of these authors could overlap in their writing was quite simply because they were all basing themselves on Menelaus, known through Ptolemy. Richard was not at all reluctant to cite Gebir's *Flores* after he had encountered it, as he had by the time he wrote his treatise *Albion*—just before becoming abbot—and in the revised version of *Quadripartitum*, to which he gave the new title *De sectore*, written during his time as abbot. When the Merton College astronomer Simon Bredon wrote a commentary on *Almagest* in 1340, he took much from Gebir, as well as from Richard of Wallingford's *Albion* and his *Quadripartitum*.[220] Regiomontanus likewise borrowed from Gebir in his important *De triangulis*. Regiomontanus's unacknowledged plagiarism of a whole section of Gebir was later noted by Girolamo Cardano, the sixteenth-century mathematician, astrologer, medical writer, and connoisseur of plagiarism. One of Gebir's most useful contributions was to reduce number of terms in the theorem of Menelaus from six to four, by suitably arranging for appropriate right angles in the basic figure.[221]

Richard's *Quadripartitum* opens with definitions of sines and versed sines, relating them to arcs and chords. They are regarded as lengths, rather than ratios, all being supposed normed on a radius of R. One could say that we take a radius of unity. The usual medieval values for R were 60 parts (a tradition older than Ptolemy) and 150 parts (ultimately from

Hindu sources). Richard of Wallingford used them both, on occasion. Thus where we should say that the sine of 30° is 0.5, he would have taken it to be either 30 or 75 parts, according to context. The context in question was often a table of sines, and he had tables from both traditions.

In the specimen examples given below, which use a notation not available to him, we shall give trigonometrical ratios an initial capital letter wherever they must be considered as lengths to a standard radius of R. He used chords, of course, as Ptolemy had done. He did not introduce the tangent function, but he had used it in his canons to Maudith's tables.

After his definitions, Richard launched into Part I of his treatise, with a series of what we may view—with the caution given above—as trigonometrical identities. We could write his first theorem as

$$\operatorname{Vers} a = R - \sqrt{(R^2 - \operatorname{Sin}^2 a)},$$

to which he adds the corollary

$$R^2 = \operatorname{Sin}^2 a + \operatorname{Sin}^2(90^\circ - a),$$

which is of course the equivalent of our familiar

$$1 = \sin^2 a + \cos^2 a.$$

Part I of the work contains a dozen propositions, and the ancestry of most of them can be guessed at. Before passing on to the second part of the treatise, we may consider the most complex of them, which is probably original. In our modern notation it can be expressed

$$\operatorname{Sin}(a-b) = \frac{2}{R} \cdot \operatorname{Sin}\left(\frac{a-b}{2}\right) \cdot \sqrt{\left\{R^2 - \operatorname{Sin}^2\left(\frac{a-b}{2}\right)\right\}},$$

where
$$\operatorname{Sin}\left(\frac{a-b}{2}\right) = \frac{1}{2}\sqrt{\left\{(\operatorname{Sin} a - \operatorname{Sin} b)^2 + (\operatorname{Vers} a - \operatorname{Vers} b)^2\right\}}.$$

This last result may be of interest, even to those who have no wish to check it, or even read it. It will not seem particularly elegant to those familiar with

$$\sin(a-b) = \sin a . \cos b - \cos a . \sin b.$$

What is so striking about the new result, however, is its sheer complexity, which prompts the sobering thought that it was obtained entirely without the aid of a convenient, modern, mathematical notation. How should we ourselves go about deriving this new pair of equations without the convenience of an analytical notation? Even if we cannot imagine ourselves doing so, we can at least try to turn the equations into words, and contemplate how we might then begin to manipulate the resulting word-equations. A little help in manipulating six-term proportionalities was obtained from simple line-drawings (Fig. 64), but it is clear that there are limits to how much the mind can hold, and that with such complex material as Richard of Wallingford was including in his text, he was near to the end of the road of verbal algebra. For a reasonably powerful symbolic language, western Europe would have to wait for another two centuries and more.

In Part II of his treatise, Richard turned to the mathematics of proportion and proportionality. Again, modern readers—armed with algebraic notations which they are ready to apply to almost anything to which a number can be attached—may find it hard to understand why so much attention was given to this subject. The foundations of the theory go back to Book V of Euclid's *Elements of Geometry*, or rather to Eudoxus, who is usually credited with it. The theory was much admired in the seventeenth century, but not until the nineteenth were its merits properly grasped—and then only after mathematicians had unwittingly retrodden its originator's path. In the middle ages, the routes by which the *Elements* had arrived on the scene were so convoluted that it was known only in a muddled state, a mixture of often inconsistent translations and comment, with interpolations by such as Plato, Aristotle, Boethius, and Adelard of Bath. The most influential version was that by Campanus of Novara, with unavoidable omissions and extraneous elements. It eventually became by far the most popular version, however, from the day it was published by Erhardt Ratdolt, in 1482.

64. Marginal line diagrams, drawn as aids to following the purely verbal proofs of propositions based on the theorem of Menelaus. Such figures had their origins in antiquity.

The key to Book V is in the definition of a ratio with which the book opens: it is 'a kind of relation in repect of size between two magnitudes of the same kind'. That seems innocent enough, but insistence on the idea that the magnitudes be of the same kind had far-reaching consequences. A quantity or magnitude was allowed to be equal to, less than, or greater than another only if both were of the same kind. It meant that a ratio of, for example, a distance to a time was illegitimate. In the Oxford of Richard's day, those who were discussing motion in general, and ratios of resistive power to motive force, and so forth, were having to contend with this embargo on ratios of mixed kinds.

What is now the most famous of the definitions in Euclid, namely the fifth, is admired because it lends itself to a definition of the equality of numbers.[222] The virtues of this definition, as a way of defining equality of ratios, were not appreciated in the middle ages, and it was still being criticised mistakenly by Galileo in the seventeenth century. Medieval mathematicians followed another route. They defined equality of ratios in terms of the equality of their so-called denominations, *denominationes*. The denomination of a rational number can be considered in some sense the size of the ratio. We could say it was the fraction in its lowest terms—which in some cases will of course be an integer—so that the denominations of 38/4 is equal to the denomination of 114/12, both being 19/2. This will not work for irrational numbers, for which there was no good and explicit medieval account, but those who used the concept of denomination, as did Richard of Wallingford, had a feeling for it as the size of the number. We shall not be far from the idea if we suppose it to have been given (to use our own vocabulary) by a terminating or non-terminating decimal.[223]

Those who studied Euclid in the universities learned the names of (legitimate) transformations of ratios from Euclid Book V. In this standard scholastic exercise, they learned to distinguish six kinds of proportionality by adjective: *conversa, permutata, disiuncta, coniuncta, eversa,* and *equa*. Campanus likened them to the types of argument the student would already have learned in logic, and indeed the names were close enough to have been potentially confusing. Where we have no difficulty in using algebra to deduce $(A+B)/B = (C+D)/D$ from the premise $A/B = C/D$, the student learned this as a standard inference with the name of 'conjunct proportion'. (No doubt there are still a few readers who recall learning the same rule in the context of school geometry.) It is therefore not surprising to find Part II of *Quadripartitum* opening with a series of propositions which to us appear similarly straightforward, but which are handled by rote. They actually owe more to another work by Campanus than to his Euclid. He wrote a tract—if indeed it is his—on the Menelaus theorem, which in turn drew from Arabic works on the same theme by Aḥmad ibn Yūsuf al-Miṣrī and Thābit ibn Qurra.[224] As specimens from this section: Richard's first proposition, would now be written $A.(C/A) = C$ while the fourth can be considered as an inference from $A/B = (C/D)/(E/F)$ to

$B/A = (D/C).(F/E)$. Armed with a bundle of such propositions he is now ready to tackle the standard *pattern* of the Menelaus theorem—not the geometrical theorem, but the proportionality needed to represent it. Let us suppose that we express this in our own notation as follows:

$$\left(\frac{a_1}{a_2}\right) = \left(\frac{a_3}{a_4}\right).\left(\frac{a_5}{a_6}\right),$$

which by an obvious convention we may represent as 123456. It will be clear enough to a modern reader that 123654 is an equivalent mode, derivable from it, as is 132456. Richard first shows that there are eighteen such possible modes, and then proceeds to prove them, drawing heavily on the Latin version of Thābit. Having done this, he gives an elaborate proof of the unacceptable character of other meaningful combinations of terms. This is all set out in a quasi-axiomatic way, and is thorough to a degree, even though no one would now regard it as exciting mathematics.

The same is true of Part III of the *Quadripartitum*, in which the foregoing work on abstract proportions, in the tradition of Euclid Book V, is at last turned into geometry, in fact into proofs of Menelaus's theorem for plane triangles. Here there are thirty-six modes, however, eighteen where the transversal cuts all three sides of the fundamental triangle externally (*cata coniuncta*), and eighteen where only one side is cut externally (*cata disiuncta*). The word *cata* was from Ptolemy's Greek, through Arabic, and was also translated into Latin as *sector*, so that when astronomers refer to the 'sector figure'—as did Richard of Wallingford in his *De sectore* version of the text—they are referring to this theorem.

Finally, in Part IV, it is the turn of the thirty-six modes of the theorem of Menelaus on a spherical surface, done with comparable thoroughness, but supplemented—after Richard came to write *De sectore*—with material from Gebir. He took from Gebir almost precisely the same material as Regiomontanus, more than a century later. At last, Richard of Wallingford reached the stage at which he could make use of the basic theorems he had proved up to this point. In this final part, with it twenty-five propositions in all, he presented us with a condensed survey of spherical astronomy. The summary rested on the astronomy of the *Almagest*, of course, and also drew from al-Battānī, but it was consciously set out as an application of what had gone before, rather than as a text book in itself. It included some calculations based on the *Toledan Tables*, and it has to be said that Richard's arithmetical talents appear not to have matched those in his geometry.

Viewed from the perspective of the historian who can see two thousand years at a single glance, Richard of Wallingford's *Quadripartitum* cannot astonish us in the way that a Eudoxus or a Ptolemy can. Viewed, however, as a product of its age and culture, it represents a considerable step forward. It was the first substantial tract of its kind in western science, and for all its faults, it gives evidence of a strong feeling for the need to axiomatise mathematics, and to provide rigorous proofs. Quite incidentally, it tells us something about an obstacle to science that was being tackled at the time it was written. It shows us just how great was the intellectual barrier standing in the way of the mathematisation of physical concepts, the 'denominations' of the ratios of different kinds of term, such as length and time. It was all very well for people like Grosseteste and Bacon to say, in effect—as Galileo was later to say—that mathematics is the language of Nature, but not until this problem of ratios was cleared out of the way would much progress be possible. The astronomers were familiar with equivalent problems; as mentioned earlier, they could handle complex problems involving accelerated motions, but when the same scholars picked up their Euclid Book V, they did not show as much independence of spirit as they might have done. Finally, we should remember that *Quadripartitum* was written early in Richard of Wallingford's career, and is all the more remarkable for that. From the additional fact that long afterwards, when he was abbot, Richard wrote his revised version *De sectore*, it is clear that his first great enthusiasm stayed with him throughout his life, and that millstones, clockbuilding, and abbey administration did not occupy all of his time.

Material Instruments

John of Whetehamstede was an Oxford theologian of some distinction who had studied at Gloucester College. He was twice abbot of St Albans, having stepped down for a time on the grounds of ill health. During his first abbacy, between 1420 and 1440, he composed a work with a title which was a pun on his own name, *Granarium*, and in this granary he lauded his great predecessor Richard of Wallingford, while including some brief remarks on a number of different instruments. He naturally included those designed by the former abbot—the rectangulus, albion and *horologium astronomicum*. He might well have had the *Albion* treatise in hand to prompt him as he wrote, for in the preamble to its third part Richard names some of the same instruments, including the horologe. The difference is that Richard is boasting that his albion can do all that can be done with the others, and more besides. Whetehamstede's slightly fuller and more impersonal list makes a useful starting point from which to examine this particular facet of medieval astronomy. Without going into much detail as to the actual workings of the instruments named, we should

65. Detail of a woodcut from Oronce Fine, *De solaribus horologiis et quadrantibus* (Paris, 1542). Urania, muse of astronomy, is shown instructing the author. Fine holds up an astrolabe, showing the hour-lines on the upper half of the back and the shadow square below them. On the ground lies an astrolabe quadrant (a 'new' quadrant).

at least point out that they fall into two broad groups: some were for common use, while others required a certain expertise, and were likely to have belonged only to a practising astronomer. The fact that all of them required astronomical knowledge for their design is irrelevant to that distinction.

Whetehamstede's list starts with the plane astrolabe, which belongs to the more advanced class. The back of it was generally equipped with what amounted to a separate instrument, to find the hour on the basis of the altitude of the Sun (Fig. 65). This might well have been used by those of slight education, but the standard uses of the astrolabe—three dozen, or more were taught—would have required some knowledge of the astronomy of the quadrivium. Whetehamstede distinguished the plane astrolabe from Ptolemy's 'spherical' or 'solid' astrolabe. (The word 'solid' here refers to its three-dimensional character, not to its incorporation of a solid ball.)

While the horary instrument on the back of the astrolabe was probably rarely used, at least by those capable of using the main instrument, it had edged into the western European astronomical tradition at least as early as the 1140s—it was in the treatise on the astrolabe by Raymond of Marseilles. There was, however, a related class of instruments, the horary quadrants, on which the same sort of lines stood alone, and the circumstances under which they appeared on the scene are not fully known.[225] Whetehamstede went on to name one such instrument, which he mentioned as carrying a cursor, and he associated it with the name of Sacrobosco, who indeed wrote a very competent treatise on it. It is now

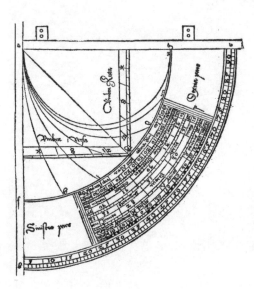

66. Woodcut illustration (Strasbourg, 1539) of an 'old quadrant', or 'quadrant with cursor', for finding the hour of the day. The ungraduated circular arcs are the unequal-hour lines, uppermost being that for noon. The cursor slides in a groove, and carries scales on which is set the Sun's place in the zodiac for the day in question. It is not usually so centrally placed: its position is decided by the observer's geographical latitude. The plumbline, with its sliding pearl bead, would have hung from the upper left corner.

67. The construction of the new, or astrolabe, quadrant of Profatius. From a late sixteenth-century book of instruments (Venice, 1597). From this a rough idea may be had of the relationship of the lines on the quadrant to those on a conventional plane astrolabe. They are clearly very different from those on the old quadrant.

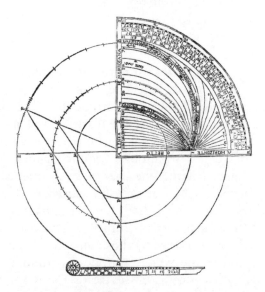

usually known as the 'old' quadrant, although 'quadrant with cursor' would be a safer description, and the cursor was a rather hefty affair, carrying as it did zodiacal scales (see Fig. 66). The cursor's aim was to simplify the task of setting a seed-pearl marker on the plumbline, this bead being what ultimately indicated the hour.

68. The front and back (below) of the sundial known as the 'ship' or 'little ship of Venice'. This example, illustrated from the *Gentleman's Magazine* for January 1787, is now lost. The sights are in the poop and forecastle. On the back are a shadow square and unequal-hour diagram, irrelevant to this side. Before observing, the mast is set with its projection below the keel at the appropriate day's reading on the solar (zodiac) scale, and a bead on the thread is positioned with the help of the other half of the same scale. The plumbline should hang from an adjustable slider on the mast, already apparently lost when the drawings were made.

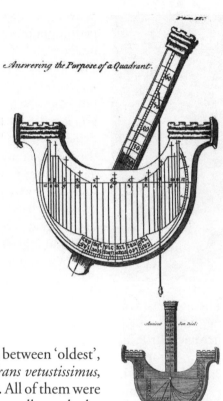

It is now customary to distinguish between 'oldest', 'old', and 'new' quadrants—*quadrans vetustissimus*, *quadrans vetus*, and *quadrans novus*. All of them were originally meant for observation as well as calculation: two sighting pinnules were fitted to one edge of the quadrant, and the angular altitude of the celestial object observed was indicated on the peripheral scale by a plumbline hanging from the corner of the instrument. The oldest form, *vetustissimus*, is close to the horary instrument on the back of an astrolabe, and was seemingly known in Baghdad by the ninth or tenth century. The origins of the quadrant with cursor are obscure, but it certainly entered western Europe from Spain before Sacrobosco's time, and might be ultimately of eastern origin. The last of the trio was essentially a folded astrolabe, although with no moving parts, and it operated on entirely different principles from the other two. It belonged emphatically to the professional class of instruments (Fig. 67), and in this respect also was different from the older forms. In addition to the text by Profatius, Whetehamstede alluded to that by the Danish scholar Peter Nightingale.[226]

69. Set on a level surface, or suspended so that it hangs perfectly vertically, the dial when in use should have a perfectly horizontal gnomon. This is directed so that it heads in the direction of the Sun, in which case its shadow will fall vertically. The tip of the shadow cast on the body of the cylinder then gives the time, as judged by the grid of hour lines on the barrel. (That grid is not universally valid, but is designed for a particular geographical latitude.) Before the dial is used, the gnomon must be set correctly for the time of year. The knob from which it protrudes is twisted round until the gnomon is opposite the correct date on a graduated band wrapped round the cylinder. The band is graduated, as finely as space allows, with the signs and degrees of the zodiac, and the months and days of the month.

The next instrument on Whetehamstede's list is the computus disc (or ring, *annulus*) of John of Northampton, which seems out of place here, since it was little more than repository of calendrical information, manipulated on revolving discs. He went on to mention a sundial known as the *navis* (ship), obviously referring to what is now better known as the *navicula* (little ship), also known as the 'little ship of Venice'. Here he ascribes its invention to a monk of Glastonbury named Peter of Mucheleyo—presumably Muchelney, where there was a Benedictine monastery, less than twenty miles from Glastonbury. There are several surviving tracts dealing with the little ship, the earliest with an English connection. Together with the ship-like form of what is in effect a portable sundial, they give a deceptive impression that this was a simple device. It would have been found easy enough to use, following a rule book, but few readers would have found it easy to understand why it indicated the hour as it did. It stood in a tradition of sundial design to which Vitruvius and Ptolemy contributed, but whether it was based on an early design, or was an English design based on an old astronomical principles, is not known.[227]

After mentioning the little ship, John Whetehamstede made some remarks about plane sundials, remarks which suggest that he was unaware of ancient contributions to this subject. He mentioned Albategni and Azarchel for dials showing unequal hours, and a St Albans monk, Robert Stikford, who is said to have first put equal hours on a mural dial.

(Albategni was the Latin name for al-Battānī.) There are four references to Robert in the index to the *Gesta abbatum*, showing that he lived at the end of the fourteenth century, but nothing else is known of him, as far as can be seen. Stickford is a village north of Boston in Lincolnshire.

Having repeated Richard of Wallingford's claim about all that could be done with the albion, Whetehamstede finally listed instruments about the inventors of which he was uncertain, for some reason even including here the rectangulus. The others were the equatorium (he was thinking of a version prior to the albion); the directorium (an astrological device not unrelated to it); the cylinder sundial; the armillary sphere; the saphea (a universal astrolabe plate adopted by Richard of Wallingford); and the turketum (turquetum or torquetum), which Richard's rectangulus was designed to replace. With the exception of the cylinder dial (Fig. 69), all of these would have fallen squarely into the class of instruments meant strictly for expert astronomers. The cylinder dial, or chilindre (from the Latin word *chilindrum*) was one which any poor scholar might have fashioned for himself out of turned wood, while at the other extreme of society a prince might have owned an example in precious metals.[228]

In a marginal note, Whetehamstede later supplied items of missing information about the various inventors, but only tentatively, ascribing his previous hesitation to his uncertainty. Some are said to believe that Azarchel invented the cylinder dial and saphea, Campanus the equatorium, Ptolemy the armillary, and Richard the rectangulus and astronomical clock. Considering the materials at his disposal in the monastery, these were quite reasonable conjectures. What is more, they confirm our feeling that by the early fifteenth century at the latest, astronomy was being looked upon as a subject with a strong identity of its own. As viewed by Whetehamstede, who was an educated, informed, but not expert, scholar, it found its identity in its apparatus, each item of which offered visible testimony to genius.

The Rectangulus

The most basic, and virtually unavoidable, calculations the astronomer needed to perform were all related to the geometry of the sphere, and it was natural that astronomers should seek ways of simplifying the relevant tasks. At the heart of most problems was the need to work in three different coordinate systems, and to move between them. The most obvious frame of reference to which observations may be referred is that based on the horizon plane: (angular) altitudes are measured above the horizon and azimuths are measured round from some point on it, such as the direction of north. The daily rotation of the heavens around the poles recommends to

Compotus manualis ad vsū Oxoniēsiū.

70. The woodcut frontispiece to the short *Computus manualis* printed by Charles Kyrfoth in Oxford in 1519-20, for the use of arts students there. Note the armillary sphere, for teaching purposes, below the wall clock, the hanging hour glass, and the astrolabe. The styles of dress and ornament have changed since the university's foundation, and the method of teaching the computus, but not the computus itself.

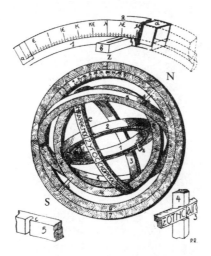

71. The armillary sphere described by Ptolemy in his *Almagest*, as reconstructed by the engineer P. Rome for the edition by his brother (the Abbé A. Rome) of the commentary by Pappus on the *Almagest* (Vatican City, 1931-33). The labelled ring *e* is the ecliptic ring. The ring inside it, and at right angles to it, carries the adjustable sights (see also the upper figure). *N* and *S* are the north and south celestial poles (note the pivots on which the inner rings all turn with the daily motion), and *z* is the zenith.

us a second coordinate frame, based on the celestial equator (or equinoctial) and some point on it (usually taken to be the head of Aries, where the ecliptic crosses it). The third important frame is that based on the Sun's annual path, the ecliptic, and some point on it, usually the same 'first point of Aries', the 'vernal point'. Historically, some of these coordinate systems were combined, but by late antiquity the three were all clearly distinguished, and have remained so to our own day. Ptolemy clearly recognised that there would be a great saving of time if he could measure the ecliptic coordinates he wanted (ecliptic latitude and longitude) directly, avoiding calculation. To do this he designed his armillary sphere (*astrolabon*, occasionally called a 'spherical astrolabe', although this has other meanings). A simple armillary (from *armillae*, rings) was a common teaching device in the middle ages, and is frequently shown in classroom use (Fig. 70). Ptolemy's instrument, however, was a much more serious affair, accurately built of bronze, on a large scale, mounted in the outdoors, and equipped with sighting pinnules (diopters) so that he could make direct measurements of ecliptic latitude and longitude (Fig. 71). This is the instrument to which Whetehamstede was referring, when he spoke of the armillary.

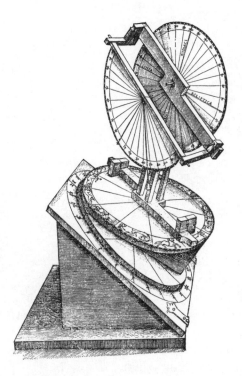

72. A torquetum, from Petrus Apianus, *Introductio geographica* (Ingolstadt, 1533). Apian reused the woodcut in other works.

It will be obvious from the figure that the difficulties of making an accurate instrument of this kind would have been prodigious. To avoid them was the ambition of those who built the torquetum, to which Whetehamstede referred. This type of device is well known from the published writings of Regiomontanus and Peter Apian; Apian's clear woodcut illustration of it, first published from his own Ingolstadt press in 1533, is often reproduced (Fig. 72). The torquetum was already known in Europe at least as early as the thirteenth century, however, when it was described in a Latin work by Franco of Polonia.[229] At first sight it may appear to have no connection with Ptolemy's armillary, but they have the same purpose, and function in closely related ways. To see the connection it is necessary only to observe that the torquetum is built out of three basic units. Each has two pivots at right angles. A single basic unit comprises a circular scale at the centre of which is pivoted an arm, able to move around another circular scale (or partial scale) in a plane at right angles to the first. As a result, the arm can be pointed first to any chosen longitude, and can then be pivoted around the other axis until it points to any chosen latitude. Ptolemy's design has these three units, ideally, at least, sharing a common centre. In the torquetum, they are stacked one above another. The final sighting is done from the uppermost unit, which is for ecliptic latitude (the far disc) and longitude (where the lower sighting rule is). The lowest unit does not quite fit my description of a 'basic unit', but that is because the plane of the equator, the square inclined plane at its base (with the polar axis at right angles to it), is fixed, for any given geographical latitude. At the top, the half-disc hangs free, and in a vertical plane, so that the plumbline registers altitude against it. There is no easy azimuth reading. Stacking the three units in this way would not be at all acceptable if angular measurements were being made of objects close at hand—by a surveyor, for instance—but offsetting them is of no consequence when viewing astronomical objects.

The torquetum, like the armillary, had a double function: it was possible to use it for observation, giving the results in any of three chosen systems of coordinates, without undue calculation. There are several medieval records of its having been used for observation. Alternatively it could be used for calculation alone, that is chiefly for switching between coordinate systems, relatively painlessly. To every ecliptic latitude and longitude a declination and right ascension correspond. Finding them was a problem Richard of Wallingford dealt with in his *Quadripartitum*, but was there no easier way? One might have chalked the two coordinate grids on a ball, for instance, in different colours. Following in Ptolemy's footsteps, one might have built the celestial sphere out of pivoted

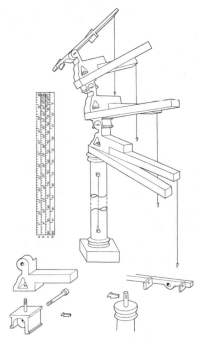

73. The rectangulus. Each of the three pairs of arms opens like a pair of scissors, while the upper two pairs, and the sighting arm on top, all pivot at right angles to the plane of the pair below. Plumblines hang from four of the arms as shown. The surrounding figures are details of the main structure. The strip with scales shows how the arms are to be graduated so as to allow the angles between them to be measured with the help of the threads, in many cases pulled so as to fall at right angles to the partner arm.

hoops of metal or wood, with graduated rings. Unfortunately, perfect metal spheres, even metal rings, are difficult to construct and graduate accurately. Throughout life Richard of Wallingford made attacks on the general problem of coordinate transformation, as fragments in the Ashmole manuscript from St Albans make clear, using a board with trigonometrical scales, for instance, to provide quick answers. It was this same problem which led him to design his rectangulus, although that had an observing function as well.

In Part One, we saw that in 1326, towards the end of his second Oxford period, Richard composed his work on the rectangulus, and that it was later carefully examined and edited by Simon Tunsted, the provincial minister of the Franciscans. Tunsted died in 1369, and interest in the work seems to have faded by the end of the fourteenth century. Mechanically speaking, it was an extremely ingenious system of rods, pivoted in three dimensions so as to cover all of the movements of the standard torquetum, indeed, to cover them more fully than was done on that instrument. Plumblines were added to indicate the required angles, by the graduations at which the lines crossed the scales on the rods. Our drawing (Fig. 73) will give a better idea than any short description. The scales were therefore not uniformly divided, but it was easy enough to graduate them by dropping perpendiculars from the degree marks from a circle on to a diameter. One disadvantage would have been that accuracy was bound to be low with small angles. This, indeed, was the greatest weakness of the rectangulus.

There were also practical problems. The idea of replacing a spherical instrument with one comprising straight jointed rods was not at all easy to put into practice. There were very few professional instrument-makers in European centres of learning before the mid-fourteenth century, and all but the richest of scholars who wanted to own an astronomical instrument would have been thrown on their own resources. Making the complicated joints of the rectangulus was something for a skilled worker in metal, and certainly not for ordinary clerks, unversed in the ways of the workshop. The torquetum may look more imposing, but it would have been considerably easier to make, and a passable instrument could even have been made in wood. It would also have been easier to make it mechanically stable.

Finally, there was the idea behind it. The design of the rectangulus showed a mastery of three-dimensional intuition, but there is good reason for thinking that most ordinary mortals—perhaps even most of those who studied astronomy—found three dimensions one too many. Richard's first excursion into instrument design did not have a large following. No example of a medieval rectangulus has survived to the present day. We may be reasonably certain that Simon Tunsted owned one, and it goes without saying that the son of the Wallingford smith would not have left his idea untried.

22

Albion

RICHARD OF WALLINGFORD'S greatest achievement in astronomy was in his design of the second instrument he described at length before finally leaving Oxford, the albion. He tells us that he wrote his treatise on the albion, like that on the rectangulus, in 1326. While the appearances of the two instruments were utterly different, they did have certain qualities in common. Both, if they were to be used with understanding, made great demands on the user's ability to rationalise the moves being made. Both allowed people of lesser ability to follow a rule book and achieve the desired result without having any real understanding of what they were doing. The main difference between the two instruments was that, whereas the rectangulus was primarily a tool for solving problems of spherical astronomy, the scope of the albion was much wider, taking in planetary astronomy too.[230] Part of the albion's historical importance—which could not have been foreseen by its designer—lay in the fact that it made astronomers look at certain mathematical questions in new ways, for many generations to come, to the benefit of western mathematical science beyond the mere realm of computation.

Early Equatoria

Richard's *Treatise on the Albion* (*Tractatus albionis*) describes the construction and use of an instrument which it is convenient to describe as an equatorium. We recall that the word refers to a mechanical device for calculating as quickly and painlessly as possible the positions of the Sun, Moon, and planets in the heavens, for any moment of time, past, present, or future. (It is related to the word 'equation', which we encountered in connection with the calculation of planetary longitudes.) From the derived positions, the instrument allowed eclipse prediction, and much more besides. The information it yielded was of a type needed by every practising astrologer, but that was not the only justification for the treatise. Astronomical theory had long been an end in itself, presenting intellectual challenges which were enough of a justification for studying it, at least in

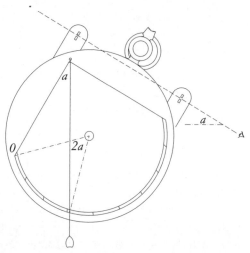

74. The method of observing altitudes on the albion, using a scale with double the accuracy that would be attained with a plumbline issuing from the centre of the scale. The altitude of the object observed is *a*, and by a well known theorem of Euclid, the angle at the centre is 2*a*. The scale is shown divided at intervals of 10° (altitude). The same technique was used on the Merton College equatorium, dating from mid-century (Fig. 76).

the minds of a select group of scholars. Equipped as the albion was with sighting vanes and a plumbline, it could also be used for observational purposes. It might not have satisfied the needs of those who aimed to revise the parameters of a theory, but it would have given better altitudes than most other hand-held instruments: not only was it larger than most quadrants and astrolabes, but it used an ingenious method for doubling the accuracy of measured altitudes (Fig. 74). The few scholars who were privileged to set eyes on it could not fail to have been impressed, whether or not they understood it. The social parallels with modern computers are obvious.

Some such device was needed, since even with the best collections of tables, the calculation of planetary positions remained a slow business. Consider the case of a person who wished to calculate a horoscope and place the Sun, Moon, and all the planets on it. Many scholars who would dearly have liked to do it—while not thinking of themselves as astrologers in any special sense—would have been quite incapable, despite the smattering of astronomy they had been forced to imbibe from their compulsory university courses in astronomy. Others might have taken days to work out the answers. A practised expert with a talent for practical arithmetic—something that few indeed had—would have taken perhaps an hour or two, or perhaps more. A substitute for those accurate but slow and painful techniques was desperately needed, and the equatorium was the instrument meant to provide it. The earliest equatoria were ostensibly simple arrangements of graduated discs of metal, wood or parchment, with epicyclic discs moving around a larger disc (say on the end of an arm pivoted slightly off-centre), the idea being to simulate the various circles (epicycles, carrying circles, and so forth) of a planetary theory of the Ptolemaic type. Having turned Ptolemaic diagrams into material discs, they could be set correctly for the required time, using scales marked on them, so that the positions of the Sun, Moon, and planets, could be immediately read off the zodiac scale, without the need to calculate and combine their mean motions and equations.

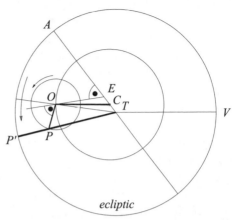

75. The Ptolemaic model for the relatively simple case of the planet Venus (compare Fig. 61), to illustrate the chief constraints when turning the geometrical model into a mechanism. The two angles marked with black spots are those which increase at a constant rate, and the thick lines are those of constant length.

Simple as this idea may seem, it presented many problems, the chief of them being mechanical. It is not unduly difficult to graduate a circle uniformly, that is, at degree intervals. One such circle will be needed for the ecliptic (or zodiac), at the centre of which is the Earth, and others will be needed at the various centres of uniform motions stipulated by Ptolemaic planetary astronomy. Unfortunately, there are many such centres, often inconveniently close to the Earth, and they themselves may move. It is easier to speak glibly of discs pivoted on pivoted rods than it is to create such an arrangement, true to the Ptolemaic model. Consider the case in very general terms: the planet P lies on an epicycle with centre O, and O moves round a deferent circle (with centre at C) which is eccentric to the Earth, T. (Fig. 75 is a modified version of one used in our earlier account of planetary theory.) The movement of O is regular, but only when seen from the equant centre E, which in not at T. We can centre a regular equant scale at E, and have a rod or thread stretching out over the graduations of that scale as far as O, at which point another scale can be put for the regular ecliptic rotation. Unfortunately, however, OE is not of constant length. Here is just one of the mechanical problems astronomers encountered, when they tried to simulate the Ptolemaic model. One rather inelegant solution offered at a later date was to fit the arms of variable length with sliding joints, controlling the length OE by ensuring that the distance OC (not OT) was kept constant.

Much of the history of the simpler type of equatorium is the history of solving this general mechanical problem, but there is another aspect of equatorium design which should not be overlooked. It is characteristic of the planetary models in the *Almagest* that each planet has a model to itself, and that—despite the physical system advocated in Ptolemy's *Planetary Hypotheses*—each model will produce answers (ecliptic positions) more or less independent of the rest. All that matters, as far as the overall geometry is concerned—as opposed to the velocities—is that the proportions within each model, the diameters and distances, be correct. This immediately opened up a whole range of possibilities to any astronomer who wished to integrate all planetary models as economically as possible. Size did not matter. To illustrate the folly of supposing otherwise, taking an extreme modern parallel: anyone who today tried to represent the motions of the planets between Mercury and Saturn, adopting an equatorium on which they moved around ellipses drawn to scale, would be faced with a situation where a reasonable orbit for Mercury, say 10 cm across, would require a disc of nearly 2.5 metres diameter to accommodate Saturn. It was invariably recognised by Ptolemaic makers of equatoria that scale was what mattered most, and that any convenient base measurement would suffice, as long as the diameters of epicycle, deferent, and other measurements were in the right proportions.

There were three classic cases of early equatoria, all produced in al-Andalus, and they had various deferent sizes for the different planets, but chosen only for convenience in fitting on to a given base plate—not for their relative sizes, which were unimportant. In later examples, from the fourteenth century, their designers used a single deferent circle and chose the scale of the epicycles accordingly; and there were other similarly late European instruments where a common equant scale was shared. They all have their advantages and disadvantages, but this is not the place to pursue such minutiae. What is more to the point is the historical sequence of events in the early history of the equatorium, whereby one idea led to another.

The first reasonably comprehensive equatorium—we shall return shortly to a simple precursor, to illustrate a different point—was devised by

Abū-l-Qāsim Ibn al-Samḥ, an Andalusian astronomer who died in 1035. Known in the West by the first part of his name, Abulcasim, Ibn al-Samḥ lived at the end of a period of Muslim political unity in the Iberian peninsula, and one might have imagined that intellectual affairs would have gone into decline. In fact the period that followed was marked by a distinctive, and often brilliant, style of astronomy, more or less independent of what had gone before it in eastern Islam. Ibn al-Zarqālluh, Azarchel, who died in 1100, produced a second, and rather more sophisticated, equatorium; and Abū-l-Salt (c. 1067-1134) designed a third.[231] Ibn al-Samḥ's equatorium had a set of plates, one for each planet, and a single epicyclic plate to serve all. He stacked them all inside a 'mother' in the way that was done with the plates of an astrolabe. For ease of use, he engraved tables of mean motions in the empty spaces on the discs themselves.

Azarchel and Abū-l-Ṣalt achieved still greater economy, placing as much as possible on a single base plate. They arranged their graduated equant scales close to the corresponding deferent circles, and nested them as well as possible, so as to conserve space on a single base disc. They did not use metal linkages, but silk threads, epicycle plates being carried on a thread through the eye of a pin at the equant centre. The epicycles were placed by eye, rather than by mechanical means, with their centres on the corresponding deferent.[232] Planetary tables were still needed for the mean planetary motions, that is to say, to supply the angles marked with black spots on our Fig. 75. The procedures to be followed would have been intuitively obvious to those who had studied planetary astronomy, and with a fairly large and well-made instrument, an accuracy of better than a degree (judged against a calculation using tables) should have been attained easily enough.

The Azarchel instrument possesses one highly unusual property which deserves mention, even though we have not here included a detailed account of the Ptolemaic model for the planet Mercury, since it is much more complex than those for the other planets. The deferent of Mercury is not a circle, but is created out of two circular motions, according to rather complicated rules. As far as almost all astronomers were concerned, that was the end of the matter. Azarchel, on the other hand, compounded the motions so as to produce a single curve for the Mercury deferent. It was oval, egg-shaped, nipped in slightly at the waist. It should not be seen as in any sense an anticipation of Kepler's ellipses, which followed half a millennium later, but it represents a doubly interesting historical step. First, it shows an astronomer who is at home with the idea of compounding motions, in a way some natural philosophers were finding it hard to do many centuries later. Second, it implied rejection of the philosophical sanctity of the circle, issuing from ancient Athens and still countenanced by Copernicus in the sixteenth century. We are reminded, perhaps, of the oval wheel of Richard of Wallingford's clock: while it has no connection with Azarchel's 'pine cone' it shared one property with that. It was the result of mathematical need, not of aesthetics or Aristotelian theory. In fact Richard of Wallingford had another oval, this time in his albion, as we shall see shortly.

The Andalusian instruments, or at least the ideas behind them, seem to have made their way into Latin Europe by the thirteenth century, and into eastern Islam by the fifteenth. Their transference to Europe is not well understood, although the warm reception the equatorium idea was given can hardly be doubted. The treatises on the first two Andalusian instruments were put into Castilian by order of that great patron of astronomy, Alfonso X, and yet the most popular of all texts on an equatorium in the Latin west had apparently been composed even before that happened. The text made up a significant fraction of the *Theorica planetarum* treatise composed by Campanus of Novara at some time between 1261 and 1265. His instrument is simpler than Azarchel's: each planet had its own plate and epicycle, the latter not on a thread but in a concavity in another disc which turned on top of the first. Mechanics apart, this might all have been derived from a lost translation from the Arabic of one of the Andalusian works already mentioned, but it raises the question of whether the idea behind all of these equatoria was not so obvious that independent invention would not have been improbable. Those who seek the origins of all good things in eastern Islam do not need this hypothesis, but there is another alternative. Ptolemy himself may well have sown the seeds of the idea in the course of writing the text accompanying his *Handy Tables*, for there he gives instructions for finding the longitude, first of the Moon, then of Mercury and the other planets, by a series of graphical construction. Drawings of the eccentrics and epicycles are to be done within a permanent circular reference scale, on some sort of board.[233] The next step, to moving discs, would surely have been obvious enough. Perhaps Ptolemy's idea reached Campanus from a Byzantine source, since the *Handy Tables* had most currency in the Greek-speaking world of the middle ages.

Part of the popularity of the Campanus text in the world of Latin scholarship can no doubt be put down to the fact that it arrived first on the scene, but—as its title suggests—it had the merit of ranging over far more than the instrument described in its pages. While Campanus does not use the word 'equatorium' of his instrument, others did so, and many copyists excerpted only that part of the work which dealt with the planetary instrument. Demand for such a simple device was almost certainly being driven by a wish to simplify astrological calculation. When one finds an unqualified reference to an equatorium in a writer of the later middle ages, the chances are that it is to the Campanus treatise.[234]

Equatoria, considered purely as calculating instruments, did not have to be taken out of doors, and could be made of less durable material than an astrolabe, for example. Many would have been of parchment or paper, pasted on wood. They could be made quite large. Geoffrey Chaucer, who was almost certainly the author of a treatise on such an instrument dating from the 1390s, recommended a board six feet in diameter, which would have allowed him to take readings down to about four minutes of arc. Dating from the mid-fourteenth century there is a large brass equatorium—although lacking its epicyclic scale—in Merton College, where it has been more or less continuously since it was bequeathed to the college by Simon Bredon in 1372 (Fig. 76). It is a fine specimen of Oxford instrument-making from a period shortly after Richard of Wallingford's death. In one small respect—the technique for doubling the accuracy of angular measurements when using it as an observing instrument—it follows Richard of Wallingford's lead (Fig. 74), but in all other ways it follows in a more elementary style—not quite that of the Andalusian astronomers, but making the planets all share a common deferent circle. (This meant that a single epicycle centre could be shared by all planets, although of course their radii were different. See Fig. 77.) The first European equatorium to deal with all planets on a single instrument seems to have been that by Peter Nightingale, the Danish Parisian master mentioned by John Whetehamstede. John of Lignères, another Paris master who was a close contemporary of Richard of Wallingford, followed in the same general tradition, and there were many others after him.[235]

With the qualifications made earlier about the different ways of placing the circles on the discs of the instrument, intuitively simple equatoria, of simulation-type, continued to be designed well into the seventeenth century. In 1540, the Ingolstadt scholar-printer Peter Apian published a sumptuous printed book containing paper equatoria. This, his *Astronomicum Caesareum*, gives to each planet its own instrument; but as far as longitudes are concerned, they remained in the Campanus tradition. (Apian made use of Richard of Wallingford, but in another connection.) There are post-Copernican equivalents, and after Galileo had discovered the satellites of Jupiter, he designed a 'Jovilabium' along similar lines.[236] Despite a superficial similarity to all such instruments—at least in the eye

76. The main plate of a finely-made but simple fourteenth-century equatorium from Merton College, Oxford. This is on the back of a fairly conventional astrolabe. The scales include the zodiac and equant circles for the planets. A separate plate (carrying the epicycles) has been lost. For a reconstruction of it, see Fig. 77.

of the person to whom all engraved circular discs are alike—Richard of Wallingford's albion worked on very different principles.

Tacit Geometry

The albion, when used to calculate planetary longitudes, did not reproduce, so to speak, the moving diagrams of planetary motion in the traditional and visually obvious way. Instead it reproduced the trigonometrical steps taken in a normal calculation made on the basis of planetary tables,

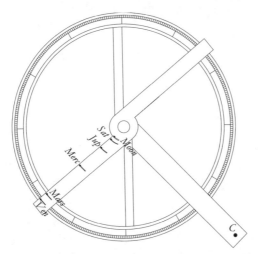

77. The probable form of the disc of epicycles, now missing from the Merton equatorium. Point C is the deferent centre; the centre of the scale is the centre of each epicycle; and the marks on the rotating rule correspond to the positions of the different planets, scaled according to the sizes of their epicycles in relation to their deferent circles on the instrument.

and to this end it made use of various ingenious graphical procedures. As with the rectangulus, no concessions were made to a weak imagination; but, as with the rectangulus, a person following the rule book could have operated the instrument without understanding its underlying principles. It might not have been so easy to learn, nor could it achieve the same degree of precision as calculation with tables, but, once learned, it was very much faster than they.

It was not the first instrument where this approach was taken, but the only earlier example I found was extremely crude. Describing itself as *astrorum speculum generale* ('a general mirror of the stars'), it contained a date of 1319 and was designed for the Paris meridian. It was in disc form, with an astrolabe on one side and the equatorium on the other, but it also included a computus disc and a wind-rose. This strange concoction contained numerous mistakes, and the Latin text was extremely cryptic and perfunctory, but the idea behind it was straightforward: just as when using tables, one is to find the ecliptic place of each planet by determining separately the mean motion and the equations of argument and centre, and then to combine them by addition—or, when necessary, subtraction.

Addition and subtraction with the help of a circular scale is scarcely different in principle from addition and subtraction with the help of a

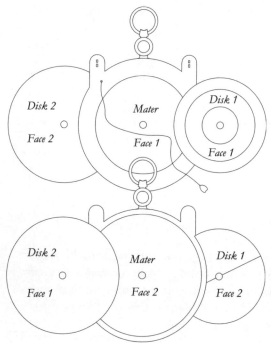

78. The arrangement of the discs of the albion, providing for its numerous solar, lunar, and planetary scales and subsidiary instruments. There are seventeen items on the first disc's first face alone. The first face of the second disc has the spiral of thirty-one turns, but even this has four parts (thirteen turns for the mean argument of the Moon, one turn for its mean motion in hours, and so forth). Parts of the mater, front and back, were hidden during some operations, but could be used when the covering discs were removed.

straight ruler, a simple matter, even if it is not one to which many people are likely to have given much thought. With straight rules, having marked off the interval to be added to a first reading, say on a card, or on a second ruler, one then moves it so that it starts at the endpoint of the other interval on the main ruler. Its other end then registers the sum of the lengths on that main ruler. This is exactly how logarithms are added on a linear logarithmic slide rule. Subtraction should need no further explanation. With circular scales, the angle to be added can be transferred either using a second circular scale or—with more difficulty—by holding threads or rules set with reference to the main scale and rotated until the end-point of the one angle coincides with the first point of the second. Handbooks occasionally tell the user of brass instruments to mark the scale with a spot of ink to assist in this quite trivial operation.

Richard of Wallingford opted for the same mathematical approach as that of the anonymous 1319 treatise—although it seems unlikely that he drew anything much from that particular source. He was able to keep the number of mechanical movements to a minimum, all rotations being around a single centre, the axis of the main disc. The angles which were to be added and subtracted with the help of threads drawn out from that common centre were found from some of the numerous scales engraved on the discs. Fine silk threads allowed more accurate readings to be taken than the usual pivoted metal rules of the period. There was a small parchment disc for lunar calculations, but the main metal discs were three in number, and on their six faces upward of seventy graduated scales were engraved, some of them spiral, some of them oval, their graduations often non-uniform. (For the overall plan of the discs, see Fig. 78.) As for graduated spirals—one such on the albion had thirty-one turns: its advantage was simply that the spiral could carry much more information than a single circular scale.[237] It was in the graduation of the scales, however, that most of the subtlety of the instrument was to be found.

New Ways with Old Theory

The *Albion* treatise is in four parts, the first of them laying down some of the astronomical theory which was needed at a later stage. Here Richard drew on the *Almagest*, but also on an *Abbreviated Almagest* and a work of Albategni. At first sight, the account offered is merely a summary of standard doctrines, but on closer examination it is much more. For example, he provided geometrical constructions for various parameters, and for the mammoth task of drawing and dividing the deferent circles on the instrument, on the basis of pre-existing astronomical tables. These mathematical techniques were, of course, not needed by the user of the instrument, who could have remained entirely oblivious to methods used. In choosing suitable parameters, Richard had no empirical evidence to guide him between Ptolemy, Albategni, and Azarchel. For the most part, his was all a geometrical exercise in transferring astronomical hypotheses to non-uniformly graduated scales, but there were many places in which he displayed an exceptional mathematical ability.

One example may be mentioned briefly here, with somewhat more detail in Fig. 79. It will be easily appreciated from such a Ptolemaic diagram as Fig. 61 that the angle subtended by the epicycle at the Earth will vary as it goes round the deferent circle. In the course of calculating a planetary longitude from standard tables, one step in the argument requires us to interpolate between the maximum and minimum values of the angle subtended at the Earth. Richard of Wallingford needed the intermediate value for his own purposes, and he presented a (valid) construction for it with an extremely interesting logical pattern. He presupposed that if a certain construction holds in two limiting cases, and there is no reason to suppose that the quantity (the angular size of the epicycle) would change in anything but a continuous manner between its two extremes, then the same construction could be used for it at any intermediate point. This is certainly

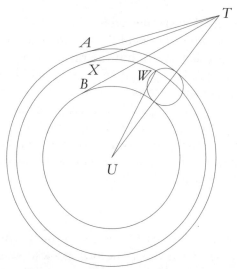

79. In *Albion* I.11 it is supposed that the maximum and minimum values of the angle subtended by an epicycle at the Earth are known. The aim is to find the intermediate value of the angle subtended, corresponding to a known value of the so-called 'true centre' (γ in Fig. 61). The construction begins with concentric circles drawn so that from an external point T the half-angles, maximum and minimum, are as shown here, ATU and BTU. A circle is then drawn touching the first two, as here, and a point W is found on it so that angle TUW is equal to γ. The theorem is that a circle centred at U passing through W will subtend the required half-angle (UTC). (It will be seen that there are two solutions in each case.)

not a Euclidean approach to mathematics, and throughout most of history would have been thought trivially intuitive and dubious, but support for the approach can be found in modern mathematics.

Part II of *Albion* deals with the practicalities of manufacture of the metalwork, with that of the plates recessed into a mater, and a clever pin for holding them together. It goes on to the astronomy behind the engraving of the scales, which of course were of paramount importance. Part III, where step-by-step instructions are given for the use of the various parts of the instrument, includes very little of this theoretical sort. Often it is only by taking the two parts together that we can follow Richard of Wallingford's train of thought. Part IV is an extremely important part of the work, for it contains a series of tables by which the scales on the albion are to be graduated or—if they have been graduated independently—verified. There are eighty-nine chapters contained in the last three parts of the work, many of them marked by touches of genius. The first example we shall consider embodies an idea which Richard borrowed from astrolabe design and brilliantly extended, an idea which was taken over by others to some effect in later centuries. The most we can do in a short space is explain it in very general terms, beginning with its source.

On the back of most conventional astrolabes there was a scale from which the position of the Sun in the ecliptic (zodiac) could be found for each day of the year. There were two scales, one of 365 days and one of 360 degrees, and using a radial rule, each date could be correlated with an ecliptic position. (For simplicity's sake, we shall take it that the year has 365 days.) The Sun, however, does not move uniformly round the ecliptic, so both scales cannot be uniform. Ptolemy's solar model tells the astrolabist the way. It was simply to draw an eccentric circular scale around (or inside) the concentric calendar scale, choosing the same eccentricity as in the standard solar model. The whole point of the model was to find a circle on which the Sun moves uniformly, so dividing that eccentric scale into 365 equal parts will provide daily stages in the Sun's motion. Our radial rule will correlate dates with ecliptic angles quite correctly. Another arrangement was sometimes adopted, however. The eccentric scale could be used as a dispensable intermediary. A third scale, could be drafted, now concentric with the degree scale, its divisions laid off from the 365-division eccentric, which could then be dispensed with. The doubly-concentric combination of scales of date and degree was sometimes felt to be more elegant. In the case of the albion, with its dozens of scales, there was plainly more than elegance at stake.

Richard of Wallingford used this type of procedure twice over, for each planet. The first application of the method each time was used to transfer the divisions of an equant circle (around which the line through the epicycle centre passes) on to a deferent scale. The latter, so divided, thus carried gradations marking the epicycle centre's places at equal time intervals—for example, day intervals. Richard then took the idea further, and transferred these divisions from a different centre (T, in our previous notation) on to another scale, providing graduations of so-called 'true centre' (γ). This final circle, doubly non-uniform, he called by the planet's name—'circle of Saturn', and so forth. (This might be confusing if it were thought to have given the planet's final position, but of course that was not the intention.) When he came to 'the circle of Mercury', he had to deal with a situation where the deferent was not a simple circle but, as we have seen, an oval. This he knew full well, possibly from Azarchel. He did not draw the oval, but he drew its graduations, which were all that he needed as an intermediate step. Like all the intermediate construction curves, these were meant to be engraved very lightly and then polished away after they had been used as stepping stones to the final scales.

The importance of the step Richard of Wallingford had taken at this stage should not be overlooked. He had shown how, graphically, one variable may be expressed as a function of three others: γ is a function of deferent position, which is a function of equant position, which is a function of time. He used this idea of continued functional dependence over and over again, as we shall see, with many different graphical constructions.

The next step on the path to planetary longitudes was that for which Richard needed a technique for working out the angle known as the equation of the argument (ζ in Fig. 61). His solution was ingenious, but too complicated for a short account.[238] Suffice it to say that a thread with a sliding bead (as marker, *almuri*) is drawn out over the angle γ found from the scale just described. The bead is then set at a position where the thread crosses a certain curve on the same disc, known (perhaps confusingly) as 'the eccentric of the epicycle'. The thread and bead are then manipulated further, leading finally to the planet's place. There was one 'eccentric of the epicycle' for each planet, and drawing them all was again testimony to a very fertile mind. The result summarised (but not proved) in our Fig. 79 was called for, and we shall not return to consider it any further, but one thing that is well worth noting is Richard's handling of the special case of Mercury.

The 'eccentric of the epicycle of Mercury', as he called the appropriate curve, is not a circle, since the epicycle moves round an oval deferent. The curve his method demanded was another kind of oval, not similar to Azarchel's, although related to it. Today we should no doubt evaluate its analytical form and plot it in some suitable coordinate system.[239] Such techniques were not open to him, and he therefore broke the curve up into four circular arcs (Fig. 80). This is strongly reminiscent of the procedure he adopted when designing his oval contrate wheel of 331 teeth for the St Albans clock. He defended his albion procedure on the grounds of its simplicity and its accuracy, at one point in his account making a statement to the effect that the curve would not have been appreciably in error had the instrument been even as great as 60 cubits (about 100 feet). Was this an idle boast, or did he proceed by trial and error, after drawing a very large specimen curve and checking it against tables? There is unfortunately no means of knowing how he arrived at his assessment, but other astronomers before him had paid much attention to the question of instrument size.[240]

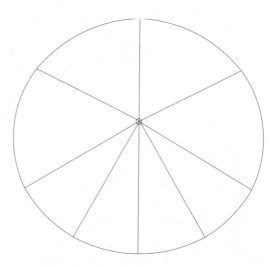

80. The so-called 'eccentric of the epicycle of Mercury', which takes into account the non-circular form of the Mercury deferent. This highly accurate curve is compounded out of eight circular arcs (four pairs).

Of the numerous scales still unexplained, we might mention one for the 'equation of days', closely related to our 'equation of time', the sort of thing often found on sundials to give the difference between true (observed, solar) time and mean solar time (clock time). Richard gave no geometrical construction here for what was a standard piece of doctrine, deriving from Ptolemy or his predecessors. The albion parameters for the mean motions, used on the appropriate scales, came from the Toledan Tables, since Richard did not yet have the Alfonsine.

There were other scales which were directly engraved from standard astronomical tables—for example, from tables of ascensions—which were included in the text, but do not merit especial attention, since they were not original. They do, however, link the work to Oxford, and to a table by John Maudith; and a copy of the work associated with Simon Tunsted has a table of oblique ascensions for Tynemouth. Another moderately standard addition was a plate with the character and lines of an ordinary astrolabe plate for Oxford. There was no rete: the stars were engraved on the plate, which therefore had to serve both purposes. (There was the necessary short star catalogue for 1327 included in the *Albion* text.) One could simply describe the result as a Profatius quadrant restored, unfolded, so as to fill a full circle once more. In place of the rotation of a rete, movements were achieved mentally, with the help of a thread and bead, angles of rotation being read off a scale on the outer rim. This simple instrument was remarkable in one respect: as a part of its star sphere (its quasi-rete) it had parallels of ecliptic latitude and lines of longitude, making for a mesh covering only the zodiac, with its conventional 12° width. This would have

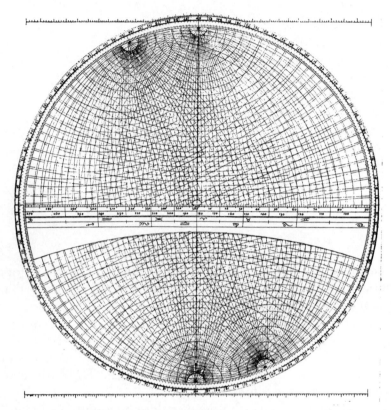

81. A seventeenth-century example of the use of the universal astrolabe ('saphea') projection. In this example it is used twice over, and projections of the equatorial and ecliptic coordinate systems are superimposed. The albion was engraved with an instrument of saphea type. (Museum of the History of Science, MS Radcliffe 74. This is a paper instrument with movable disc.)

made the device superior to an ordinary astrolabe, where a wide ecliptic ring obscures the plate underneath it. Astrolabe makers occasionally tried to achieve the same effect with a pierced rete, with effects that were more beautiful than useful.

The idea of constructing an astrolabe from a single plate, needing no more moving parts than a thread and bead, very probably owed something to another instrument from the astrolabe family, which Richard of Wallingford borrowed for the middle region on the back of the chief albion plate. This was an instrument with a history already nearly three centuries old, and he was in no way responsible for its design. It was a star map combined with a universal astrolabe—universal in the sense that it could in principle be used for all geographical latitudes. Fig. 81 shows the general character of the projection adopted on it. Such a universal instrument was chiefly known in the West as a *saphea Azarchelis*, or 'Azarchel plate'. It was type of instrument of which Richard was especially fond, and he used the underlying principle often, in the course of his long fascination with the problem of converting from one system of celestial coordinates to another. We have touched on this problem on more than one occasion, especially as it relates to moving between three systems, one where positions in the heavens are specified with reference to the equator, the second where they are related to the ecliptic, and the third where they are related to our local situation, where we specify a point in the sky by altitude and azimuth.[241] The universal astrolabe, with its superimposed sets of coordinate lines, was well designed to speed up the conversion from one system to another. In the case of the albion, all three coordinate sets were inscribed in the recess on the back of the main plate. (Adding permanent altitude and azimuth lines makes it non-universal in that respect, but increases the speed with which it can be used.)

Azarchel composed his treatise on the saphea in 1048-49, for a future king of Seville, who was then only eight or nine years old. It is an instrument by which very many of the problems of the astronomy of the celestial sphere may be solved, but intellectual standards are not what they were. It is now best considered as an instrument for experts, and not something to be explained in anything less than a short monograph. Even Richard of Wallingford's later revisers, Simon Tunsted and John of Gmunden, thought that he had been too brief, for in their recensions of the text they added long explanatory passages on the saphea.

Sun, Moon, and Eclipse

Much of the treatise on the albion was concerned with solar and lunar astronomy, and the theory and calculation of eclipses, both in timing and magnitude. Part I, where the theory of eclipses was set out in considerable detail, was broadly Ptolemaic in plan—but so indeed is much of the modern analysis of eclipses, solar and lunar. There are three basic entities here: the solar and lunar discs on the celestial sphere, together with another invisible disc, namely that corresponding to the shadow of the Earth (as cast by the Sun) at the lunar distance. These three discs are regularly passing and repassing one another as they circle the ecliptic (in the case of the Sun and the Earth's shadow), or the apparent lunar orbit inclined at a little over five degrees to the ecliptic (in the case of the Moon). The Moon, of course, moves against the stellar background more than twelve times as fast as the

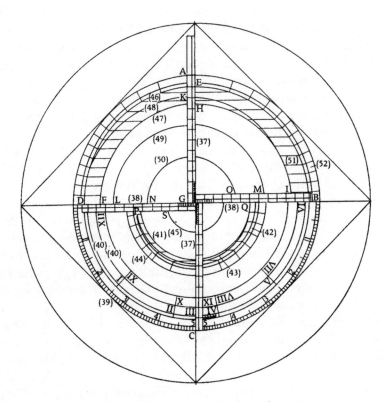

82. The general pattern of the lunar and solar eclipse instruments on the albion. (They were put on the second face of the first disc.) The numbers are keys to my printed edition of the text. Johannes Schöner's early printed versions of these eclipse instruments are shown separately in the next two figures.

other two. We note that this problem of relative velocities on two intersecting lines was more intricate by far than anything discussed in the 'kinematics' classrooms. Whether the Moon disc will enter the shadow disc (entailing an eclipse of the Moon), and whether the Moon and Sun discs will overlap (resulting in a solar eclipse, the Moon being nearer to us) depends on a variety of factors. The chief of them concern the closeness of the discs when they pass through the two main crossing-points of the ecliptic and the Moon's path. These so-called lunar nodes are the head and tail of the Dragon which we met in connection with the St Albans clock. In the case of solar eclipses, the observer's position on the Earth is relevant, but this was not taken fully into account until many centuries later.

Richard of Wallingford's eclipse instruments were far from simple affairs: they contained, for instance, thirteen separate lines and scales of relevance

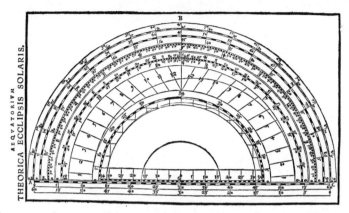

83. Johannes Schöner's solar eclipse instrument. (Compare the next figure.)

to the calculations. They are illustrated here (Figs 82, 84, and 83), but to explain them would take us too far afield.[242] They make use of a certain important technique, however, which will be explained shortly in another connection. It should be recognised that, in order to use the eclipse instruments, information was needed as to the positions of the Sun (in longitude) and Moon (in longitude and latitude), and that this was to be taken from elsewhere on the albion. The solar instrument was simple enough, but the Ptolemaic lunar model needed to be almost as complex as the Mercury model. Those devices for longitude fall into a category we have already discussed. There was one further device, however, which was needed

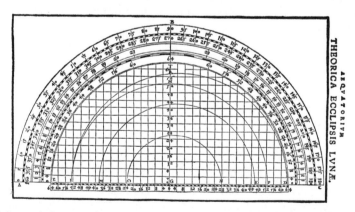

84. Schöner's lunar eclipse instrument. Like the solar instrument, this was borrowed from Richard of Wallingford's treatise on the albion. Both woodcuts are found in Schöner's *Aequatorium astronomicum* of 1521, and both were re-used in his *Opera mathematica*.

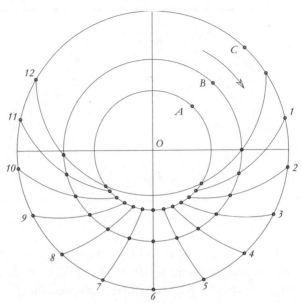

85. The unequal hour lines as drawn on the plate of an ordinary astrolabe. They are drawn for the night hours, when the Sun is below the horizon line. (The hours of the day can be found by symmetry.) The Sun (on the rete) rotates with the daily motion clockwise around O, its distance from O depending on its place on the ecliptic. It cannot be closer than A (summer solstice) or further than C (winter solstice), and is at B at the equinoxes. The hour lines are arcs of circles, drawn through points marked with black spots, these dividing the Sun's nightly paths at the appropriate seasons into twelve equal parts. The assumption made is that the arcs divide the Sun's nightly paths into twelve equal parts no matter what the season, that is, no matter what its distance from O on the astrolabe. This is a good approximation to the truth.

to provide data for eclipse calculation. This was an instrument for determining the precise times of true conjunction of the Sun and Moon. The conjunction instrument actually contains within it another instrument, a curve representing the lunar velocity, 'the equator of lunar velocity' (the heavy eccentric line in Fig. 86). We shall refer to this again, although without explaining how it is obtained. Lunar velocity, as it happens, was something which astrologers also wanted to know, since the power of the Moon was supposedly related to its velocity. On the other hand, one cannot help wondering how many astrologers there were who would have been able to find its value from the albion.

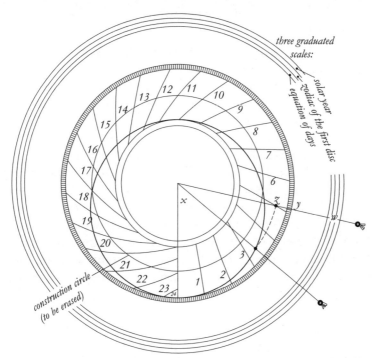

86. The instrument for determining the true times of conjunctions of the Sun and Moon on the albion. For the method of constructing the curves, see the text. In use, the thread is stretched out from the centre to position A at a certain angle (x) provided from a previous calculation. Next, the sliding bead (almuri) on the thread is moved to the 'equator of lunar velocity' (the heavier eccentric circle). The thread is then moved round to a reading on the scale corresponding to another predetermined angle (y). Finally, the first desired quantity (z) is read off from the spiralling lines, and a second quantity, w, from the middle of the three outermost scales. For simplicity's sake, the bead is here shown actually on one of the spiral lines (line 5), but in general it will not be so, and it will be necessary to interpolate. (One has to imagine the broken line to be continued round the circle, so that is creates a scale with 24 divisions at the points where it crosses the spiralling lines.)

It is not necessary to know the meaning of x, y, z, and w, in order to appreciate Richard of Wallingford's general technique, but for those who want to pursue the question, the angles are these: x is the lunar argument; y is the true elongation of the Moon from the Sun at the time of mean conjunction; z is the time interval between true and mean conjunctions, and w is the longitude interval between true and mean conjunctions, now counting one sign (30°) on the outer scale (called 'zodiac of the first disc') as one degree of longitude.

Curves for Functional Relationships

The conjunctions instrument is well worth considering further, even if only in broad outline, since it illustrates one of Richard of Wallingford's greatest scientific achievements. Before explaining the graphical technique he used in the device, however, it might be helpful to consider what was almost certainly his source of inspiration for it, namely the lines of unequal hours on an ordinary astrolabe plate. An explanation of the way in which those lines are constructed is given under Fig. 85. It is not necessary to draw the rete of the astrolabe in that figure, as long as it is understood that on any particular day the Sun has a definite position on the rete—on the ecliptic ring, of course—and that as the rete is turned clockwise, to represent the daily rotation of the heavens around the pole, the Sun can be considered to keep the same radial distance as it crosses the hour lines on the plate. The whole idea of an unequal (or seasonal) hour is that it should be a twelfth part of the day (by day) or of the night (by night). In other words, the hour lines should be such that they divide the arc swept out by the Sun, from western to eastern horizon, into twelve equal parts, twelve equal time intervals. (When astronomers called the hours 'unequal', it was meant that they varied with season, or as between day and night, not that the twelve were unequal during a given day or a given night.) If we were still to measure hours in this way, and wanted to draw the lines accurately, we should no doubt evaluate the forms of the separate lines on theoretical grounds, and plot them, using any suitable coordinate system. The early astrolabists simply assumed that circular arcs would suffice, as long as they satisfied the required condition—dividing the Sun's arc into twelve equal parts—in the special cases of solstices and equinoxes. In our case, the accurately plotted lines would not be circular arcs, but they would not deviate very far from them. The idea of plotting curves from an algebraic equation, justified on theoretical grounds, was of course not available—but more of that in due course—and circular arcs are in any case easy to draw.

The problems Richard of Wallingford wished to solve through his conjunctions instrument were very much more complex than this, but the idea of using circular arcs as good approximations to curves which effectively correlate one variable with another—an hour angle with an unequal hour, in the case of the astrolabe lines—was exactly what he needed for it. To explain his method, using the full battery of terminology required for a full understanding of the *astronomical* problem, would for most people certainly obscure his strategy. Instead, we shall refer anonymously to five different variable quantities, a radial distance r, and four different angles, x, y, z, *and* w. Here z and w are what we are aiming to find, while x and y are supposed known, having been found from other instruments on the albion. The astronomical meanings of the four angular variables are given, for those who want them, in the caption to Fig. 86.

As explained in the caption to the figure, the thread is stretched out from the centre at angle (x), and the radial distance of the bead (r) is set with the help of a certain eccentric circle. The quantity r and the predetermined angle y are then used with the spiral curves to provide the desired quantity z. This last step is exactly analogous to the procedure for unequal hours on an astrolabe, although the curves themselves are quite different. On an astrolabe, the Sun's polar distance is the analogue of r, its hour angle is the analogue of y, and the unequal hour that of z. What Richard of Wallingford has done is introduce more variables into the chain. In our language, r is a function of x; z is another function of r and y; and w is a third function of z (here using one of the outer scales). He even continues this sequence of functions, using another of the outer scales on the disc, to correct w—which is a certain time—for the equation of days.

How were the spiralling lines, used to express a variable as a function of two others, actually drawn in the first place? They register hours of lunar motion, and are correlated with the lunar velocity. Richard of Wallingford had tables which related the two, and geometrical models from which the tables ultimately came, and he could plot angle against hour for every possible velocity. In fact he plotted just three points for each curve, and joined them with a circular arc—which, as it happens, is extremely accurate. In short, he followed the lead of the astrolabists to that extent, but then went very much further in exploiting the method. What he did should be admired for a different reason. The hour lines on an astrolabe have an obvious meaning. They mark out quite simply the divisions over which the Sun, in its daily course, may be considered to pass. The conjunctions instrument, and others designed for the albion, lend themselves to no such interpretation. They simulate nothing that can be readily pictured as happening in the heavens. However they might have been described by Richard to his contemporaries, to us they are graphical representations of functional relationships based on Ptolemaic theory.

The Fortunes of Albion

As mentioned earlier, the treatise on the albion was much copied and edited by later scholars. Despite the fact that at least seven variant editions were produced, however, only two metal albions are known to survive from the middle ages, and they only in part (for one example, see Fig. 87). The albion had a poor survival value, since even with the basic treatise it

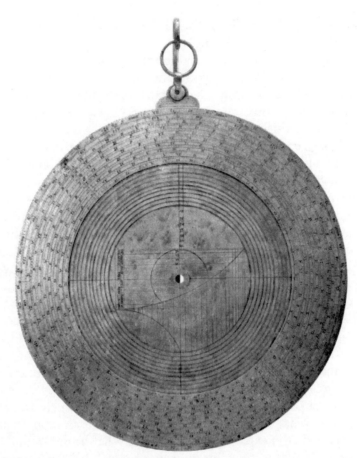

87. Part of the brass albion now in the Astronomical Observatory of Monte Mario, Rome. This is the first face of the mater. There is a central disc extant, but here removed.

would have been hard to understand, and without the treatise impossible. It is a sad reflection on sixteenth-century scholarship in England that the mathematician Robert Recorde could think that 'Simon Tunstede's rulers' (the rectangulus) were derived from the albion. Richard of Wallingford did not live to see anyone of note take up the lead he had given in his treatise on it. It was natural enough that all should judge it as an instrument of calculation—as we are inclined to do. That was its purpose, after all. From this point of view, we might say that, if very carefully engraved, it would probably have been capable of giving positions to within at least a third of a degree of those positions derived from traditional tables—much better than equatoria of the traditional type. On the other hand, we might despair, with Chaucer, of obtaining accuracy with any such instrument: 'For wel woot every astrologien that smallist fraccions ne wol not be shewid in so small an instrument as in subtile tables calculed for a cause.'[243] On a more positive note, we might stress, not its accuracy, but its universality, as did its designer when he named it. There was scarcely any standard calculation that could not be done on it, in principle if not in practice. We might list them all, adding that even horoscopes could have been cast with its aid. This is rather like saying that the abbot's crozier would have made a good walking stick. If the albion was by far the most sophisticated instrument of its kind in its day, this was not because it covered such a large variety of uses, but for the mathematical ingenuity to which almost every detail of it testified.

Whether or not those who read *Albion* had any plans to make the instrument as instructed, if they worked conscientiously through the entire treatise they would have acquired new insights which they would not have found easily anywhere else. Its instruments, with contours in the form of circular arcs, did indeed strike a chord in the most influential German instrument designers of the late-fifteenth and sixteenth centuries. In considering some of the ways in which this happened, we must avoid one historical danger in particular—the temptation to persuade ourselves that we are in the terrritory of Cartesian geometry before Descartes. Too many false claims to this effect have been made in the past. The idea has been claimed for antiquity, with the rectangular grids of the Roman surveyors, for example, or with the use of stellar coordinates by Hipparchus. It has been claimed for the Carolingian middle ages, with their graphs of planetary movement in latitude; and for Nicole Oresme's use of 'latitude of forms' to represent the motions of bodies. We should resist the temptation to get excited at the sight of squared paper or polar coordinate meshes, without some further evidence of motive. But what motive might we hope to find? The problem is largely one of definition.

Algebraic geometry is the correlation of what had been regarded as 'geometrical' concepts and objects—for geometry came first—with certain algebraic concepts and objects. The subject permitted the former class of objects to be shown to have such and such properties, by using only algebraic methods. As it is often more briefly put, the essence of the analytical method in geometry is the study of geometrical loci by means of their equations; but this is a narrower definition, and the primary concern of those who use it is with geometry. An important stage in the evolution of algebraic geometry is reached—and was reached in antiquity by

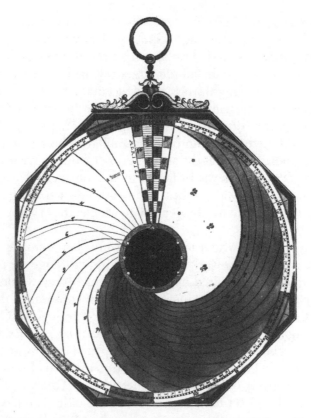

88. One of the instruments (following *enunciatum* 20) from Peter Apian, *Astronomicum Caesareum* (Ingolstadt, 1540). This follows the lead given by Richard of Wallingford in his conjunctions instrument.

Menaechmus—when the new algebraic relationship is looked upon as standing proxy for the geometrical object, for example a curve or a surface, so that the connections between it and other geometrical objects may be deduced by the manipulation of their algebraic equivalents. A third stage comes when the algebraic relationship is considered to be, as it were, that in which is encapsulated *all* that can be said about the curve, once it is properly interpreted. Fermat and Descartes reached this mental state.

Richard of Wallingford stood in an entirely different tradition. We have seen how he perceived the idea of functionality through graphical representation. Consider the *Astronomicum Caesareum* of Peter Apian, lavishly published in large folio at Ingolstadt in 1540, as a present for the emperor and his brother. We mentioned that Apian followed in the simpler

Campanus tradition when he designed his paper instruments for planetary longitudes. He had the eclipse instruments of the albion, in a heavily disguised form, but added nothing essentially new to them. Just as in some manuscript versions of the original eclipse instruments, the semicircles were backed by a square mesh, giving the appearance of our modern squared paper. Its purpose here was quite different from ours, for it was simply to assist the user in dropping perpendiculars to the scales, as the canons required. As so often in the history of 'coordinate geometry', appearances are deceptive. In several other instruments, however, Apian made use of precisely the same principle as Richard of Wallingford had used in the conjunctions instrument on the albion. How he developed the theme further is of some importance.

To take, first, two instruments which were more or less directly inspired by Richard of Wallingford's, those in Apian's *enunciata* 20 and 21: Apian here faced similar problems, for he wanted to determine small changes in the aspects of the Moon with the planets, or of a planet with other planets.

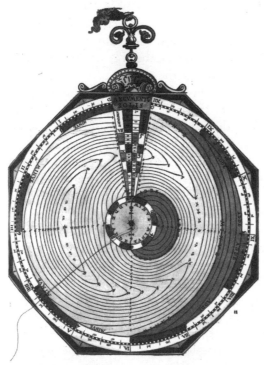

89. Another of Apian's many instruments (here following *enunciatum* 26) in the same tradition, but now adopting non-circular contour lines.

As on the albion, one variable was evaluated as a function of two others, the dependent variable being given by spiralling circular contour-lines, the independent variable by polar coordinates (Fig. 88). Following his explanation in *enunciatum* 26, Apian presented yet another instrument of the same general character, to relate the time to the hourly changes in lunar and solar argument at syzygy (Fig. 89). In this case the style was subtly changed, as it was in a remarkable series of instruments which followed it in the later parts of the book. All were still of the same general form, that is, they expressed a variable as a function of two others, and all were used in the same way as those already explained.[244] On them, however, the loci which on the albion and immediately derivative instruments had been circular arcs were now curves of a much more intricate form. We need not go deeply into the algebraic equivalents of Apian's contour lines of constant time difference, which he himself would certainly not have been able to provide, but there are two important points to be noticed. One is that these curves were all ultimately dependent on certain strings of Ptolemaic procedures, geometrical algorithms for which we could, if we tried, find modern analytical equivalents. The second point is that, in Apian's case, the curves were found in quite another way: they were plotted from previously calculated tables. Only in a very remote sense did those tables contain empirical data, but even had they done so, the way of turning them into graphical form would have been much the same.

Apian was not the first to work in the albion tradition. In his instruments for lunar and planetary latitude, for instance, we can see that he had in front of him latitude instruments designed by Sebastian Münster. Münster's work was published in his *Organum Uranicum* (Basle, 1536), where again the link with Richard of Wallingford is plain to see. The contours there are still arcs of circles. By studying Münster we can see why Apian was forced to change them: because he changed the graduation of a certain radial scale, from what was effectively a non-uniform to a uniform one, he was obliged to produce the non-circular contours.[245] Of course he had a choice, and the step was a bold one.

Apian's latitude instruments had a more influential career than their publication in his sumptuous book alone would have guaranteed them. They were adapted, for example, by Jacques Bassantin in his work *Astronomique discours* of 1557, having already been given a more public form on the face of the extraordinary astronomical clock by Philip Imser of Tübingen, dated 1555. It is hard to believe that many of those who contemplated the intricately engraved discs of contour lines on Imser's clock knew either their source or the means of interpreting them. The technique exploited by

Richard of Wallingford clearly lent itself to enormous generality, but there was a ceiling to what could be expected of those who followed in his footsteps. The talents of Münster and Apian, Bassantin and Imser were rare. The instruments which comprised the albion would have found a more appreciative audience among practitioners of what is now a virtually forgotten subject, nomography. This was a loosely-knit collection of graphical techniques, much used by nineteenth-century engineers, for example, to perform rapid calculations in standard situations. Theirs was precisely the situation in which medieval astronomers had found themselves. But is all this to count as 'geometric algebra'? The great advances in algebraic notation which came in the late sixteenth and early seventeenth centuries were unavailable to the astronomers we have been discussing, but, even without them, those earlier scholars were undoubtedly conscious of the fact that they were using their graphs as substitutes for underlying algorithms, and that the graphs were in a sense representations of the working-out of those algorithms. This was geometric algebra, in a mannner of speaking, even though it was not analytical geometry in the Cartesian sense.

23

Epilogue

> The seeds of knowledge may be planted in
> solitude, but must be cultivated in publick.
>
> <div align="right">Samuel Johnson, <i>The Rambler</i>, no. 168.</div>

RICHARD OF WALLINGFORD never achieved the fame of a Grosseteste or a Bacon. His name was rarely mentioned in later centuries, when their reputations were still green, but that was the penalty for writing above the heads of those who make reputations. It is probably true to say that it was through his theoretical writings, the fruit of his Oxford period, that his influence would be felt most forcibly—but only by like-minded posterity. The fact remains that, in his life, he was committed to the service of the church, and within his convent he was not forgotten until that was itself dissolved. When he expressed regret at the time he had spent in scientific pursuits, it was not a mere figure of speech. Christian belief was the stable core of the elaborate intellectual network which animated the universities. There are those today who spend their lives writing about intellectual affairs, carefully filtering out all theological components, only to end by making the well-worn generalisation that neither philosophy nor science was able to make progress before it was secularised, freed from theological bondage. Does the case of Richard of Wallingford support or run counter to this sweeping claim? To the extent that pure astronomy and mathematics were very rarely anything but secular, his case is hardly relevant. To the extent that he regarded himself a student of God's creation, his career at least shows that these secular studies were not a full-time activity, but what of their interaction with his religious belief?

His formal academic exercises are lost, and we are never likely to find any explicit evidence for the interplay of his religious belief and his scientific work. What survives of that work, however, reveals no religious influence, either supporting or hindering. One might say that his wish to serve the church was what drove him to design his clock or to write on the calendar, but that scarcely qualifies as bondage to theological principle. For the most part, his work is theologically neutral. We have seen a few examples of natural philosophers for whom this could not be said, men whose science was centred on—even if it did not spring from—their commentaries on Peter Lombard's *Sentences*. There is no doubt that many excellent scholars of the middle ages refused to countenance what seemed to be rational conclusions, when they seemed to conflict with religious othodoxy. (The movement of the Earth around the Sun is a notorious example, but there are many others.) The same continued to be the case, however, long after the supposed coming-of-age of science: witness the perennial debate on evolution, for example. The best minds have always found ways of achieving compromise, and in many subjects, including those with which Richard of Wallingford was most concerned, compromise was rarely difficult.

The very fact that we know nothing of Richard of Wallingford's academic theology is bound to distort our view of him. That his monks could not share his intellectual interests means that they cannot help us. They knew how, as clockmaker, he had brought God's universe into the church. They knew how, as a Christian believer, his thoughts and words were of heaven. They knew that his heart had for many years been with mathematical astronomy, and that this filled him with guilt. They judged him to be a great abbot, and they could see how he was helped by his analytical skills; but as abbot, his feet were firmly on the ground, and it required no great acumen to recognise his worldly merits. Those merits mattered a great deal to monks whose first duty—after they had made sure of the salvation of their souls—was instant obedience to their abbot. He, however, lived in two alien worlds. He lived a life of great power and privilege; and there can have been few members of the convent with who he could share his scientific ideas. He must often have felt intellectually isolated, after leaving Oxford.

It would have been interesting to read his lost commentary on the *Sentences*; but even if it were to be found, it is unlikely that it would show theology acting as a constraint on his scientific thinking. His confessed neglect of theology suggests as much. The medieval sciences often assumed the role of servant to theology, but of an expert and even wilful servant, with one eye on what the master wanted, but with freedom in plenty when out of sight. Christian astronomers did not feel themselves at liberty, for example, to defend the idea of an uncreated and eternal universe, but when they put an age to the world, or estimated when it would end, the theologians could only stand by and marvel—apart from those who were themselves expert in such matters. Apart from one or two grand theses, for instance those concerning the creation and end of the world, there was rarely any obligation felt to turn the argument around, and use scripture or church tradition to prove difficult scientific points. The few who tried to do so were rarely in the first scientific rank. Even such notable theologians as Albertus Magnus and Thomas Aquinas consciously resisted the temptation. The very fact that university theologians were thoroughly trained in the quadrivium and natural philosophy often led them to borrow the methods they had learned there when discussing God and his creation, but this again was a form of applied science, and not the kind of science one has in mind when speaking of the failure of science to advance. It was certainly not science in the tradition in which Richard of Wallingford's known writings stood. Medieval natural philosophy was likewise much freer from constraint than is often supposed, the Tempier condemnations notwithstanding. In an examination of 310 questions drawn from treatises by four leading fourteenth-century Parisian natural philosophers, Edward Grant found fewer than a third with traces of theological sentiment, and only ten with a lengthy discussion of God or the faith. This is all the more surprising, in view of the fact that Aristotle's works—*On the Heavens, On the Soul, On Generation and Corruption,* and *Categories,* for instance—offered them so many opportunities to introduce such a theme into their discussions. The one notable medieval natural philosopher positively to advocate intermingling theology and natural philosophy was Roger Bacon, of the previous century, but in practice even he tended to avoid doing so.[246]

The most serious threat to the independent scientific spirit was surely the doctrine of God's absolute power, together with the acceptance of miracle and—on a lower plane—magic. To those who have no philosophically ingenious way of by-passing the problem, such ideas weaken the very foundations of the sciences, empirical and rational alike, since one of the first requirements of the sciences is universality. Here lurked one of the chief dangers in the Tempier prohibitions of 1277. On the other hand, there is plenty of evidence from fourteenth-century Oxford that even the doctrine of God's absolute power could generate a valuable scientific discussion. It led to the investigation of alternative hypotheses about what God could do with the world—the more the merrier, as long as scholars did not require God to violate the laws of logic, the law of non-contradiction.

In a letter to Jean Dalembert, Madame du Deffand commented famously on the story that St Denis had walked two leagues carrying his severed head in his hands. The distance was nothing, she said: it was the first step that counted. Historians writing on the scientific revolution of the seventeenth century are often inclined to take a similar view of science. They would have us believe that the first scientific step was taken then, or at least in the early modern period. Blinded by that miracle, many are unable to see beyond it. To suggest replacing theirs with a medieval miracle would be to commit a similar fallacy. The first scientific step is as old as human existence—which is not to say that the intellectual web spun throughout human history has been an uneventful continuum. Scientific genius—the genius of a Galileo, a Kepler, or a Newton, or the clustering of such genius—has been favoured to different degrees at different times, by changing social and cultural circumstance. It is not only the first step that counts, but it is true that not all steps count equally. Oxford in the first half of the fourteenth century was a place where some exceptional steps were taken.

The Oxford of that time, galvanised by ideas that we may still trace back through two thousand years of earlier history, was a place more stimulating intellectually than it would be again for another two centuries. Richard of Wallingford took from, and contributed to, the scientific spirit of that exceptional community. He was doubly fortunate, able as he was to draw sustenance from his position within the Benedictine order. His life bears witness to a series of remarkable episodes without which later history would have been very different. He was without doubt the most talented English astronomer before the seventeenth century. His history, like that of medieval western science as a whole, is a story of the acquisition of ideas from elsewhere—from the distant past, and from medieval Muslim and Jewish learning—but it is also the story of the extension of those ideas. He and his fellow Oxonians multiplied their inheritance, to the great benefit of early modern science. The seventeenth century was not without those who knew this, knew that they were standing on the shoulders of giants, even when they were unable to name them—the metaphor itself was one they borrowed from the middle ages. Whether they knew it or not, one of those giants was Richard of Wallingford.

Notes

The main sources used in this work are my *Richard of Wallingford: An Edition of his Writings, with Introductions, English Translation and Commentary,* 3 vols (Oxford, 1976), and H. T. Riley (ed.), *Gesta abbatum monasterii Sancti Albani a Thomas Walsingham ... compilata,* 3 vols, Rolls Series (London, 1867-69). I refer to them here only when a precise reference will not be easily found. For the main account of Richard's rule in the *Gesta abbatum,* see vol. ii, at pp. 83-299. To restrict their number, these notes do not contain chapter and verse for historical material of a general character, except when it is likely to be difficult to locate, unfamiliar, or controversial.

1. The word he used was *mathesis,* which had that double meaning.
2. Alban's martyrdom is now widely accepted as authentic. The traditional date was 303. John Morris argued that it took place as early as the year 209, but Charles Thomas, in his *Christianity in Roman Britain to AD 500* (London, 1981), opts with good reason for the middle of the third century.
3. They were eventually combined into one, as far as the organisation of the Benedictines was concerned, but that was only in 1336, the year of Richard of Wallingford's death.
4. Such contests depend on how the measurement is taken: here it is from the west door to the centre point of the crossing, under the main tower. The only French rival was the cathedral at Bourges, but that is a church without transepts.
5. See Eileen Roberts, *The Hill of the Martyr: An Architectural History of St Albans Abbey* (Dunstable, 1993), p. 26. This work presents a well-balanced history of the architecture of the church in general.
6. This was detected by E. W. Tristram in 1931, and an artist's impression of it now hangs on the wall at the north side of the tower arch.
7. For more detail, see p. 221.
8. St Albans School moved to the gatehouse in 1870, after being housed for more than three centuries in the Lady chapel.
9. The decay, restoration, and reinterpretation of the abbey from the Reformation up to the present day are well described in Roberts, *The Hill of the Martyr,* pp. 151-262. The Victorian age was by far the most active. She rightly defends Sir George Gilbert Scott (1811-78) against his many later critics; but Scott died before his schemes came to fruition. Of Sir Edmund Beckett, first Baron Grimthorpe (1816-1905), the wealthy lawyer and amateur architect who took centre stage after Scott's death, she cannot avoid mentioning his arrogant bad taste and lack of historical sense, but she is kinder than most when she concludes that 'his strengths and weaknesses were on an heroic scale'. There was another side to the character of this notoriously abrasive man that is now often forgotten: he wrote a work on horology that saw many editions, and he designed over forty large clocks. In the 1850s, in association with Sir George Airy, astronomer royal, and the clockmaker Edward John

Dent, he designed the most famous clock of its day, that in the Victoria tower of the Houses of Parliament. In an episode characteristic of his biography as a whole, the designer of Big Ben (the bell rung by that clock) successfully sued him for libel.

10   The town has now been transferred to the county of Oxfordshire. Even at Domesday, much of the land in Wallingford belonged to Oxfordshire.

11   The original, as given in A. R. Myers, ed., *English Historical Documents, 1327-1485* (London, 1969), p. 1055, reads as follows: 'Swarte-smeked smethes, smatered with smoke,/ Drive me to deth with den of here dintes:/ Swich nois on nightes ne herd men never,/ What knavene cry and clatering of knockes!/ The cammede kongons cryen after Col ! Col !/ And blowen here bellewes that all here brin brestes./ Huf ! Puf ! seith than on, Haf ! Paf ! that other./ They spitten and sprawlen and spellen many spelles,/ They gnawen and gnacchen, they grones togidere,/ And holden hem hote with here hard hamers./ Of a bole hide ben here barm-felles,/ Here shankes ben shakeled for the fere-flunderes./ Hevy hemeres they han that hard ben handled,/ Stark strokes they striken on a steled stocke./ Lus, bus, las, das ! rowten by rowe./ Swich dolful a dreme the Devil it todrive !/ The maister longeth a litil and lasheth a lesse,/ Twineth hem twein and toucheth a treble./ Tik, tak, hic, hac, tiket, taket, tik, tak,/ Lus, bus, lus, das ! Swich lif they leden,/ Alle clothemeres, Christ hem give sorwe !/ May no man for brenwateres on night han his rest?'

12   There were houses of Augustinian canons at St Frideswide's and Oseney, Benedictine nuns at Godstow, and a secular community centred on the chapel of St George in the castle. For more on this early history see R. W. Southern, 'From Schools to University', in J. I. Catto (ed.) , *The History of the University of Oxford*, i, *The Early Oxford Schools* (Oxford, 1984), pp. 1-36.

13   R. W. Hunt, edited and revised by Margaret Gibson, *The Schools and the Cloister : The Life and Writings of Alexander Nequam, 1157-1217* (Oxford, 1984).

14   The worst of all riots was on St Scholastica's Day 1355, when sixty-three members of the university died.

15   For further reading on this central figure see A. C. Crombie, *Robert Grosseteste and the Origins of Experimental Science, 1100-1700* (Oxford, 1953); J. McEvoy, *The Philosophy of Robert Grosseteste* (Oxford, 1982); R. W. Southern, *Robert Grosseteste: The Growth of an English Mind in Medieval Europe* (Oxford, 1986).

16   It was said that when the Oxford title was changed to that of chancellor, in 1214, Grosseteste was denied its use. Whether or not he is to be counted as Oxford's first chancellor is of less importance than the fact that the significant change in nomenclature occurred at about this time, with him or immediately afterwards.

17   On the social forces operating in the early university, see M. B. Hackett, 'The University as a Corporate Body', in J. I. Catto and Ralph Evans, eds, *The History of the University of Oxford*, i, *The Early Oxford Schools* (Oxford, 1984), pp. 37-95.

18   The canons of Lincoln even tried to have Grosseteste canonised, and there were revivals of his memory by Wycliffe, the Lollards, and even the anti-Lollards up to the mid-fifteenth century. For his place in the history of the church, see Southern, *Robert Grosseteste*, especially chapter 12.

19   For an account of Grosseteste's relations with the university in general, see C. H. Lawrence, 'The University in State and Church', in J. I. Catto (ed.), *The History of the University of Oxford*, i, *The Early Oxford Schools* (Oxford, 1984), especially pp. 100-2. On his opposition to the *Sentences*, see p. 101. For differences between

Grosseteste's and Peter Lombard's readings and emphases, see Southern, *Robert Grosseteste*, especially pp. 194-204.

20 See Lawrence, 'The University in State and Church', pp. 114-15.

21 Six Middle English manuscripts containing the work have been found and described by Linda Ehrsam Voigts. See her 'The *Declararacions* of Richard of Wallingford: A Case Study of a Middle English Astrological Treatise', in I. Taavitsainen and P. Pahta, eds, *Medical and Scientific Writing in Late Medieval English* (Cambridge, 2004), pp. 195-208. Three of the six texts seem to be copies of the same translation, but there are no fewer than three different versions in all, showing how strong was the appeal of the simpler sorts of astrology.

22 The basic text of Tempier's prohibitions was edited by H. Denifle and A. Chatelain in 1889; they were printed in a more logical order by P. F. Mandonnet in 1908. For a recent printing, with their sources, see Roland Hissette, *Enquête sur les 219 articles condamnés à Paris le 7 mars 1277* (Louvain, 1977) or David Piché, *La condemnation Parisienne de 1277: texte latin, traduction et commentaire* (Paris, 1999). An English translation (1963) of many of them from Mandonnet's version by E. L. Fortin and P. D. O'Neill was printed in R. Lerner and Muhsin Mahdi, *Medieval Political Philosophy: A Sourcebook* (New York, 1963), pp. 337-54. For the historical context see J. M. M. H. Thijssen, *Censure and Heresy at the University of Paris, 1200-1400* (Philadelphia, 1998).

23 See Lawrence, 'The University in State and Church', p. 117, and his references to F. van Steenberghen and R. Zavalloni on this point. Decima L. Douie, *Archbishop Pecham* (Oxford, 1952), pp. 280-301, gives an excellent account of the condemnations from a Pecham perspective.

24 On the relationship of the 1277 condemnations to Aristotelianism and Muslim interpretations of it, see Edward Grant, 'The Condemnation of 1277', in his *Studies in Medieval Science and Natural Philosophy* (London, 1981), ch. XIII; and Cees Leijenhorst, Christoph Lüthy and Johannes M. M. H. Thijssen, eds, *The Dynamics of Aristotelian Natural Philosophy from Antiquity to the Seventeenth Century* (Leiden, 2002).

25 For more on Bradwardine's ideas about God's occupation of what would otherwise be a void, see p. 316 below.

26 The matter was greatly complicated by several ongoing disputes, and the fact that the archbishop of Canterbury, John Pecham, was a former Franciscan and a partisan supporter of that order in its long-running competition with the Benedictines. He was disturbed at the thought of a spread of Thomism at Oxford, and had a low regard for the intellectual level of the monks, whom on one occasion he described as 'idlers, dunces and blockheads'. The whole question of a Benedictine house of studies was mingled with another, concerning reform of the liturgy, and especially the shortening of the divine office. See Douie, *Archbishop Pecham*, pp. 164-69.

27 As an order of friars the Carmelites could not sell their possessions, so the bishop of Norwich acted as intermediary, in a rather devious manoeuvre. The most important collection of early documents relating to Gloucester College was added by V. H. Galbraith to H. E. Salter, *Snappe's Formulary and Other Records* (Oxford, 1924), pp. 337-86b. Another valuable source is Andrew Clark (ed.), *Survey of the Antiquities of the City of Oxford Composed in 1661-6, by Anthony Wood*, ii (Oxford, 1890), pp. 246-63.

28 Anthony Wood at one point remarks that St Albans also sent novices to St Alban Hall, but he does not give any reference or date, and the context is of fifteenth-century developments. See Andrew Clark (ed.), *Survey* , ii, p. 255. This might have been true at some period, but Wood elsewhere explains the origin of the name of the hall in a way that shows it to have no connection with St Albans Abbey. Its founder was a Robert de St Alban, a burgher of Oxford, who early in the reign of Henry III gave St Alban Hall and the adjacent Nun Hall to the nuns of Littlemore. The former hall eventually absorbed the latter, and was itself by degrees acquired by neighbouring Merton College. When Merton College was founded, Nun Hall was used for the education of the founder's kin. See George C. Brodrick, *Memorials of Merton College* (Oxford, 1885), pp. 313-14.

29 Chaucer's description of the Friar (lines 208-69) follows that of the Monk (lines 165-207) in the General Prologue. For both see L. D. Benson (ed.), *The Riverside Chaucer*, 3rd edition (Boston, 1987), pp. 26-27. For the relevant incident later in Richard of Wallingford's life, see p. 105 below.

30 Averroes (or Ibn Rushd), who lived in Muslim Spain and north Africa in the twelfth century, was the leading medieval commentator on the works of Aristotle. The Averroists—or better 'Latin Averroists', when European scholars are being described—were those who studied Aristotle through the medium of his teaching. They were often resented because they seemed to have little regard for potential conflicts with Christian orthodoxy. Notoriously controversial were their views on the eternity of the world and the oneness of the intellect for all men. Baconthorpe's Averroism was far from extreme.

31 Cobham died in 1327, before the library was finished, and the newly founded Oriel College, having settled some of his debts, took possession of the library and the books he had intended for it. Not until 1410 was the dispute settled. Cobham's book rested there from 1410 to 1480, when Duke Humfrey's Library was finished, so laying the foundations of what became the Bodleian Library. On the church in general, see T. G. Jackson, *The Church of St Mary the Virgin* (Oxford, 1897).

32 The fact that the canonical age for ordination as priest was twenty-five strengthens our arguments for placing Richard's date of birth around 1292.

33 Thomas of Lancaster had died a rebel's death in 1322, at the hands of the king's late father. The fourteen-year-old Edward III had been crowned on 1 February 1327, and Thomas's brother Henry of Lancaster had become chief of the council of regency. Henry lost little time in restoring much of his brother's property to the family, and as for the letter carried by Burley, it was dated 28 February. See C. Martin, 'Walter Burley', in R. W. Southern, ed., *Oxford Studies Presented to Daniel Callus* (Oxford, 1964), pp. 194-230, especially pp. 214-17. See also p. 55 below.

34 For brief remarks on the astrolabe, see pp. 64-67 below. For fundamental observations (not a medieval speciality) only a large astrolabe would have been of use, and there were better ways open to astronomers; but the question is of peripheral interest here.

35 For the later fortunes of the instrument in a European context, see the section beginnning on p. 373.

36 Edward II had to put an armed guard around the priory to keep off weeping crowds. The building of a chapel was soon begun on the place of Thomas's execution (the collection of funds was authorised on 8 June 1327 in the name of Edward III), and this, and a tablet in St Paul's church in London, became focal points for the devo-

tions of those who saw him as a second Simon de Montfort, one who had helped bring the king to heel in 1311 with a list of forty-one demands (the so-called Ordinances). It was also in the name of Edward III that rather ludicrous requests were made in 1327, 1330, and 1331 to have Thomas of Lancaster canonised. See May McKisack, *The Fourteenth Century, 1307-1399* (Oxford, 1959), pp. 12-30, 62-67, 98-102. Richard of Wallingford must have seen the first of the three attempts at close quarters, as we saw at p. 55.

37 For some remarks on famine in the early century, see p. 132.
38 The first Avignon papacy, with which Richard of Wallingford had dealings, is often wrongly confused with the Great Schism. There was an antipope in Richard of Wallingford's time (see p. 84), but the major crisis came only after Gregory's action in 1378, which led to the cardinals of the Sacred College selecting a second pope, who assumed the vacant Avignon seat. For more than a century after this there were two parallel series of popes (popes and antipopes). The Great Schism was not resolved until 1417. One important consequence was the encouragement of conciliarism, the idea that a general council of the church has greater authority than the pope and may even depose him.
39 Her statue is said to have drifted into the port on a crewless ship in the seventh century. After many vicissitudes, just the hand of the statue now survives, and is preserved in a reliquary in the church.
40 In all Roman strictness, therefore, he was abbot by papal provision only from 1 February 1328, not 29 October 1327, the original election in England having been disallowed.
41 *Gesta abbatum*, ii, p. 370.
42 For more on their status, see p. 116.
43 Judging by attendances, the boroughs of Hertfordshire for some reason became virtually disenfranchised after this time.
44 Two of the twenty-two tenants were obedientiaries. The cost of each knight was about £11 per annum (or the same for the substitution of two squires). The tenants could in principle nominate from among themselves, although the abbot could hire strangers directly if he thought it preferable. See L. F. R. Williams, *History of the Abbey of St Alban* (London, 1917), p. 123.
45 There was a manor called Butler's in the neighbourhood of that town two centuries later, and there is still a Butler's Farm, perhaps connected with the Aignel claim. The claim itself was not intrinsically unjustifiable. Lavish expenditure on the festivities connected with the enthronement of bishops and abbots was entirely usual. Their chief vassals expected substantial perquisites in return for supervising the preparations and performing the ceremonial tasks of butler, chamberlain, pantler, and cup-bearer.
46 The Latin text of the constitutions of St Julian's hospital, as formulated by Richard of Wallingford's successor, Michael of Mentmore, are in the *Gesta abbatum*, ii, pp. 483-510. For an English translation, see Michelle Still, *The Abbot and the Rule: Religious Life at St Albans, 1290-1349* (Aldershot, 2002), pp. 281-91. The charter of the house for women is in the *Gesta abbatum*, i, pp. 202-04.
47 The temporalities of the monastery were secular sources of income, such as land, manorial rights and dues, profits from farming, and leases of property. Spiritualities, by contrast, included church patronage, and income from subordinate churches generally, pensions, and (in the case of the fortunate few) oblations at a

shrine. The shrine of St Alban was a profitable commodity, not only spiritually speaking.

49 The details are not in the *Gesta abbatum*, but see the *Calendar of Papal Letters*, ii, pp. 381, 509.

50 The question was a long-running one which had achieved some notoriety after a bull promulgated by Alexander IV in 1288 confirmed the friars' rights. Their supporter, Archbishop Pecham, on that occasion described the friars' enemies as 'barking dogs rising like sulphurous stink from the abyss'. See Douie, *Archbishop Pecham*, p. 228. The papal bull of 1300, *Super cathedram*, stipulated that for a friar to hear confessions he had merely to be licensed by the bishop of his diocese. The orders were expected to name those who could be licensed. Boniface's motives were thinly disguised. He had delusions of grandeur and wanted to be considered emperor as well as pope. He hoped to recruit mendicant support for his ambitions by these measures, and his actions helped to consolidate the system of indulgences in general. He even declared 1300 a year of plenary indulgence for all who made a pilgrimage to Rome, visited certain churches there, and performed various pious acts. This year was the first Christian 'jubilee', and was meant to be repeated every hundred years, although the period became progressively shorter, and is now quite arbitrary. See T. S. R. Boase, *Boniface VIII* (London, 1933).

51 Not in the *Gesta abbatum* but in the *Calendar of Patent Rolls, 1327-30*, for 6 February 1329.

52 It was also of a respectable standard in works of the kind studied in the universities, including Aristotle's works on physics and astronomy. Richard of Wallingford's works would have been part of the abbot's library until after his death, but there were some works of high science in the abbey in the thirteenth century. The evidence for the holdings of the abbey in general is of course sparse, but see R. W. Hunt, 'The Library of the Abbey of St Albans', in M. B. Parkes and A. G. Watson (eds.), *Medieval Scribes, Manuscripts and Libraries: Essays presented to N. R. Ker* (London, 1978), pp. 151-77.

53 On the requests to the pope to interfere, see the *Calendar of Papal Letters*, 1332 (509) and 1333 (381).

54 All resting places for the body of Edward I's queen, on its last journey from Lincolnshire to Westminster in 1290, had such a cross. That at St Albans was in such disrepair that in 1700 it was carted away. It was later replaced by a plainer market cross.

55 Changes in the law under Edward I were by far the most important between the Conquest and the end of the Wars of the Roses, but their effect on the economic status of the villeins was not rapid. The greatest changes came after the Black Death, when services were largely replaced by money rents. See G. A. Holmes, *The Estates of the Higher Nobility in Fourteenth-Century England* (Cambridge, 1957).

56 For some remarks on famine in the early century, see p. 132 below. On the withdrawal of labour in the St Albans case, in 1246, 1265, the 1270s, 1309, and 1318-27, see R. Faith, 'The Class Struggle in Fourteenth-Century England', in Raphael Samuel, ed., *People's History and Socialist Theory* (London, 1981), pp.53-54. She notes that the same names crop up repeatedly in court cases: fourteen out of forty-five names occur in more than one year in the 1320s. Michelle Still gives many examples of dues, customs, fines, and so on, from the court book for the manor of Park, which covers the period 1237-1460. See her *The Abbot and the Rule:*

Religious Life at St Albans, 1290-1349 (Aldershot, 2002). See also Williams, *History of the Abbey of St Alban*, p. 130. Tallage was a tax exacted by a lord on his unfree tenants; merchet was a fine paid by a villein on marriage; leywrite was a fine paid by a villein guilty of immorality; chevage a fine paid by a villein for permission to reside outside the manor; heriot was the gift of the best beast from a deceased tenant to the lord. There were also many local taxation customs, for instance concerning the selling of wardships and marriages, and obtaining permission to sell certain animals.

57 In this, the wheel is small, and horizontally mounted, with angled vanes as in a turbine, into which the water falls. Since its axle is vertical it can drive the moving millstone directly, without any gearing. The design was almost always crude. As with the overshot wheel, this arrangement requires a reasonably good head of water such as might be provided by a mountain stream.

58 There has been much debate among economic historians in recent years on the question of the economic viability of fulling mills, and whether they were much used in our period. Since any decision rests on the relative frequency of the more costly but more efficient overshot type, and since statistics for this are very uncertain, I leave the question open. It seems to be widely agreed that fulling mills were less profitable than corn mills.

59 Every year, he required the bailiff of the liberty to present in writing at the abbot's treasury his accounts of crown pleas, gaol delivery, chattels of felons, 'and green wax'. Green wax is here said to be strictly reserved to the abbot's treasurer—who might have been his adviser. See *Gesta abbatum*, ii, pp. 206-7. The list is continued in a different hand, giving an outline of his scientific writings and then a sentence about his legal skill.

60 Not in the *Gesta abbatum*. See the *Calendar of Close Rolls*, 12 October 1330.

61 Hertfordshire shared a sheriff with Essex until the reign of Elizabeth I.

62 Trailbaston (in the *Gesta* 'traylebaston') was a special judicial commission that handled both criminal and quasi-criminal business. Edward I issued the first trailbaston commissions in November 1304. It addressed problems of trespasses involving violence, and was surprisingly bold in its attacks on the upper levels of society. See Amy Phelan, *A Study of the First Trailbaston Proceedings in England, 1304-1307* (unpublished Ph.D. dissertation, Cornell University, 1997).

63 *Calendar of Close Rolls*, 13 April 1332.

64 An inquisition found for the monks, and their adversaries were gaoled; the judgement was then reversed; and then on appeal upheld. The lawyers grew richer and the St Albans women poorer.

65 See n. 21, above.

66 The editor of the *Gesta abbatum*, ii, p. 280, thinks that the mill *de Mora* was the place of that name near Rickmansworth, where there was a priory and where three rivers meet. It is conceivable, however, that the reference is to Redbourn, where there were mills and where there is a lake, still called 'the Moor'.

67 For a comprehensive study of the problem, see William Chester Jordan, *The Great Famine: Northern Europe in the Early Fourteenth Century* (Princeton, 1996).

68 The text does not make it clear whether the result was by design or serendipity. Peter Newcome, *The History of the Abbey of St Alban from the Founding Thereof in 793 to its Dissolution in 1539* (London, 1793), suggests that the abbot had been trying to create a four-acre copse by planting with acorns, but the text of the *Gesta abbatum*, ii, p. 281, has 'et fecit plantari croftum ibidem de glandibus et fowis, quia

prius annis quattuor seminatum nec de semine respondebat'. The croft would have been the cultivated land surrounding the house at St German's. The word *glans* could mean acorns, but also, more generally, mast, the material from the woodland floor (especially acorns and beech seed) on which pigs grazed. The passage surely suggests the idea of adding compost to unproductive land. It is hard to see why ground planted with acorns to create a copse would have been overplanted with wheat. Newcome's carefully prepared book made an important contribution to the history of the abbey and is still of value, although it was written before his source materials had been edited and published. He was rector of the nearby parish of Shenley. A very slender supplement to the volume was also published under his name: *Various Extracts from the History of St Alban's Abbey* (Barnet, 1848). This rare book is nothing more than a few extracts, selected and supplemented by E. W. Edgell. It of no great importance, unless it is for its closing sentence, blaming the rise in the number of Dissenters in Virginia for the loss of the American colonies.

69   For the Ripoll alarm, see Francis Maddison, Bryan Scott, and Alan Kent, 'An Early Medieval Water-Clock', *Antiquarian Horology*, 3 (1962), pp. 348-53. The manuscript, mentioned again below in Part Four (see p. 232), is presently MS Ripoll 225 in the Archivo de la Corona de Aragón in Barcelona. For an edition of it, see José Millás Vallicrosa, *Assaig d'història de les idees físiques i matemàtiques a la Catalunya medieval*, i (Barcelona, 1931), pp. 316-18. For the illustration in the moralised Bible, see C. B. Drover, 'A Medieval Monastic Water-Clock', *Antiquarian Horology*, 1 (1954), pp. 54-58 and 63.

70   See my 'Monasticism and the First Mechanical Clocks', in J. T. Fraser and N. Lawrence, eds, *The Study of Time* (New York, 1975), pp. 384-85, reprinted in my *Stars, Minds and Fate* (London, 1989), pp. 174-75; and *Richard of Wallingford*, ii, pp. 368-69.

71   The word, derived from the Greek, was originally used of the risings of stars, and hence for a type of clock that shows those risings. See Vitruvius, *De architectura*, ix, viii.8. For a critical account of the texts, see André W. Sleeswijk and Bjarne Huldén, 'The Three Waterclocks Described by Vitruvius', *History and Technology*, 8 (1990), pp. 25-50; and for a survey of literature on the subject, A. J. Turner, 'The Anaphoric Clock in the Light of Recent Research', in Menso Folkerts and Richard Lorch (eds), *Sic Igitur ad Astra: Studien zur Geschichte der Mathematik und Naturwissenschaften* (Wiesbaden, 2000), pp. 536-47.

72   The earlier Roman scholar Marcus Tarentius Varro (116-27 BC), in his *De agricultura* describes a large three-dimensional revolving astronomical framework which he compared with the horologe in the Tower of the Winds in Athens. The latter was known by the Athenians as the House of Kyrrhestian, after its designer, the Macedonian astronomer Andronikos of Kyrrhos. For a conjectured reconstruction of the anaphoric mechanism formerly held within it, see references in the previous note, and also J. V. Noble and D. J. de Solla Price, 'The Water-Clock in the Tower of the Winds', *American Journal of Archaeology*, 72 (1968), pp. 345-55.

73   There are actually no fewer than sixteen ways in which the astrolabe dial may be made. See my *Richard of Wallingford*, ii, p. 342. On the Salzburg fragment the signs of the zodiac circulate in the same (anti-clockwise) direction as on an astrolabe. Strictly speaking, on the Grand fragment only dates in the Roman calendar are marked, but they have to follow the signs of the zodiac, and as advocated by Vitruvius the order is here the reverse of Salzburg's.

74 Despite its diameter of about 120 cm, the Salzburg clock had its holes 2 degrees apart, so that here the peg was to be moved to the next hole at intervals of two days. The diameter of the Grand plate was only a third as great, but it had the full complement.

75 The passages are in *Paradiso*, canto X, lines 139-48 and canto XXIV, lines 10-18. In paraphrase—which must naturally lose much of the poetic meaning—the first reads as follows: 'Like a clock which calls us at the hour when the Church rises to sing mattins to Christ, ... in which one part pushes or pulls the other, sounding *ting! ting!* with notes so sweet that the well-disposed spirit swells with love, so did I see the glorious wheel move and render voice to voice within the movement [*or* in harmony, *in tempra*] and in sweetness that cannot be known where there is no everlasting joy'. The second passage reads: 'Those joyous souls made themselves spheres on fixed axes, flaming in the guise of comets as they turned. And as wheels revolve within the movements of clocks [*or* within the harmonious arrangements within clocks, *in tempra d'oriuoli*], so that to those who watch them the first seems still and the last seems to fly, so those carols, some fast some slow, made me judge of their riches'.

These passages seem to point to more extensive trains of wheels than are needed in a simple water-clock, wheels in a harmonious mechanical arrangement. The sound (*tin tin sonando*) does not seem to indicate a large clock, and perhaps Dante was thinking of a circle of small bells when he wrote of the 'glorious wheel' (*la gloriosa rota*, compare Fig. 22).

76 Isaac ben Sid (or Isḥāq ibn Sīd, also called Rabi Çag of Toledo) was responsible for many vernacular (Castilian) treatises on the construction of instruments and timekeepers. He wrote his own notes in Arabic, but in Hebrew characters.

77 See A. T. Lantink-Ferguson, 'A Fifteenth-Century Illustrated Notebook on Rotary Mechanisms', *Scientiarum Historia*, 29 (2003), pp. 3-66.

78 In a marginal note to the edition by J. S. Brewer this is mistakenly described as a planisphere, an astrolabe. Bacon's reference is to Ptolemy's *Almagest*. Ptolemy does describe planispheres elsewhere, but in the *Almagest* he describes an armillary sphere and how it can be used for purposes of demonstration and observation. Bacon therefore has a three-dimensional model of the heavens in mind.

79 In his *De universo creaturarum*. See Lynn White, Jr, *Medieval Technology and Social Change* (Oxford, 1962), p. 132.

80 This idea has an interesting later history. The Copernican William Gilbert, in his great work on the magnet published in 1600, tried to modify the idea to allow for the various motions of the Earth. Galileo, a great admirer of his, nevertheless criticised him here; and this in turn delighted, quite irrationally, certain Jesuit critics of Copernicanism in general and of Galileo in particular.

81 See Chapter 22 below.

82 The word 'planetarium' is not medieval, but in its later use it describes precisely the medieval devices under discussion.

83 See my 'Opus Quarundam Rotarum Mirabilium', *Physis*, 8 (1966), pp. 337-72; reprinted in my *Stars, Minds and Fate*, pp. 135-170, with an endnote pointing out that the astrolabe dial was in the common (southern) projection. Cf. p. 151 above on this question.

84 He also proposed that a missing link in the European history of the clock be filled with an idea transmitted somehow from China, perhaps via Islam. The chief argu-

ment against this is not that influences from that quarter were never felt, but that the western mechanical escapements have nothing whatsoever in common with the Chinese device he had in mind.

85 Derek J. de Solla Price, 'On the Origin of Clockwork, Perpetual Motion Devices and the Compass', *United States National Museum Bulletin, 218* (1959), pp. 81-112; D. S. Landes, *Revolution in Time: Clocks and the Making of the Modern World* (Cambridge, Massachusetts and London, 1983), pp. 54-56.

86 The word *foliot* is often said to relate to *folier*, to play the fool, but the Old French noun *folier* could also mean a press (as for wine or olives), and this carried a cross-beam on a vertical support, as does a verge and foliot escapement.

87 *Le Paradis d'amour: L'Orloge amoureus*, ed. Peter F. Dembowski (Geneva, 1986).

88 One copy is now manuscript 551 in the Jagellonian Library in Cracow.

89 The second is to a great extent a condensed fair copy of the first. Note that, despite their names, these are multiple volumes, not single codices.

90 See Codex Atlanticus, 397v-b. (This manuscript is now in the Biblioteca Ambrosiana, Milan.) In Codex Madrid (Biblioteca Nacional, Madrid, MS 8937) the most relevant pages are 7r (one strange device reminiscent of the strob device and one strob, both to produce a linear oscillation, perhaps to ring a bell); 8v (another resembling the first), 9r (opposed pallets), 27r (cf. the first), 27v (full strike), 61v (two opposed, one of them a virtual pendulum), 93r (mechanism related to a strob, reversed to raise a weight), 115v (opposed pallets).

91 The first professionally constructed models using my analysis of Richard of Wallingford's text were made by Mr Peter Haward, first for the escapement alone, as exhibited in 1973 by Asprey and Company of New Bond Street, London, and later for a full-scale reconstruction of the entire St Albans clock. The latter was made for Mr Seth Atwood's Time Museum in Rockford, Illinois, and was exhibited for a short period beforehand at the Science Museum, South Kensington, in 1980. With the rest of the Atwood horological collection—the most remarkable of its kind outside Europe—it was acquired by the City of Chicago in 1999. There have since been several replicas made, mostly on a smaller scale. One such replica .

92 The drawing and partial explanation are on the upper half of fol. 12 r. He starts from a pawl-and-ratchet device, the ratchet teeth being on the outside of a wheel with teeth also internal to it. A separate lever moved at hourly intervals presses on the pawl, being resisted by a spring (a strip of steel, not a spiral spring). Richard of Wallingford alludes to a spring of this type in his striking mechanism.

93 For the Italian method of time-reckoning, see p. 147 below.

94 For his arguments, and those of the next paragraph, see Gerhard Dohrn-van Rossum, *History of the Hour: Clocks and Modern Temporal Orders*, trans. from the German by Thomas Dunlap (Chicago, 1996), pp. 108, n. 176, and pp. 127-34.

95 The bells found today in a fifteenth-century clock tower in the town have the stamp on them of a London bell-founder (either William or Robert Burford), which allows them to be dated at between 1371 and 1418. It is not unlikely that their striking was controlled by a mechanical clock soon after the installation of the bells; but true or not, a clock was certainly installed before 1485, in which year repairs were ordered to it.

96 See my *Richard of Wallingford*, ii, p. 339 for the Latin text.

97 See p. 217 below.

98   It is natural to choose a hard material for a bearing—a ninth-century horizontal mill at Tamworth is known to have had a bearing of high quality steel—but a mixture of materials is preferable. In mills, wrought iron was often used for pivots and bearings, but by the thirteenth century oaken journals in wrought iron bearings were in use in England, and the reverse in north-western France and northern Italy. For these and other examples, see John Muendel, 'Friction and Lubrication in Medieval Europe: The Emergence of Olive Oil as a Superior Agent', *Isis*, 86 (1995), pp. 373-93, especially pp. 375-76, 384-85. By the thirteenth century, if not earlier in England, undershot and overshot grain mills were using brass and (better still) bronze for bearings or pillows, weighing as much as 16lb, under secondary vertical axles or spindles. Brass, working against wrought iron, has a low coefficient of sliding friction (below 0.14). Muendel notes that the brass bushings in the surviving Salisbury clock were formerly assumed to be late additions, but I am sure he is right to suppose that they were original (*c.* 1385), not because his reference to Dondi makes this likely, but because mill practice would have taught the technique. His article discusses lubricants—suet and lard in the fourteenth century and olive oil in Tuscany, in regular use by 1300.

99   The larger wheel would be a bevel of zero angle. Bevel gears are today commonly used with four broad classes of pinion: worm, spiroid and hypoid and simple bevel (where the line of its shaft intersects with the shaft of the larger gear). The teeth of both wheels can be straight or spiral, even in the last case. Offsetting the shaft has the advantage that it can if necessary be arranged with support at both ends. The advantage of the hypoid gear in the differential gear of a motor car is that it makes a lower drive shaft possible—an irrelevance in a clock.

100   See Jane Geddes, 'Leighton, Thomas of (*fl.* 1293-1294)', *Oxford Dictionary of National Biography* (Oxford, 2004).

101   First described in summary form in Catalan by L. Camós i Cabruja in 1936, it was described in English by Cecil Beeson in 1970, finally edited by him—with the help of Francis Maddison and others—and published posthumously in1983.

102   At the most famous of all medieval valuations, the Ecclesiastical Taxation of Pope Nicholas IV (1288-92), St Albans was assessed at nearly £850. In 1086 it had been £284. See Williams, *History of the Abbey of St Alban*, p. 121. By the time of the dissolution it exceeded £2,000. Monetary equivalents are difficult to express at this period. In 1356 the Perpignan florin (first minted in 1346) was worth about three-quarters that of Florence, on which comparisons are best made. French currency was very unstable, as a result of the Hundred Years War, while English currency remained extremely stable throughout the thirteenth and fourteenth centuries, at about six Florentine florins to the pound. In Perpignan reckoning, the sum stated could have been worth about £370 sterling. In livres tournois, it might have been as much as £420 or as little as £120. For further information, see Peter Spufford, with the assistance of Wendy Wilkinson and Sarah Tolley, *Handbook of Medieval Exchange* (London, 1986), pp. 148, 176, and 200.

103   An alternative phrase was *machina mundana*, 'world machine'. See my 'Macrocosm, Microcosm and Analogy', in Martin Gosman, Arjo Vanderjagt and Alasdair Macdonald (eds), *Groningen Studies in Cultural Change* (forthcoming).

104   See p. 164 above for more about this important device.

105 Emmanuel Poulle took this step. See his *Les Instruments de la théorie des planètes selon Ptolémée: équatoires et horlogerie planétaire du XIIIe au XVIe siècle* (Geneva/Paris, 1980), pp. 680-81.
106 Poulle, *Les Instruments*, p. 684 and Fig. 196.
107 British Library, MS Cotton Julius. D.vii. It has also been ascribed to Abbot John de Cella (d. 1213), but this is probably a mistake.
108 John Whetehamstede, *Registrum*, i, p. 457; Amundesham, *Annales*, ii, p. 260.
109 For an explanation of this point, see p. 151 above.
110 It may help to perform a rough calculation of the difference between the solar day and the stellar day. The Sun's annual movement through the stars implies that they complete a full circle in slightly less than a day. In a solar year of 365¼ ordinary solar days, each of 24 mean solar hours, the stars will have performed an extra revolution, 366¼ in all. From this it is a short step to calculating that the sidereal day is about 4 minutes less than 24 hours. Astronomers such as Richard of Wallingford knew a much more accurate value for this, as we shall see.
111 For more about the lines of unequal hours on an astrolabe plate, see p. 372.
112 See, for example, Fig. 57, p. 298.
113 See below, Part Four, p. 377.
114 British Library, MS Sloane 1697, fols 25r-32r: 'Tabula eclipsis lunaris secundum diametros Ricardi Abbatis de Sancto Albano.' The copy is interesting chiefly because it is late. I did not know of it when preparing *Richard of Wallingford*, but described the text on the basis of earlier manuscripts. See iii, appendix 34. The Sloane copy was possibly owned by John Dee. The volume contains many personal horoscopes, some of them for the mayor of London, Henry Billingsley, Dee's close friend.
115 See p. 11 above.
116 *Gesta abbatum*, ii, pp. 201, 280-83.
117 John Stevens, *The History of the Ancient Abbeys, Monasteries, etc.*,... a supplement to William Dugdale, *Monasticon Anglicanum*, ii (London, 1723), pp. 196-97.
118 Oxford, Bodleian Library, MS Savile 29, fol. 3v. Somewhat later than this were the extracts mentioned in British Library, MS Sloane 1697, as mentioned earlier.
119 We still have Boethius's translations of Aristotle's *Categories, De interpretatione, Prior Analytics, Topics*, and *Sophistici elenchi*. Porphyry's *Isagoge* was usually prefixed to Aristotle's logical works.
120 For this source see Bruce S. Eastwood, *The Revival of Planetary Astronomy in Carolingian and Post-Carolingian Europe* (Aldershot, 2002).
121 See my 'The Western Calendar—"Intolerabilis, horribilis et derisibilis": Four Centuries of Discontent', in G. V. Coyne, M. A. Hoskin and O. Pedersen (eds), *Gregorian Reform of the Calendar : Proceedings of the Vatican Conference to Commemorate its 400th Anniversary, 1582-1982* (Vatican City, 1983).
122 The manuscript, Ripoll 225, now in the Archivo de la Corona de Aragón in Barcelona, was described in various places by José Millás Vallicrosa. For an account in English, see his 'Translations of Oriental Scientific Works to the End of the Thirteenth Century', in G. S. Métraux and François Crouzet (eds), *The Evolution of Science* (New York, 1963), pp. 128-67, at pp. 138-44. Millás edited the manuscript in his *Assaig d'història de les idees físiques i matemàtiques*, i, pp. 269-335. For the water-driven alarm, see pp. 316-18; also p. 148 above.

123 Alfonso I of Aragón was also pretender to the throne of Castile after 1109, and is therefore sometimes known also as Alfonso VII of Castile. Petrus made an excellent informant on Muslim and Jewish religious practice. Through his influence, both of the kings he served showed a tolerant attitude to Jews in general. He was a true convert, who used his fluency in Hebrew and Arabic to draw on rabbinic and cabbalistic material for his anti-Jewish writings. See Y. Baer, *A History of the Jews in Christian Spain*, trans. L. Schoffman, i (Philadelphia, 1966), p. 52; also C. Roth, *A History of the Jews in England* (Oxford, 1941), p. 6.

124 For a comprehensive introduction to recent Adelard studies, see Charles Burnett (ed.), *Adelard of Bath: An English Scientist and Arabist of the Early Twelfth Century* (London, 1987).

125 For a general introduction to the fate of Aristotelian texts over a very large span of time, from a mainly philosophical standpoint, see F. E. Peters, *Aristotle and the Arabs: The Aristotelian Tradition in Islam* (New York, 1968).

126 Dante, *Inferno* iv.144. For this and more of Commentator's lavish praise, see the old but classic study by Ernest Renan, *Averroès et l'Averroisme* (Paris, 1867), pp. 54-57.

127 The best short survey of his work is in the article by Lorenzo Minio-Paluello, 'Moerbeke, William of', in C. C Gillispie et al., eds, *Dictionary of Scientific Biography*, ix (New York, 1974), pp. 434-40.

128 Tzvi Langermann, *The Jews and the Sciences in the Middle Ages* (Aldershot, 1999), p. 3.

129 There are some who are so anxious to dwell on enlightened cooperation—such as when Maimonides was physician to the khalif in Egypt, and Jewish astronomers worked for the Christian king Alfonso X—that they forget the sporadic persecutions of Jews in Spain by Muslims and Christians. These, however, were largely governed by local circumstance. Christian scholasticism was not especially friendly to the Jewish community, and it had much responsibility for the evils of the Catholic Inquisition, but not for the worst of them. The guidelines set down by the universities for seeking out heresy had little to do with the empirical or mathematical sciences, and in any case the Inquisition at its worst simply disregarded them.

130 Abraham said that the scriptures are compatible with the sciences, but that they do not tell us everything, and that to learn the sciences we may turn to the Greeks, although reading his own works would provide another way. In theology, he greatly influenced the early fourteenth-century Franciscan exegete, Nicholas of Lyra. For much on the secularisation of theology, in Christian and Jewish culture, see Amos Funkenstein, *Theology and the Scientific Imagination from the Middle Ages to the Seventeenth Century* (Princeton, 1986), especially pp. 213-19. A recent and more selective study is Shlomo Sela, *Abraham Ibn Ezra and the Rise of Medieval Hebrew Science* (Dordrecht, 2002).

131 *The Guide*, chapter 24. See Moses Maimonides, *The Guide for the Perplexed*, trans. M. Friedländer (London, 1904; repr. New York, 1976); or Moses Maimonides, *The Guide of the Perplexed*, trans. Shlomo Pines, with Introductory Essay by Leo Strauss (Chicago, 1963).

132 This they did in spite of Maimonides' denial of the immortality of the individual human soul (as opposed to the actual intellect, which he thought lacks all individuality and could survive death). Here he followed Alexander of Aphrodisias and other Aristotelians. The *Guide* later influenced Jean Bodin, and in due course Spinoza and Leibniz.

134 The three were, respectively, Don Vidal Menahem ben Solomon Me'iri, Abba Mari ben Moses, and Solomon ben Adrat.
134 The ineffectual character of the Interdict of 1305 is underlined by Gad Freudenthal, 'Les sciences dans les communautés Juives médiévales de Provence: leur appropriation, leur role', *Revue des Etudes Juives*, 152 (1993), pp. 79-80. It was not helped by the expulsion of the Jews from France by Philippe le Bel in 1306.
135 Gersonides, also known as Ralbag, Leon de Bagnols, or Leo Hebraeus (1288-1344). For a highly significant Jewish treatise of the late 1320s on Aristotelian physical cosmology by Jedaiah Ha-Penini from Béziers, see Ruth Glasner, *A Fourteenth-Century Scientific Philosophical Controversy: Jedaiah Ha-Penini's 'Treatise on Opposite Motions' and 'Book of Confutation'* [in Hebrew] (Jerusalem, 1998). Reviewed by Gad Freudenthal in *Isis* and *Early Science and Medicine*. The first printed edition of Jedaiah ben Abraham Bedersi (Ha-Penini), Behinat Olam, with commentaries by Moses ibn Habib and Joseph Frances, was published in Ferrara by Samuel Gallus, 1552.
136 For further references see B. R. Goldstein, 'An Anonymous Zij in Hebrew for 1400 AD: A Preliminary Report', *Archive for the History of Exact Sciences*, 57 (2003), pp. 151-71. My allusions are to Abraham bar Hiyya (twelfth century, Barcelona), Levi ben Gershom (Orange, c. 1344), Immanuel Bonfils (fl. 1350, Tarascon), Jacob ben David Bonjorn (fl. 1360, Catalonia), Judah ben Verga (fl. 1455-80, Lisbon), and Abraham Zacut (Salamanca, d. 1515).
137 For this division, and a general survey, see especially Gad Freudenthal, 'Science in the Medieval Jewish Culture of Southern France', *History of Science*, 33 (1995), pp. 23-58, and his richly illustrated 'Les sciences dans les communautés Juives médiévales de Provence: leur appropriation, leur role', *Revue des Etudes Juives*, 152 (1993), pp. 29-136.
138 Joseph Jacobs gave many examples of influences on the English Jewry, chiefly from France, especially twelfth-century Orleans. Few of these influences were in any deep sense scientific. Joseph Jacobs, *The Jews of Angevin England: Documents and Records, from the Latin and Hebrew Sources, Printed and Manuscript* (London, 1893), pp. 403-6. He even asks whether the loss of Normandy to the English crown in 1206 meant more to the English Jews than to the English as a whole.
139 Of this dynasty, the translator Moses ibn Samuel fathered Judah ibn Moses. The most famous member of the generation of Judah, however, was without doubt Jacob ibn Machir. All took the side of the supporters of Maimonides in the latter half of the thirteenth century. The name Profatius comes through a Provençal form of the Hebrew *mehir*.
140 José Millás Vallicrosa, *Estudios sobre historia de la ciencia española*, i, new edn by J. Vernet (Madrid, 1987; 1st ed. 1949), pp. 103-5. This refers to several other works by Millás on the subject. For a more recent account of the saphea and its principles, see Roser Puig Aguilar, *Los Tratados de construcción y uso de la azafea de Azarquiel* (Madrid, 1987).
141 Lawrence, 'The University in State and Church', pp. 123-24.
142 The statement was repeated by Sāʿīd al-Andalusī, in ch. 8 of his *Book of the Categories of Nations*, translated and edited by S. I. Salem and A. Kumar (Austin, Texas, 1991), p. 30.
143 Galen, *Selected Works*, trans. P. N. Singer (Oxford, 1997), p. 138.

145 See D. A. Callus, 'The Introduction of Aristotelian Learning to Oxford', *Proceedings of the British Academy*, 29 (1943), pp. 238-52.

145 This is to follow the account in Southern, *Robert Grosseteste*, pp. 120-40.

146 Consider this apology to the theory he presents in his tract *On Light (De luce)*: 'It is my opinion that this was the meaning of the theory of those philosophers who held that everything is composed of atoms, and said that bodies are composed of surfaces, and surfaces of lines, and lines of points.' See Clare C. Riedl, *Robert Grosseteste on Light* (Milwaukee, Wisconsin, 1942), p. 12. The pamphlet contains a translation of the *De luce*. The basic edition is in Ludwig Baur (ed.), *Die philosophischen Werke des Robert Grosseteste*, in Clemens Baeumker's *Beiträge zur Geschichte der Philosophie des Mittelalters*, 9 (1912).

147 Bradwardine asked whether his assumption of the truth of geometry did not tacitly amount to a denial of atomism at outset, and he went on to show that some forms of atomism are denied, but others not, by common (Euclidean) geometry. See my 'Stars and Atoms', in R. Fox (ed.), *Thomas Harriot: An Elizabethan Man of Science* (Aldershot, 2000), pp. 186-228.

148 The phrase was first used by Clemens Baeumker, in his studies of the natural philosophy of Witelo, and was applied to a cluster of Neoplatonic doctrines that Ludwig Baur—Grosseteste's editor—subsequently found in Grosseteste and traced back to St Augustine and St Basil.

149 Born in Basra in Iraq, in 965, he worked mostly in Egypt, where he died in or around 1039.

150 He was devious here, for he took the eye to be so beautifully symmetrical that any ray passing at right angles through the outer surface (the cornea) would cross or meet other critical surfaces at right angles. In other words, his was a completely idealised eye, a geometer's eye. The best modern edition of the work is Ibn al-Haytham, *The Optics*, translated with introduction and commentary by A. I. Sabra (London, 1989).

151 'Perspective' was still in common use in English in the late seventeenth century. In some Latin versions, the title of the work of Alhazen was translated as *De aspectibus*.

152 For more details of the dependence, see David C. Lindberg, *Studies in the History of Medieval Optics* (Aldershot, 1983), paper III, pp. 339-41 (reprinted from *Isis*, 1967).

153 Adam of Buckfield's work was not unknown in Paris. It is not, however, of the same quality as that of Albertus Magnus. On Adam's natural philosophy, see S. Donati, 'Per lo studio dei commenti alla *Fisica* del XIII secolo', *Documenti e studi sulla tradizione filosofica medievale*, 2 (1991), pp. 361-442.

154 See the reference to Simon Tunsted in my 'Natural Philosophy in Late Medieval Oxford', in J. I. Catto and Ralph Evans, eds, *The History of the University of Oxford*, ii: *Late Medieval Oxford* (Oxford, 1992), p. 100. Note also references (at pp. 100-2) to the Oxonian treatment of questions of a photometric character.

155 George Molland, *Thomas Bradwardine, Geometria Speculativa: Latin Text and English Translation, with an Introduction and a Commentary* (Wiesbaden, 1989). On the Bacon and Alhazen references, see pp. 155-56. The first of the four texts which Molland considers as candidates for the lost arithmetic is not Bradwardine's (although it was printed under his name in 1495) but Simon Bredon's. See Keith Snedegar, 'The Works and Days of Simon Bredon, A Fourteenth-Century Astronomer and Physician', in Lodi Nauta and Arjo Vanderjagt, eds, *Between Demonstra-*

tion and Imagination: Essays in the History of Science and Philosophy (Leiden, 1999), p. 291.

156 Kilwardby, a classifier *par excellence*, was nevertheless capable of restraint. Faced with the countless arts, he scorned those who made them out to be only seven in number in order to correlate them with the seven liberal arts. See Robert Kilwardby, *De ortu scientiarum*, ed. Albert G. Judy (London and Toronto, 1976), pp. 127-37.

157 For the background to this question, see Steven Marrone, *The Light of Thy Countenance: Science and Knowledge of God in the Thirteenth Century* (Leiden, 2001). Franciscans, led by such as Bonaventure and Matthew of Aquasparta, gave great prominence to the theory in defending it. For a somewhat similar doctrine, even near the heart of seventeenth-century science, consider the claim of Descartes that there are truths, ideas, which are naturally in our souls, ideas of which the ultimate source is God.

158 Cited by O. F. Anderle, 'Theoretische Geschichte', *Historische Zeitschrift*, 185 (1958), p. 28.

159 There has been much discussion of this point in connection with his treatment of one of Aristotle's favourite examples of scientific demonstration, the lunar eclipse, which may or may not happen at full moon. I will not discuss this case, except to say that there has been confusion over the place of hypothesis in the Grosseteste commentary on Aristotle. If the initial conditions are right, he says in effect, then the eclipse will take place. This is not the hypothetico-deductive stance of the empiricist, who says that if the theoretical premisses are accepted (and the initial conditions are right) then the conclusion will follow, and who adds that the conclusion must be tested (perhaps together with others) by observation.

160 See Jeremiah Hackett, ed., *Roger Bacon and the Sciences: Commemorative Essays* (Leiden, 1997), for references to recent literature. For a summary of Bacon's achievement see A. C. Crombie and J. D. North, 'Bacon, Roger', in *Dictionary of Scientific Biography*, ed. C. C. Gillispie et al., i (New York, 1970), pp. 377-85, where references are made to earlier Bacon literature. The study in T. Crowley, *Roger Bacon: The Problem of the Soul in his Philosophical Commentaries* (Louvain and Dublin, 1950) is still of the first importance.

161 See Ernest Moody, 'Buridan, Jean', in *Dictionary of Scientific Biography*, ed. C. C. Gillispie et al., ii (New York, 1973), pp. 603-8.

162 For a discussion of this point, see H. Carterton, 'Does Aristotle have a Mechanics', in J. Barnes, M. Schofield, and R. Sorabji (eds), *Articles on Aristotle*. I: *Science* (London, 1975), pp. 161-74. On the role of the concept of vacuum in medieval thought, see Edward Grant, *Much Ado About Nothing: Theories of Space and Vacuum from the Middle Ages to the Scientific Revolution* (Cambridge, 1981).

163 See Grant, *Much Ado About Nothing*.

164 We are being generous to the extent that—in time-honoured fashion—Bradwardine simply took specimen illustrative ratios of the velocities, such as double and half. He never considered the possibility of an irrational quotient, for example. Modern writers who talk of his logarithmic law are being rather too generous.

165 *The Works of Aristotle Translated into English*, ed. David Ross, i (Oxford, 1928), chapter 8 (10b 34-36), trans. E. M. Edghill. The context is important. He lists ten categories, all derived from corresponding linguistic entities—*substance* from words like 'man', *quantity* from a phrase like 'two cubits long', *quality* from an ad-

jective like 'white', and so forth. The first category, substance, is given a special rank, for it is presupposed by the rest, which are all present in a subject in some sense. One of the characteristics of substance is that it has no contrary and no degrees, but it may be qualified in contrary ways or in different degrees. He passes on to a psychology of perception, and the qualities appropriate to each of our senses.

166 This idea is sometimes traced back to a scale of degrees for health and sickness proposed by Galen, but he was in debt to Aristotle for that.

167 For a fuller account, with further references to the main literature, see my 'Natural Philosophy in Late Medieval Oxford', especially pp. 82-95.

168 Nicole Oresme, around 1350, transformed the arithmetical language into a more easily visualised geometrical analogue. For the chief text, see M. Clagett, *Nicole Oresme and the Medieval Geometry of Qualities and Motions: Tractatus de configurationibus qualitatum et motuum* (Madison, Wisconsin, 1968). Oresme's diagrams may be described (although somewhat anachronistically) as graphs in rectangular coordinates, the intensity of the moving body's velocity being represented by the ordinates and the time by the abscissa. He saw that the area under the curve representing the body's motion (a straight line for constantly accelerated or constantly decelerated motion) then represents the distance travelled. The mean speed theorem appears almost trivial in this light; but this type of geometrical representation was of course capable of handling far more complex problems of non-uniform acceleration ('difformly difform motion', in the jargon of the time).

169 The natural philosophers first called the range of degrees in which the intensities of a quality might lie its 'latitude' (*latitudo*). The word was later applied to the concept of an arbitrary specific intensity. It is quite possible that the astronomical uses of that word, all close to our own, influenced the natural philosophers. On the other hand, it does seem that there were two social groups forming. The writer of the Oxford *Theorica planetarum*, possibly Simon Bredon, comments on the different approaches of the natural philosophers, for whom uniform motion meant covering equal spaces in equal times, and the astronomers, for whom it meant covering equal angles with respect to any geometrical centre. See the passage quoted by Olaf Pedersen in 'The Problem of Walter Brytte and Merton Astronomy', *Archives internationales d'histoire des sciences*, 36 (1986), p. 245.

170 For more on this controversial question see my '1348 and All That: Science in Late Medieval Oxford', ed. L. Berggren and B. Goldstein (Copenhagen, 1987), pp. 155-65; reprinted in my *Stars, Minds and Fate*, pp. 361-71.

171 Averroes' position should be understood in the context of an attack on philosophy by al-Ghazālī, to which he was responding in his *The Incoherence of the Incoherence* trans. by S. van den Bergh (Oxford, 1954). He argued that theologians are unable to reach the highest levels of demonstrative knowledge, and are therefore powerless to give correct interpretations of divine law. He therefore needed to separate philosophy as far as possible from theological argument. The doctrine of two truths grew out of this move of his.

172 They turn out to be highly impressionistic, but the attempt to represent planetary motions in a partially quantitative way is interesting. For an assessment of these and other diagrams, and the historical context, see Eastwood, *The Revival of Planetary Astronomy*. The same historical period is covered by Stephen C. McCluskey, *Astronomies and Cultures in Early Medieval Europe* (Cambridge, 1998). On the stereographic character of a type of ninth-century manuscript illustration associated

with the Latin Aratus, see my 'Monasticism and the First Mechanical Clocks', at pp. 387-90; reprinted in my *Stars, Minds and Fate*, pp. 178-81.

173 For a classic work in the field, although weak on the middle ages, see A. von Braunmühl, *Vorlesungen über Geschichte der Trigonometrie* (Leipzig, 1900). See also his 'Die Entwicklung der Zeichen- und Formelsprache in der Trigonometrie', *Bibliotheca Mathematica* (3rd series), 1 (1900), pp. 64-74. On Regiomontanus and his correspondence with Giovanni Bianchini, see Armin Gerl, *Trigonometrisch-Astronomisches Rechnen kurz vor Copernicus* (Stuttgart, 1989).

174 The original Arabic version of the Arzachel the canons is lost, but two versions have survived in Latin, together with many variants. The standard work on the Toledan Tables in general is now Fritz S. Pedersen, *The Toledan Tables: A review of the MSS and the Textual Versions with an Edition*, 4 vols (Copenhagen, 2002). The Savasorda tables mix Jewish, Islamic and Ptolemaic material. The epoch is the equivalent of 21 September 1104, and the year is not Jewish but the old Egyptian-Ptolemaic-Battānī year of 365 days. He uses al-Battānī's meridian (al-Raqqa), a very strange convention for an astronomer from Barcelona.

175 The now standard English translation of the text is in G. J. Toomer, *Ptolemy's Almagest* (London, 1984). For a step-by-step guide to the entire work, see Olaf Pedersen, *A Survey of the Almagest* (Odense, 1974). For a survey of Ptolemy's philosophical outlook, see my 'Ptolémée', in *Le savoir grec*, ed. Jacques Brunschwig and Geoffrey Lloyd (Paris, 1996).

176 The convention was to express the chord of a circle (a function of the angle at the centre) as a sexagesimal fraction of the radius of the circle, the latter usually taken to have sixty parts. Ptolemy's table, with 240 entries at ½° intervals, is based on a value of the chord of 1°, which he evaluates by a clever approximation procedure. To take a specimen entry: an arc of 151° is said to correspond to a chord of 116 parts 10 minutes of a part (sixtieths) and 40 seconds of a part (sixtieths of minutes of a part). It is now customary in historical commentary to write this as $116;10,40^p$. The figure is correct to the nearest minute, as are almost all the entries in this remarkable table. (Allowing for scribal error is a troublesome factor when making any such judgement, however.) Sines, which came from Indian trigonometry, were eventually found to be more convenient than chords, and were coming into use in England in Richard of Wallingford's day. They represent no great mathematical advance, being easily derived from chords. (The chord of an angle is twice the sine of half the angle.)

177 Menelaus (fl. AD 95-98) made observations in Rome, where he was known to Plutarch. His book on 'spherics' is known only through its Arabic translation. Book 3 has the famous theorem. He too had a table of chords. See Thomas Heath, *A History of Greek Mathematics*, 2 vols (Oxford, 1921), ii, pp. 260-73. The relations between Menelaus and earlier writers in the same tradition is obscure. His debts very probably went back to Hipparchus. See Nathan Sidoli, 'Hipparchus and the Ancient Metrical Methods on the Sphere', *Journal for the History of Astronomy*, 35 (2004), pp. 71-84.

178 The work is edited and translated in Lynn Thorndike, *The 'Sphere' of Sacrobosco and Its Commentators* (Chicago, 1949). Thorndike includes much ancillary material, for example the commentary by Robertus Anglicus. For what little we know about Sacrobosco, see Olaf Pedersen, 'In Quest of Sacrobosco', *Journal for the History of Astronomy*, 16 (1985), pp. 175-221, supplemented by Wilbur R. Knorr,

'Sacrobosco's *Quadrans*: Date and Sources', *Journal for the History of Astronomy*, 28 (1997), pp. 187-222. Knorr discusses Sacrobosco's work on the quadrant, which was also influential, and argues—I think convincingly—that he died in 1256, appreciably later than was previously thought.

179 See the unpublished doctoral thesis by Bruce Robert Maclaren, *A Critical Edition and Translation, with Commentary, of John Pecham's Tractatus De Sphera* (University of Wisconsin, Madison, 1978).

180 The standard biography of Pecham is Douie, *Archbishop Pecham*. She places his birth between 1230 and 1235, probably at Patcham, near Lewes.

181 On this book, *Jerarchie*, see Douie, *Archbishop Pecham*, p. 52. She notes that the analogies drawn between the nine orders of angels and the orders of royal officialdom are conventional, but that the three inferior celestial ranks are likened to the French seneschals, bailiffs, and provosts, and not to the English justices and sheriffs. As for the king's relations with the new archbishop, they were soon generally good, his initial objection having been more to papal interference than to Pecham. His own candidate, Robert Burnell, was notoriously corrupt. See Douie, pp. 110-12.

182 Lindberg, *Studies in the History of Medieval Optics*, paper X, reprinted from *Speculum*, 46 (1971), pp. 66-83. Lindberg tentatively concludes that Witelo, who wrote his optical tract at Viterbo in the period 1270-78, probably at the earlier end of it, influenced Pecham. He concludes that Pecham's *Perspectiva communis* (one of his two books on the subject) dates from the period 1269-79.

183 John Pecham makes a fairly precise statement about the equinoctial arc covered by the head of the sign of Cancer, putting this effectively at 240°, 'or very nearly'. See the Maclaren edition, p. 208, for the text. There were several latitudes used for Paris, but all give a truly excellent fit to this statement. Thus the commonly accepted 48°40' gives a figure of 239°26'. Oxford, by comparison, gives an answer over 7° high, and Viterbo and Rome give figures which fall more than 13° short of 240°. I think this tells us something more about Pecham, namely that he was a respectable calculator at a time when few scholars were.

184 The earliest extant Oxford statutes (from before 1350) do not mention the *Theorica*, but on the other hand, neither do those of 1409, and we have manuscripts of it traceable to Oxford from the thirteenth century. On the problem of authorship of the commonest text, see Olaf Pedersen, 'The Origins of the *Theorica planetarum*', *Journal for the History of Astronomy*, 12 (1981) pp. 113-23. He argues for an unknown thirteenth-century author. Graziella Federici Vescovini, 'Michel Scot et la *Theorica planetarum Gerardi*', *Early Science and Medicine* 1 (1996), pp. 272-82 suggests that it was translated by Gerard of Cremona or someone from his circle. Another Gerard whose name often appears in manuscripts is Gerard of Sabbioneta (thirteenth century). A fuller incipit is: 'Circulus eccentricus vel egresse cuspidis vel egredientis centri dicitur qui non habet centrum suum cum centro mundi ...' The late Oxford revision begins in almost the same way, with the result that the two are often confused: 'Circulus eccentricus et egresse cuspidis et egredientis centri idem sunt ...' The latter has long been associated with a supposed Merton astronomer named as Walter Brit, but almost everything in this tradition is questionable. See Olaf Pedersen, 'The Problem of Walter Brytte and Merton Astronomy'.

185 For a more extensive introductory account see my *Fontana History of Astronomy and Cosmology* (London, 1994).

186 For the early history of the tables in Paris and England, see my 'The Alfonsine Tables in England', in Y. Maeyama and W. G. Saltzer (eds), *Prismata: Festschrift für Willy Hartner* (Wiesbaden, 1977), pp. 269-301; reprinted in my *Stars, Minds and Fate*, pp. 327-59. The most comprehensive guide to what is known of the original work is José Chabas and Bernard R. Goldstein, *The Alfonsine Tables of Toledo* (Dordrecht, 2003).

187 Translation from Arabic into Latin was well established at Toledo, the ancient Visigothic capital, at the courts of the archbishops. Alfonso established a school that included Christian and Jewish savants, as well as a Muslim convert to Christianity. He presided over the group in some sense, revising their work and writing parts of the introductions to it. The names of fifteen collaborators are known from the complete collection of Alfonsine books, which includes much important important astronomical material.

189 For the sake of brevity, I have not commented on the fact that the aux line of Venus is in the same direction as that of the Sun, or that the mean motus of Mercury and Venus are the same as the Sun's. These facts slightly shorten the working when a whole set of longitudes is needed. For more detail, see my *Richard of Wallingford*, iii, appendix 29.

189 They are the result of essentially simple sexagesimal addition, with the occasional need to perform simple interpolations, but it should be noted that in practice a whole series of tables was needed. To take a somewhat simplified example: we suppose that we have tables where the radices of the motions are for midnight commencing 1 January 1100, and the aim is to calculate the mean motus (measured from the vernal point V) at 3:25 a.m. on 7 February 1327. We might have a small table listing motions at century intervals, another for motions at intervals of scores of years, one for single years, one for (named) months, one for tens of days, and one for single days, another for hours, and one for minutes. The time difference having been worked out, we shall enter the first table with 2 (for the centuries), the second with 1 (for the scores of years), the third with 7, and so forth, ending with 25 in the table of motions for minutes. Adding the results to the root value of the mean motus (that provided, for the radix date), we shall very probably end up with a result much greater than 360° (or 12 signs, if our tables gives angles in 30°-signs and degrees), and in this case we shall discard multiples of 360°. We shall then repeat the entire exercise with tables of mean argument. There may be complications. If our tables are for a Muslim calendar, we shall need additional tables for calendar conversion, and the lengths of years will not be the same as in the Christian calendar. It is also worth noting here that whereas the old Toledan tables used signs of 30°, the Alfonsine tables usually took signs of 60°, in a purer sexagesimal style which made for slightly easier calculation.

190 The latter were known as *signa physica*, physical signs. By 'adapting to a new meridian' we refer to the need to take into account the time difference between places. The planets will move by small amounts in the time interval between noon at Oxford and noon at Paris, for example.

191 S. A. Pantin and W. T. Mitchell, *The Register of Congregation, 1448-1463* (Oxford, 1972), p. 32.

192 After teaching in Paris, Henry of Langenstein (often called Henry of Hesse) spent the last years of his life in Vienna, and played an important part in the reorganisation of the university there. His 'Treatise Refuting Eccentrics and Epicycles' (written in 1364 in Paris) is much given over to petty academic criticism of the standard *Theorica planetarum*, but it found a sympathetic readership among the cosmologists. See Michael H. Shank, '*Unless You Believe, You Shall Not Understand*': *Logic, University, and Society in Late Medieval Vienna* (Princeton, 1988).

193 See Willy Hartner, 'Medieval Views on Cosmic Dimensions', in *Mélanges Alexandre Koyré* (Paris, 1964), pp. 254-82, where he concluded that there was a crucial passage missing from the published text. Bernard R. Goldstein located the missing part, which he also found in a Hebrew translation. (Its omission from the known text had been an unfortunate consequence of the death of one of the editors of the Leipzig edition, earlier in the century.) For the text, with translation and commentary, see Goldstein's *The Arabic Version of Ptolemy's Planetary Hypotheses* (Philadelphia, 1967).

194 Completed and printed privately by his pupil Regimontanus (Nuremberg, *c.* 1473), the *Theoricae novae planetarum* was reprinted at least a dozen times before the end of the sixteenth century. See several relevant papers by Willy Hartner, reprinted in his *Oriens-Occidens* (Hildesheim, 1968), especially pp. 480-95.

195 Edward Grant, *Planets, Stars, and Orbs : The Medieval Cosmos, 1200-1687* (Cambridge, 1994), pp. 97-103. Giles of Rome took the firmament to be a continuous orb, with channels through which the planets moved.

196 Grant, *Planets, Stars, and Orbs*, chapter 14.

197 For a brief account of the earlier history of the notion, see Gregor Maurach, *Coelum Empyreum: Versuch einer Begriffsgeschichte* (Wiesbaden, 1968).

198 Graziella Federici Vescovini, ed., *Il 'Lucidator dubitabilium astronomiae' di Pietro d'Abano: Opere scientifiche inedite* (Padua, 1988), pp. 200-1.

199 Dante Alighieri, *Divine Comedy*, Paradise, cantos xxx-xxxiii.

200 See my 'Celestial Influence'.

201 For more detail of Bradwardine's general position on these questions, see Gordon Leff, *Bradwardine and the Pelagians: A Study of His 'De causa Dei' and Its Opponents* (Cambridge, 1957), pp. 27-34, 159-60.

202 The best guide to Bradwardine on the extracosmic void space, and some of its later manifestations, is Grant, *Much Ado About Nothing*, pp. 133-45, 345-46; see also his *Planets, Stars, and Orbs*, pp. 173-76, and compare his account of Robert Holcot's argument for a present-day vacuum beyond the existing world, where a body could exist but does not. Walter Burley argued along similar lines. See pp. 170-73. (I omit them here since they do not directly concern God's dwelling.) The Augustine reference is to *Confessions*, book 7, chapter 5.

203 For the many different mathematical techniques, see my *Horoscopes and History* (London, 1986).

204 On great conjunctions as a cause of religious change, see my 'Astrology and the Fortunes of Churches', *Centaurus*, 24 (1980), pp. 181-211; reprinted in my *Stars, Minds and Fate*, pp. 59-90; also, on Chaucer, my *Chaucer's Universe* (Oxford, 1990). For the examples of Robert of Chester and Robert Grosseteste, see Southern, *Robert Grosseteste*, pp. 105, 107.

205 K. V. Snedegar, 'John Ashenden and the Scientia Astrorum Mertonensis' (unpublished Oxford D.Phil. thesis, 1988), 2 vols. Ashenden was from Northumbria, and

what he knew of English winds, for example, did not fit well with the views of Ptolemy, Albumasar, and Alcabitius.

206 The texts chiefly read were apparently Ptolemy's *Quadripartitum* (*Tetrabiblos*), the pseudo-Ptolemy *Centiloquium*, and Alcabitius's *Introductorium*—all dyed-in-the-wool astrological works, with no astronomical content. For the supplications, see Pantin and Mitchell, *The Register of Congregation, 1448-1463*, pp. 101, 129, 153, and 224. On Bredon's version of Ptolemy's work, see Snedegar, 'The Works and Days of Simon Bredon, pp. 290, 295.

207 Isidore, *Etymologiae*, ed. W. M. Lindsay (Oxford, 1911), III.27. Robert Kilwardby, *De ortu scientiarum*, p. 32. Roger Bacon, *Communia mathematica*, in *Opera hactenus inedita*, ed. R. Steele. (Oxford, 1905-40), I.3.5.

208 John Baconthorpe, *Quaestiones in quatuor libros sententiarum et quodlibetales*, 2 vols (Cremona, 1618), ii, pp. 253-54.

209 Lynn Thorndike, 'A New Work by Robert Holkot', *Archives internationales d'histoire des sciences*, 10 (1957), pp. 227-35. This *quaestio* was later incorporated in a longer work of his, *Opus quaestionum super Sententias*. This was printed in part (Lyons, 1497 and 1518), but without the relevant passage.

210 Hilary Carey, *Courting Disaster: Astrology at the English Court and University in the Later Middle Ages* (London, 1992), pp. 265-66 (Bolingbroke) and pp. 155-56 (Stacy and Blake). For much on Lewys, see the index to my *Richard of Wallingford*. See also my 'Scholars and Power: Astrologers at the Courts of Medieval Europe', *Actes de les VI trobades d'història de la ciència i de la tècnica* (Barcelona, 2002), pp. 13-28; and for William Brouncker, my *Horoscopes and History*, pp. 182-84.

211 His two chief works are the *De impressionibus aeris* (*On Forces Affecting the Weather*) and the *Correctorium alchymiae* (*A Work Correcting Alchemy*). They are known by various other titles. The alchemical work seems to have been later redrafted by a certain master Bernard, possibly Bernard of Treves (1406-c. 1490), under the title *Correctio fatuorum* (*Correction of Fools*), and was printed in both versions (Nürnberg, 1541; Bern, 1545; Strasbourg, 1659; Geneva, 1702). There is a German translation dating from 1485.

212 Snedegar, *John Ashenden*, i, pp. 55-59, 142-43. For John Walter, see my *Horoscopes and History*, pp. 84, 126-31, 158.

213 Leopold's origins are obscure. He is often 'duke of Austria', or 'son of the duchy of Austria', but nothing is known of him beyond his writings, which were much copied and then printed by Ratdolt (Augsburg, 1489). The meteorological section of his astrology influenced Firmin of Belleval, who wrote on astrological meteorology in the early fourteenth century. Leopold's work was translated into French for the French queen, Marie of Luxembourg (d. 1324).

214 See Fig. 61, and the explanation opposite (on p. 305), for the meaning of these terms.

215 Royal College of Physicians, MS 384, fols 94v-98v. The volume contains perhaps thirty-nine separate works, several in Latin but mostly in English. The present work describes a computistical instrument.

216 See p. 56 above.

217 See Snedegar, 'The Works and Days of Simon Bredon', p. 298. Bredon observed an appulse of Venus with Regulus and an occultation of Aldebaran by the Moon, perhaps having calculated approximate times in advance. He deduced the movement from Ptolemy's day to be 18°45' in the first case and 18°04' in the second.

218 On Bianchini and Regiomontanus, see p. 307 above.
219 For the best study of his trigonometrical writings, see Richard P. Lorch, 'Jābir ibn Aflaḥ and his Influence in the West' (unpublished Ph.D. thesis, Manchester, 1970), supplemented by the same author's 'The Astronomy of Jābir ibn Aflaḥ', *Centaurus* 19 (1975), pp. 85-107, and *Thâbit ibn Qurra : On the Sector Figure and Related Texts* (Frankfurt, 2001). For further information, see Julio Samsó, *Las ciencias de los antiguos en al-Andalus* (Madrid, 1992). See also the entry on Jābir by R. P. Lorch in the *Dictionary of Scientific Biography*, vii (New York, 1973) Jābir ibn Aflaḥ is not to be confused with that other famous Gebir (or Geber), the eastern Arab writer on alchemy, Jābir ibn Ḥayyān.
220 Bredon also used Theodosius, Menelaus, and Thābit. His heavily annotated *Albion* is in British Library, MS Harley 625, and his *Quadripartitum* is in Bodleian Library, MS Digby 178. See Snedegar, 'The Works and Days of Simon Bredon', p. 296.
221 For this, and a brief explanation of the original theorem, see Fig. 55 on p. 291 above.
222 The definition reads: 'Magnitudes are said to be in the same ratio, the first to the second and the third to the fourth, when, if any equimultiples whatever be taken of the first and third, and any equimultiples whatever of the second and fourth, the former equimultiples alike exceed, are alike equal to, or alike fall short of, the latter equimultiples respectively taken in corresponding order'. This divides all rational numbers into two classes.which are in a 1:1 relationship, and so defines equal ratios in the way used by Dedekind. See Thomas Heath, *The Thirteen Books of Euclid's Elements*, 3 vols (Cambridge, 2nd edn, 1926), i, pp. 114, 120-26. Campanus has a Latin sentence which translates thus: 'Quantities are said to be in the same ratio, the first to the second and the third to the fourth, when equimultiples of the first and third are similar—exceeding, equalling, or falling short—to equimultiples of the second and fourth, taken in corresponding order'.
223 This simplified account of the medieval concept of denomination does not cover all usages, but I believe is at the core of most. Bradwardine has a particularly cavalier attitude to the term. See A. G. Molland, 'An Examination of Bradwardine's Geometry', *Archive for History of Exact Sciences*, 19 (1978), pp. 113-75.
224 In his edition of the work of Thābit, Richard Lorch deals briefly with works on the sector figure by Aḥmad ibn Yūsuf and al-Sijzī, who are especially interesting since they seem to be independent of Thābit, the source of most later knowledge of the theorem. Richard Lorch, *Thâbit ibn Qurra: On the Sector Figure and Related Texts* (Frankfurt, 2001). Jābir ibn Aflaḥ wrote a commentary on the work of Thābit, now lost in its Arabic form, but known from a Hebrew translation by Qalonymus ben Qalonymus. It is summarised by Lorch in the same volume.
225 David King, 'A *Vetustissimus* Arabic Treatise on the *Quadrans Vetus*', *Journal for the History of Astronomy*, 33 (2002), pp. 237-55, makes the useful point that the old horary quadrants, with or without cursor, are universal, in the sense that they yield the time for any latitude, albeit not to the same degree of accuracy at all latitudes. His claim that 'unequal-hour lines' and 'horary quadrant for planetary hours' are modern expressions (p. 237) should be ignored. (As Chaucer has it: 'Understond wel that these houres inequales ben clepid houres of planetes', *Astrolabe*, II.10.) He also misrepresents an old suggestion of mine that the horary quadrant and related markings, as actually found on a large number of astrolabes, were relatively worthless, and were in all probability rarely put to serious use. I was not concerned with what

was theoretically possible. The main astrolabe scales would have given a much greater accuracy than the horary quadrants actually found, the lines of which are carelessly drawn, unnumbered, and small. There are two unanswered (and virtually unanswerable) problems here: one concerns the need which would have been felt for a universal quadrant on an astrolabe; the other is a question of how aware people were of its method of use.

226 'Astrolabe quadrant' is for this reason less ambiguous than the medieval term 'new quadrant'. For the Baghdad evidence on the older quadrants, see David King, ibid. The cursor on the old quadrant (*quadrans vetus*) carries graduations of ecliptic longitudes, and needs to be set for the user's latitude. It is in turn used to allow a sliding bead to be set on the plumbline. The latter is stretched over the day mark, and the bead is moved to the mid-day (semicircular) hour line. When the instrument is tilted so that the shadow of one pinnule falls on the other, the nearest hour-line to the bead will give an idea of the time. For more details, see the basic study in J. M. Millás Vallicrosa, 'La introducción del cuadrante con cursor en Europa', *Isis*, 18 (1932), pp. 218-58. There has been a vast literature since then, much of it well illustrated. An attempt to identify Sacrobosco's sources is made in Knorr, 'Sacrobosco's *Quadrans*: Date and Sources'. For a compact guide to the various quadrants, and illustrations of some Islamic instruments, see Francis Maddison and Emilie Savage-Smith, *Science, Tools and Magic. Part One: Body and Spirit, Mapping the Universe* (Oxford, 1997), pp. 266-70. For the Peter Nightingale text on the new quadrant, alluded to by Whetehamstede, see Fritz S. Pedersen, *Petri Philomenae de Dacia et Petri de S. Audomaro opera quadrivalia*, Corpus Philosophorum Danicorum medii Aevi 10:1-2 (Copenhagen, 1984).

227 There are extant examples of the instrument in Oxford, Cambridge, Greenwich, Geneva, Florence, and Milan. The Science Museum, London, has lost one; and a sixteenth-century example formerly in the British Museum was destroyed during the Second World War. The Milan specimen (in the Museo Poldi Pezzoli) carries the name of Oronce Fine, who published two versions of the instrument in his *De solaribus horologiis et quadrantibus* (Paris, 1560).

228 The instrument is often simply known in medieval English texts as a 'cylinder' (or 'chilyndre', Latin *chilindrum*), while a pocketable example could be called a 'wayfarer's dial' (Latin *horologium viatorum*). In sixteenth-century England the name 'pillar dial' was not uncommon. There are examples in two of Holbein's paintings. See my *The Ambassadors' Secret: Holbein and the World of the Renaissance*. (London, 2002), plates 4 and 6. For more on its use see my *Chaucer's Universe*, pp. 111-16.

229 While different spellings are used, the instrument should not be confused with Ptolemy's *triquetrum*. Its medieval ancestry includes texts by Franco of Polonia and Bernard of Verdun, but they were perhaps derived from related instruments from the Islamic world. The common later spelling *turketum* (cf. *turcus* for 'Turkish') seems to stem from that belief. See my *Richard of Wallingford*, ii, pp. 296-300; and for a possible contribution from Muslim Spain to its evolution see R. P. Lorch, 'The Astronomical Instruments of Jābir ibn Aflaḥ and the Torquetum', *Centaurus*, 20 (1976), pp. 11-34.

230 Those ideas concern the graphical representation of functions of several variables. See below.

231 For a comprehensive survey of equatoria, although one that is arranged thematically and not chronologically, see Poulle, *Les Instruments*. Mercé Comes, *Ecuatorios*

Andalusíes (Barcelona, 1991) is a comprehensive study of the three Andalusian authors; and see also E. S. Kennedy, 'The Equatorium of Abū al-Ṣalt', *Physis*, 12 (1970), pp. 73-81. There is a Spanish translation of the Arzachel text in J. M. Millás Vallicrosa, *Estudios sobre Azarquiel* (Madrid-Granada, 1943-50), pp. 460-79. David King has suggested the tenth-century astronomer Abū Ja'far al-Khāzin as a potential eastern ancestor of the equatorium idea.

232 Ibn al-Samḥ had the clever idea of making the epicycle touch another circle internally, that is, a circle concentric with the deferent and exceeding it in radius by one epicycle radius. Setting the opaque epicycle is then much simplified.

233 For further references to these graphical methods, see Otto Neugebauer, *A History of Ancient Mathematical Astronomy*, 3 vols (Berlin and New York, 1975), pp. 990, 1004.

234 For the Campanus instrument, see Francis J. Benjamin, Jr, and G. J. Toomer, *Campanus of Novara and Medieval Planetary Theory: Theorica Planetarum* (Madison, Wisconsin, 1971).

235 For a short list see my *Richard of Wallingford*, pp. 259-62; for more detail, Poulle, *Les Instruments*; and for the work on the so-called 'semissa' of Peter Nightingale, see F. S. Pedersen, *Petri Philomenae ... opera quadrivalia*, and articles mentioned by him there.

236 See my 'A post-Copernican Equatorium', *Physis*, 11 (1969), pp. 418-57; and 'The Satellites of Jupiter, from Galileo to Bradley', in A. van der Merwe, ed., *Old and New Questions in Physics, Cosmology, Philosophy and Theoretical Physics* (New York, 1983), pp. 689-717.

237 The same principle was applied on the circular equivalent of the logarithmic slide rule, in common use before the advent of electronic calculators. In view of Richard of Wallingford's frequent use of spirals in various connections, it is also worth noting that well-formed spirals constructed in one of the ways he adopted on the albion were a commonplace of medieval architecture, and that there is even an instance of one inscribed by a mason at St Albans. It is in a niche, high on the north side of the tomb of Humfrey, duke of Gloucester. See Jennifer S. Alexander, 'Villard de Honnecourt and Masons' Marks', in Marie-Thérèse Zenner, ed., *Villard's Legacy: Studies in Medieval Technology, Science and Art, In Memory of Jean Gimpel* (Aldershot, 2004), pp. 67-69.

238 See my *Richard of Wallingford*, i, *Albion* I.11 and II.7; and ii, pp. 146-50 and 163-68.

239 It may be considered a curve in polar coordinates where the radius vector is proportional to the sine of the greatest equation of the argument at the chosen angle. See my *Richard of Wallingford*, ii, p. 165, and the references in the previous note.

240 For further information, see my *Richard of Wallingford*, ii, pp. 166-67.

241 This makes for six types of conversion, all told, taking the direction of the conversion into account. For our earlier mention of the problem, see pp. 345 and 349 above. Azarchel and 'Alī ibn Khalaf—both of the eleventh century—designed universal astrolabes in which the projection was stereographic (as on an ordinary, latitude-dependent astrolabe), but in which the centre of projection was the vernal (equinoctial) point and the plane of projection was that through the solstitial points and poles. See the references in J. Samsó, *Islamic Astronomy and Medieval Spain* (Aldershot, 1994), chapter 1, pp. 9-12.

242 For a full account, see my *Richard of Wallingford*, ii, especially at pp. 200-26, and the relevant materials in i and iii to which reference is made there. The eclipse instruments contain a series of eccentric (semicircular) arcs that can be seen as polar graphs of various functions of the various quantities that are needed in calculating an eclipse. The radial scale (see Fig. 82) is a linear scale recording angular dimensions, and the angular argument is the lunar argument during eclipse. The curves in question are not quite what they should be according to Ptolemaic eclipse theory, but the important thing is that Richard of Wallingford was trying to make them so. He was working within the limits set by circular arcs.

243 This realistic view on the limitations of small instruments by comparison with astronomical tables comes where he promises what he will discuss in the second part of his *Treatise on the Astrolabe*.

244 Apian had learned from *Albion* how to introduce any number of variables into the plane, by the successive use of different 'instruments' on the same page: c is a function of a and b; a is a function of c and d; g of a and f; and so forth.

245 For Münster, the radial polar coordinate (set by a seed pearl on a radial thread) was set with the help of an ancillary circular scale of degrees, a sort of epicycle. This first variable (the radius vector) was thus effectively itself a function of another (the 'true argument' in the epicycle). A peripheral scale gave the variable known in astronomy as the 'true centre', while, as a function of these two, the latitude was given by the contour on which the seed pearl fell—or, rather, the imaginary scale which would have been divided by the contours. There was no radial scale, but since the length of the radius vector had been found as a function of an epicyclic argument, the effective radial scale was non-linear. Apian created a linear radial scale of the (true) epicyclic argument. The form of Apian's contours was complicated by yet another fact: he set the 360 degrees of the peripheral scale, that of the true centre, into about 340 degrees of the instrument, leaving space in the other 20 degrees for his radial scale. His were therefore not polar coordinates as we know them. Apian realised that what others might have seen as the distortions of his coordinate scales did not matter in the slightest, if he had at his disposal tabulated information he could plot on the plane of his instrument.

246 See Edward Grant, 'God, Science, and Natural Philosophy in the Late Middle Ages', in Lodi Nauta and Arjo Vanderjagt, eds, *Between Demonstration and Imagination*, pp. 243-68, at p. 263. The four men with the 310 questions were John Buridan, Nicole Oresme, Themon Judaeus, and Albert of Saxony (see pp. 257-58).

Bibliography

Manuscripts

(For comprehensive lists of the numerous manuscripts containing Richard of Wallingford's writings, see my *Richard of Wallingford*, especially ii.)

Barcelona, Real Patrimonio Archivo de la Corona de Aragon, MS N. 2435. Accounts of the building of the Perpignan clock of 1356.
London, British Library, MS Arundel 292, fol. 72v. The poet's complaint of the smithy.
——, MS Cotton Julius D.vii (13c). Astronomy and geography by Abbot John of Wallingford, with tidal material.
——, MS Cotton Nero D.i (13c)—*Gesta abbatum*, etc. See H. T. Riley in the list of printed sources below.
——, MS Cotton Vitellius E.xii (14/15c). Extracts quoted from Richard of Wallingford's lost commentary on the Rule of St Benedict.
——, MS Sloane, 1697. Includes a late sixteenth-century copy of a work deriving from Richard of Wallingford.
London, Royal College of Physicians, MS 384. An early sixteenth-century collection of astronomical writings, mostly in English.
Oxford, Bodleian Library, MS Ashmole 1796. Uniquely includes the clock material and has other works by Richard of Wallingford.
——, MS Laud misc. 264. Works of Anselm, from the St Albans abbot's study, donated by Richard of Wallingford.
A copy of Richard of Wallingford's lost *Computus* was seen by John Leland at Wymondham. See B119.2 in the *Corpus of British Medieval Library Catalogues*, i (London, 1990).

Printed Sources

Averroes, *The Incoherence of the Incoherence*, translated by S. van den Bergh (Oxford, 1954).
Roger Bacon, *Opera quaedam hactenus inedita*, vol. i, ed. J. S. Brewer (London, 1859).
——, *Opus maius* ed. by J.H. Bridges, 2 vols (London, 1900). See part 4, dist. 4, on the application of mathematics to sacred subjects.

——, *Communia mathematica*, in *Opera hactenus inedita*, ed. R. Steele. (Oxford, 1905-40).
John Baconthorpe, *Quaestiones in quatuor libros Sententiarum et Quodlibetales*, 2 vols (Cremona, 1618).
Y. Baer, *A History of the Jews in Christian Spain*, trans. L. Schoffman, vol. 1(Philadelphia, 1966).
C. F. C. Beeson, 'Perpignan 1356 and the Earliest Clocks', *Antiquarian Horology*, 7 (1970), pp. 408-14.
——, *English Church Clocks, 1280-1850* (London, 1971).
——, (posthumous), *Perpignan 1356: The Making of a Tower Clock and Bell for the King's Castle* (London, 1982).
L. Benson, see Geoffrey Chaucer.
John Blair and Nigel Ramsay, eds, *English Medieval Industries: Craftsmen, Techniques, Products* (London, 1991).
T. S. R. Boase, *Boniface VIII* (London, 1933).
A. von Braunmühl, *Vorlesungen über Geschichte der Trigonometrie* (Leipzig, 1900).
——, 'Die Entwicklung der Zeichen- und Formelsprache in der Trigonometrie', *Bibliotheca Mathematica* (3) 1 (1900), pp. 64-74.
George C. Brodrick, *Memorials of Merton College* (Oxford, 1885).
G. Brusa, *L'arte dell'orologeria in Europa: sette secoli di orologi meccanici* (Busto Arsizio, 1978).
Jean Buridan, *Quaestiones super libris quattuor de caelo et mundo*, ed. E. A. Moody (New York, 1970).
——, see also E. A. Moody and A. Ghisalberti.
Charles Burnett (ed.), *Adelard of Bath: An English Scientist and Arabist of the Early Twelfth Century* (London, 1987).
Lluís Camós i Cabruja, 'Dietari de l'obra del rellotge i la campana del castell de Perpinya l'any 1356', in *Homenatge a Antoni Rubió i Lluch : miscel·lània d'estudis literaris, històrics i linguistics*, vol. iii (Barcelona, 1936), pp. 423-46.
Hilary Carey, *Courting Disaster : Astrology at the English Court and University in the Later Middle Ages* (London, 1992).
H. Carterton, 'Does Aristotle have a Mechanics', in J. Barnes, M. Schofield, and R. Sorabji (eds), *Articles on Aristotle*. i, *Science* (London, 1975), pp. 161-74.
J. I. Catto and Ralph Evans, eds, *The History of the University of Oxford*, i: *The Early Oxford Schools* (Oxford, 1984).
——, *The History of the University of Oxford*, ii: *Late Medieval Oxford* (Oxford, 1992).

José Chabas and Bernard R. Goldstein, *The Alfonsine Tables of Toledo* (Dordrecht, 2003).
Geoffrey Chaucer, Collected works ed. L. D. Benson, *The Riverside Chaucer*, 3rd edition (Boston, 1987).
Andrew Clark (ed.), *Survey of the Antiquities of the City of Oxford Composed in 1661-6, by Anthony Wood*, 3 vols (Oxford, 1890).
Mercé Comes, *Ecuatorios Andalusíes* (Barcelona, 1991).
A. C. Crombie, *Robert Grosseteste and the Origins of Experimental Science, 1100-1700* (Oxford, 1953).
A. C. Crombie, *Styles of Scientific Thinking in the European Tradition*, 3 vols (London, 1994).
A. C. Crombie and J. D. North, 'Bacon, Roger', in *Dictionary of Scientific Biography*, ed. C. C. Gillispie et al., i (New York, 1970), pp. 377-85.
Theodore Crowley, *Roger Bacon: The Problem of the Soul in his Philosophical Commentaries* (Louvain and Dublin, 1950).
S. Donati,'Per lo studio dei commenti alla *Fisica* del XIII secolo', *Documenti e studi sulla tradizione filosofica medievale*, 2 (1991), pp. 361-442.
Giovanni Dondi dall'Orologio, *Tractatus Astrarii*. Travaux d'Humanisme et Renaissance, no. CCCLXXI. Édition critique et traduction de la version *A* par Emmanuel Poulle (Geneva, 2003).
Gerhard Dohrn-van Rossum, *History of the Hour: Clocks and Modern Temporal Orders*, trans. Thomas Dunlap (Chicago and London, 1996).
Decima L. Douie, *Archbishop Pecham* (Oxford, 1952).
C. B. Drover, 'A Medieval Monastic Water-Clock', *Antiquarian Horology*, 1 (1954), pp. 54-58 and 63.
Bruce S. Eastwood, *The Revival of Planetary Astronomy in Carolingian and Post-Carolingian Europe*(Aldershot, 2002).
E. L. Edwardes, *Weight-Driven Chamber-Clocks of the Middle Ages and Renaissance* (Altrincham, 1965).
A. B. Emden, *A Biographical Register of the University of Oxford to AD 1500*, 3 vols (Oxford, 1958).
Euclid, see Thomas Heath.
R. Faith, 'The Class Struggle in Fourteenth-Century England', in *People's History and Socialist Theory*, ed. Raphael Samuel (London, 1981), pp. 53-54.
Galvano Fiamma, ed. C. Castiglioni, *Opusculum de rebus gestis ab Azone Luchino et Johanne vicecomitibus* (Bologna, 1938).
Oronce Fine, *De solaribus horologiis et quadrantibus* (Paris, 1560).

Gad Freudenthal, 'Les sciences dans les communautés Juives médiévales de Provence: leur appropriation, leur role', *Revue des Etudes Juives*, 152 (1993), pp. 29-136.

———, 'Science in the Medieval Jewish Culture of Southern France', *History of Science*, 33 (1995), pp. 23-58.

Amos Funkenstein, *Theology and the Scientific Imagination from the Middle Ages to the Seventeenth Century* (Princeton, 1986).

V. H. Galbraith, 'Some New Documents about Gloucester College', in H. E. Salter, *Snappe's Formulary* (Oxford, 1924), pp. 337-86.

Galen, *Selected Works*, trans. P. N. Singer (Oxford, 1997).

Gesta abbatum—see H. T. Riley.

Jane Geddes, 'Iron', in John Blair and Nigel Ramsey, eds, *English Medieval Industries* (London, 1991), ch. 7, pp. 167-88.

———, 'Leighton, Thomas of (*fl.* 1293-1294)', *Oxford Dictionary of National Biography* (Oxford, 2004).

Armin Gerl, *Trigonometrisch-Astronomisches Rechnen kurz vor Copernicus* (Stuttgart, 1989).

D. Gernez, 'Les indications relatives aux marées dans les anciens livres de mer', *Archives internationales d'histoire des sciences*, 28(1949), pp. 671-91.

Alessandro Ghisalberti, *Giovanni Buridano dalla metafisica alla fisica* (Milan, 1975).

Annelies van Gijsen, 'De Middelnederlandse vertaling van John Ashendens *Summa iudicialis de accentibus mundi*', in P. W. M. Wackers et al., *Verraders en Bruggenbouwers: verkenningen naar de relatie tussen Latinitas en de Middelnederlandse letterkunde* (Amsterdam, 1996), pp. 85-111, 291-96.

Owen Gingerich, 'Sacrobosco Illustrated', in Lodi Nauta and Arjo Vanderjagt, eds, *Between Demonstration and Imagination: Essays in the History of Science and Philosophy* (Leiden, 1999), pp. 211-24.

Ruth Glasner, *A Fourteenth-Century Scientific Philosophical Controversy: Jedaiah Ha-Penini's 'Treatise on Opposite Motions' and 'Book of Confutation'* (in Hebrew) (Jerusalem, 1998).

Bernard R. Goldstein, *The Arabic Version of Ptolemy's Planetary Hypotheses* (*Transactions of the American Philosophical Society*, new series, lvii, part 4.) (Philadelphia, 1967).

———, 'Astronomy in the Medieval Spanish Jewish Community', in Lodi Nauta and Arjo Vanderjagt, eds, *Between Demonstration and Imagination: Essays in the History of Science and Philosophy Presented to John D. North* (Leiden, 1999), pp. 225-42.

——, 'An Anonymous Zīj in Hebrew for 1400 AD: A Preliminary Report', *Archive for the History of Exact Sciences*, 57 (2003), pp. 151-71.
——, see also J. Chabás.
Edward Grant, *Much Ado About Nothing: Theories of Space and Vacuum from the Middle Ages to the Scientific Revolution* (Cambridge, 1981).
——, 'The Condemnation of 1277', in his *Studies in Medieval Science and Natural Philosophy* (London, 1981), ch. XIII.
——, *Planets, Stars, and Orbs: The Medieval Cosmos, 1200-1687* (Cambridge, 1994).
——, 'God, Science, and Natural Philosophy in the Late Middle Ages', in Lodi Nauta and Arjo Vanderjagt, eds, *Between Demonstration and Imagination* (Leiden, 1999), pp. 243-68.
Robert Grosseteste, *De luce*, trans. Clare C. Riedl in her *Robert Grosseteste on Light* (Milwaukee, Wisconsin, 1942).
Jeremiah Hackett, ed., *Roger Bacon and the Sciences: Commemorative Essays* (Leiden, 1997).
M. B. Hackett, 'The University as a Corporate Body', in J. I. Catto and Ralph Evans, eds, *The History of the University of Oxford*, i: *The Early Oxford Schools* (Oxford, 1984), pp. 37-95.
A. d'Haenens, 'La clepsydre de Villers (1267): comment on mesurait et vivait le temps dans une abbaye cistercienne au XIIIe siècle', *Sitzungsberichte der Oesterreichischen Akademie der Wissenschaften, Philosophisch-Historische Klasse*, 367, no. 3 (Vienna, 1980).
Willy Hartner, 'Medieval Views on Cosmic Dimensions', in *Mélanges Alexandre Koyré* (Paris, 1964), pp. 254-82.
——, *Oriens-Occidens* (Hildesheim, 1968).
Ibn al-Haytham, *The Optics*, translated with introduction and commentary by A. I. Sabra (London, 1989).
Thomas Heath, *A History of Greek Mathematics*, 2 vols (Oxford, 1921).
——, *The Thirteen Books of Euclid's Elements*, 3 vols (Cambridge, 1926^2).
Herrad of Hohenbourg, *Hortus deliciarum*, ed. Rosalie Green and others, 2 vols (London, 1979).
D. R. Hill, *A History of Engineering in Classical and Medieval Times* (London, 1984).
Roland Hissette, *Enquête sur les 219 articles condamnés à Paris le 7 mars 1277* (Louvain, 1977).
G. A. Holmes, *The Estates of the Higher Nobility in Fourteenth-Century England* (Cambridge, 1957).
R. P. Howgrave-Graham, *Peter Lightfoot, Monk of Glastonbury, and the Old Clock at Wells: A Poem with an Illustrated Account of the Clock* (Glastonbury, 1922).

Chris Humphrey and W. M. Ormrod, eds, *Time in the Medieval World* (York, 2001).

R. W. Hunt, 'The Library of the Abbey of St Albans', in M. B. Parkes and A. G. Watson eds, *Medieval Scribes, Manuscripts and Libraries: Essays Presented to N. R. Ker* (London, 1978), pp. 151-77.

——, , edited and revised by Margaret Gibson, *The Schools and the Cloister: The Life and Writings of Alexander Nequam, 1157-1217* (Oxford, 1984).

Isidore, *Etymologiae*, ed. W. M. Lindsay (Oxford, 1911).

T. G. Jackson, *The Church of St Mary the Virgin* (Oxford, 1897).

Joseph Jacobs, *The Jews of Angevin England: Documents and Records, from the Latin and Hebrew Sources, Printed and Manuscript* (London, 1893).

Claude Jenkins, *The Monastic Chronicler and the Early School at St Albans* (London, 1922).

John of Amundesham, *Annales monasterii sancti Albani* [etc.], ed. H. T. Riley, 2 vols (London, 1870-71).

John Ashenden [here called Johannes Eschuid], *Summa astrologiæ iudicialis de accidentibus mundi quæ anglicana uulgo nuncupatur* (Venice, 1489).

William Chester Jordan, *The Great Famine: Northern Europe in the Early Fourteenth Century* (Princeton, 1996).

J. N. D. Kelly, *The Oxford Dictionary of Popes* (Oxford, 1986).

E. S. Kennedy, 'The Equatorium of Abū al-Ṣalt', *Physis*, 12 (1970), pp. 73-81.

Robert Kilwardby, *De ortu scientiarum*, ed. Albert G. Judy (London and Toronto, 1976).

David King, 'A *Vetustissimus* Arabic Treatise on the *Quadrans Vetus*', *Journal for the History of Astronomy*, 33 (2002), pp. 237-55.

Wilbur R. Knorr, 'Sacrobosco's *Quadrans*: Date and Sources', *Journal for the History of Astronomy*, 28 (1997), pp. 187-222.

David Knowles, *The Monastic Order in England: A History of its Development from the Times of St. Dunstan to the Fourth Lateran Council, 940-1216* (Cambridge, 1963).

David Knowles and R. Neville Hadcock, *Medieval Religious Houses, England and Wales* (London, 1971).

D. S. Landes, *Revolution in Time: Clocks and the Making of the Modern World* (Cambridge, Massachusetts, and London, 1983).

Tzvi Langermann, *The Jews and the Sciences in the Middle Ages* (Aldershot, 1999).

A. T. Lantink-Ferguson, 'A Fifteenth-Century Illustrated Notebook on Rotary Mechanisms', *Scientiarum Historia*, 29 (2003), pp. 3-66.

C. H. Lawrence, 'The University in State and Church', in J. I. Catto and Ralph Evans, eds, *The History of the University of Oxford*, i: *The Early Oxford Schools* (Oxford, 1984), pp. 97-150.

A. Lehr, *De Geschiedenis van het Astronomisch Kunstuurwerk* (The Hague, 1981).

Cees Leijenhorst, Christoph Lüthy and Johannes M. M. H. Thijssen, eds, *The Dynamics of Aristotelian Natural Philosophy from Antiquity to the Seventeenth Century* (Leiden, 2002).

Leonardo da Vinci, *The Madrid Codices*, 4 vols, with facsimile editions of Codex Madrid i and ii and commentaries and transcriptions by Ladislao Reti (New York, 1974). (Published also in other countries with trivial changes to the opening pages. Original Spanish title: *Tratado de estática y mecánica en italiano. Biblioteca nacional, no. 8937; Tratados varios de fortificación, estática y geometría escritos en italiano, no. 8936.*)

David C. Lindberg, *Studies in the History of Medieval Optics* (Aldershot, 1983).

Richard P. Lorch, 'Jābir ibn Aflaḥ and his Influence in the West' (unpublished Ph.D. thesis, Manchester, 1970).

——, 'The Astronomy of Jābir ibn Aflaḥ', *Centaurus* 19 (1975), pp. 85-107.

——, 'The Astronomical Instruments of Jâbir ibn Aflah and the Torquetum', *Centaurus*, 20 (1976), pp. 11-34.

——, *Thâbit ibn Qurra, On the Sector Figure and Related Texts* (Frankfurt, 2001).

Justin McCann, ed. and trans., *The Rule of St Benedict* (London, 1952).

Stephen C. McCluskey, *Astronomers and Cultures in Early Medieval Europe* (Cambridge, 1998).

J. McEvoy, *The Philosophy of Robert Grosseteste* (Oxford, 1982).

May McKisack, *The Fourteenth Century, 1307-1399* (Oxford, 1959).

Bruce Robert Maclaren, *A Critical Edition and Translation, with Commentary, of John Pecham's Tractatus De Sphera* (Madison, Wisconsin, 1978).

Francis Maddison, Bryan Scott, and Alan Kent, 'An Early Medieval Water-Clock', *Antiquarian Horology*, 3 (1962), pp. 348-53.

Francis Maddison and Emilie Savage-Smith, *Science, Tools and Magic*. Part One: *Body and Spirit, Mapping the Universe* (Oxford, 1997).

Manilius, *Astronomica*, ed. G. P. Goold (Cambridge, Massachusetts, 1977).

Moses Maimonides, *The Guide for the Perplexed*, trans. M. Friedländer (London, 1904; repr. New York, 1976).

——, *The Guide of the Perplexed*, trans. Shlomo Pines, with introductory essay by Leo Strauss (Chicago, 1963).

Steven Marrone, *The Light of Thy Countenance: Science and Knowledge of God in the Thirteenth Century* (Leiden, 2001).

C. Martin, 'Walter Burley', in R. W. Southern, ed., *Oxford Studies Presented to Daniel Callus* (Oxford, 1964), pp. 194-230.

G. H. Martin and J. R. L. Highfield, *A History of Merton College* (Oxford, 1997).

José Millás Vallicrosa, *Assaig d'història de les idees físiques i matemàtiques a la Catalunya medieval*, i (all that appeared) (Barcelona, 1931).

——, 'La introducción del cuadrante con cursor en Europa', *Isis*, 18 (1932), pp. 218-58.

——, *Estudios sobre Azarquiel* (Madrid-Granada, 1943-50).

——, 'Translations of Oriental Scientific Works To the End of ther Thirteenth Century', in G. S. Métraux and François Crouzet (eds), *The Evolution of Science* (New York, 1963), pp. 128-67.

——, *Estudios sobre historia de la ciencia española*, i, ed. J. Vernet (Madrid, 1987; 1st ed. 1949).

Lorenzo Minio-Paluello, 'Moerbeke, William of', in C. C Gillispie et al., eds, *Dictionary of Scientific Biography*, ix (New York, 1974), pp. 434-40.

Guillaume Mollat, *Les Papes d'Avignon, 1305-1378* (10th edn, Paris, 1965).

Ernest Moody, 'Buridan, Jean', in *Dictionary of Scientific Biography*, ed. C. C. Gillispie et al., ii (New York, 1973), pp. 603-8.

George Molland, 'An Examination of Bradwardine's Geometry', *Archive for History of Exact Sciences*, 19 (1978), pp. 113-75.

——, *Thomas Bradwardine, Geometria Speculativa: Latin Text and English Translation, with an Introduction and a Commentary* (Wiesbaden, 1989).

John R. H. Moorman, *Church Life in England in the Thirteenth Century* (Cambridge, 1945).

John Muendel, 'Friction and Lubrication in Medieval Europe: The Emergence of Olive Oil as a Superior Agent', *Isis*, 86 (1995), pp. 373-93.

A. R. Myers, ed., *English Historical Documents, 1327-1485* (London, 1969).

Lodi Nauta and Arjo Vanderjagt, eds, *Between Demonstration and Imagination: Essays in the History of Science and Philosophy Presented to John D. North* (Leiden, 1999).

Otto Neugebauer, *A History of Ancient Mathematical Astronomy*, 3 vols (Berlin and New York, 1975).

Peter Newcome, *The History of the Abbey of St Alban from the Founding Thereof in 793 to its Dissolution in 1539* (London, 1793).

——, excerpted and supplemented by E. W. Edgell, *Various Extracts from the History of St Alban's Abbey* (Barnet, 1848).

Rosalind Niblett, *Verulamium: The Roman City of St Albans* (London, 2001).

J. V. Noble and D. J. de Solla Price, 'The Water-Clock in the Tower of the Winds', *American Journal of Archaeology*, 72 (1968), pp. 345-55.

J. D. North, 'Opus quarundam rotarum mirabilium', *Physis*, 8 (1966), pp. 337-72, reprinted in J. D. North, *Stars, Minds and Fate: Essays in Ancient and Medieval Cosmology* (London, 1989), pp. 135-170.

——, 'A Post-Copernican Equatorium', *Physis*, 11 (1969), pp. 418-57.

——, 'Monasticism and the First Mechanical Clocks', in J. T. Fraser and N. Lawrence, eds, *The Study of Time*, ii (New York etc., 1975), pp. 381-98, reprinted in J. D. North, *Stars, Minds and Fate* (London, 1989)

——, 'The Western Calendar—"Intolerabilis, horribilis et derisibilis": Four Centuries of Discontent', in G. V. Coyne, M. A. Hoskin and O. Pedersen (eds), *Gregorian Reform of the Calendar. Proceedings of the Vatican Conference to Commemorate its 400th Anniversary, 1582-1982* (Vatican City, 1983), reprinted in J. D. North, *The Universal Frame* (London, 1989), pp. 39-78.

——, *Richard of Wallingford: An Edition of his Writings with Introductions, English Translation and Commentary*, 3 vols (Oxford, 1976).

——, 'The Satellites of Jupiter, from Galileo to Bradley', in A. van der Merwe, ed., *Old and New Questions in Physics, Cosmology, Philosophy and Theoretical Physics* (New York, 1983), pp. 689-717.

——, 'Celestial Influence: The Major Premiss of Astrology', in P. Zambelli, ed., *'Astrologi hallucinati': Stars and the End of the World in Luther's Time* (Berlin, 1986), pp. 45-100.

——, *Horoscopes and History* (London, 1986).

——, '1348 and All That: Science in Late Medieval Oxford', ed. L. Berggren and B. Goldstein (Copenhagen, 1987) pp. 155-65; reprinted in *Stars, Minds and Fate* (London, 1989), pp. 361-71.

——, *Stars, Minds and Fate : Essays in Ancient and Medieval Cosmology* (London, 1989).

——, *The Universal Frame : Historical Essays in Astronomy, Natural Philosophy and Scientific Method* (London, 1989).

——, *Chaucer's Universe* (2nd edition, Oxford, 1990).

——, 'Natural Philosophy in Late Medieval Oxford', and 'Astronomy and Mathematics', chapters 3 and 4 in Catto and Evans, *The History of the University of Oxford*, i , pp. 65-174.

——, 'Ptolémée', in *Le savoir grec*, ed. Jacques Brunschwig and Geoffrey Lloyd (Paris, 1996).

——, *The Ambassadors' Secret: Holbein and the World of the Renaissance.* (London, 2002).

——, 'Scholars and Power: Astrologers at the Courts of Medieval Europe', *Actes de les VI trobades d'història de la ciéncia i de la técnica* (Barcelona, 2002), pp. 13-28.

——, 'Diagram and Thought in Medieval Science', in Marie-Thérèse Zenner, ed., *Villard's Legacy: Studies in Medieval Technology, Science and Art, In Memory of Jean Gimpel* (Aldershot, 2004).

Günther Oestmann, *Die astronomische Uhr des Strassburger Münsters* (Stuttgart, 1993).

J. Packe, *King Edward III* (London, 1985).

William Page, ed., *The Victoria History of the County of Hertford*, 4 vols (London, 1902-23; reprinted (Folkestone and London, 1971). (ii (1908) covers the Dacorum Hundred and the Cashio Hundred, and so the city of St Albans.)

Howard R. Patch, *The Goddess Fortuna in Mediaeval Literature* (London, 1927; reprinted 1967).

Fritz S. Pedersen, *Petri Philomenae de Dacia et Petri de S. Audomaro opera quadrivalia* (Corpus Philosophorum Danicorum medii Aevi 10:1-2), (Copenhagen, 1984).

——, *The Toledan Tables: A Review of the MSS and the Textual Versions with an Edition*, 4 vols (Copenhagen, 2002).

Olaf Pedersen, *A Survey of the Almagest* (Odense, 1974).

——, 'The Origins of the *Theorica planetarum*', *Journal for the History of Astronomy*, 12 (1981) pp. 113-23.

——, 'In Quest of Sacrobosco', *Journal for the History of Astronomy*, 16 (1985), pp. 175-221.

——, 'The Problem of Walter Brytte and Merton Astronomy', *Archives internationales d'histoire des sciences*, 36 (1986), pp. 227-48.

Carlo Pedretti, *Studi vinciani: documenti, analisi e inediti leonardeschi* (Geneva, 1957).

Petrus Peregrinus de Maricourt, *Opera: epistula de magnete, nova compositio astrolabii particularis*, ed. Loris Sturlese and Ron B. Thomson, Centro di cultura medievale, 5 (Pisa, 1995).

Petrus Peregrinus—see also Sylvanus P. Thompson

Georgius Peurbach, *Theoricae novae planetarum* (Nuremberg, c. 1473).

Amy Phelan, *A Study of the First Trailbaston Proceedings in England, 1304-1307* (unpublished Ph.D. dissertation, Cornell University, 1997).

David Piché, *La condemnation parisienne de 1277. Texte latin, traduction et commentaire* (Paris, 1999).

Emmanuel Poulle, *Les Instruments de la théorie des planètes selon Ptolémée: équatoires et horlogerie planétaire du XIIIe au XVIe siècle*, 2 vols (Geneva and Paris, 1980).
Emmanuel Poulle—see also Giovanni Dondi
Derek J. de Solla Price, 'On the Origin of Clockwork, Perpetual Motion Devices and the Compass', *United States National Museum Bulletin*, 218 (1959), pp. 81-112.
——, see also J. V. Noble.
Ptolemy—see G. J. Toomer.
Roser Puig Aguilar, *Los tratados de construcción y uso de la azafea de Azarquiel* (Madrid, 1987).
Ernest Renan, *Averroès et l'Averroisme* (Paris, 1867).
Clare C. Riedl—see Robert Grosseteste.
H. T. Riley (ed.), *Gesta abbatum monasterii Sancti Albani a Thomas Walsingham ... compilata*, 3 vols, Rolls Series (London, 1867-69).
Charlotte A. Roberts, Mary E. Lewis and K. Manchester, *The Past and Present of Leprosy: Archaeological, Historical, Palaeopathological and Clinical Approaches*, BAR Archaeopress, S1054 (Oxford, 2002).
Eileen Roberts, *The Hill of the Martyr: An Architectural History of St Albans Abbey* (Dunstable, 1993).
C. Roth, *A History of the Jews in England* (Oxford, 1941).
A. I. Sabra—see Ibn al-Haytham.
Saʿīd al-Andalusī, *Book of the Categories of Nations*, translated and edited by S. I. Salem and A. Kumar (Austin, Texas, 1991).
E. Salter, 'A Complaint Against the Blacksmiths', *Literature and History*, 5 (1979), pp. 194-215.
H. E. Salter, *Snappe's Formulary and Other Records* (Oxford, 1924).
Julio Samsó, *Las ciencias de los antiguos en al-Andalus* (Madrid, 1992).
——, *Islamic Astronomy and Medieval Spain* (Aldershot, 1994).
H. R. Schubert, *History of the British Iron and Steel Industry* (London, 1957).
Shlomo Sela, *Abraham Ibn Ezra and the Rise of Medieval Hebrew Science* (Dordrecht, 2002).
Michael H. Shank, '*Unless You Believe, You Shall Not Understand*': *Logic, University, and Society in Late Medieval Vienna* (Princeton, 1988).
André W. Sleeswijk and Bjarne Huldén, 'The Three Water-Clocks Described by Vitruvius', *History and Technology*, 8 (1990), pp. 25-50.
John Smith, *Select Discourses: As also a Sermon Preached by S. Patrick at the Author's Funeral*, ed. J. Worthington (London, 1660).
K. V. Snedegar, 'John Ashenden and the Scientia astrorum Mertonensis' (unpublished Oxford D.Phil. thesis, 1988), 2 vols.

——, 'The Works and Days of Simon Bredon, a Fourteenth-Century Astronomer and Physician', in Lodi Nauta and Arjo Vanderjagt, eds, *Between Demonstration and Imagination. Essays in the History of Science and Philosophy* (Leiden, 1999), pp. 285-309.

R. W. Southern, ed., *Oxford Studies Presented to Daniel Callus* (Oxford, 1964).

——, 'From Schools to University', in J. I. Catto (ed.) , *The History of the University of Oxford*, i: *The Early Oxford Schools* (Oxford, 1984), pp. 1-36.

——, *Robert Grosseteste: The Growth of an English Mind in Medieval Europe* (Oxford, 1986).

Peter Spufford, with the assistance of Wendy Wilkinson and Sarah Tolley, *Handbook of Medieval Exchange* (London, 1986).

Michelle Still, *The Abbot and the Rule: Religious Life at St Albans, 1290-1349* (Aldershot, 2002).

Robert Egerton Swartwout, *The Monastic Craftsman: An Inquiry into the Services of Monks to Art in Britain and in Europe North of the Alps during the Middle Ages* (Cambridge, 1932).

J. M. M. H. Thijssen, *Censure and Heresy at the University of Paris, 1200-1400* (Philadelphia, 1998).

Charles Thomas, *Christianity in Roman Britain to AD 500* (London, 1981).

Silvanus P. Thompson, *Peter Peregrinus of Maricourt, Epistle to Sygerus of Foucaucourt, Soldier, Concerning the Magnet* (London, 1902).

Lynn Thorndike, *The 'Sphere' of Sacrobosco and Its Commentators* (Chicago, 1949).

——, 'A New Work by Robert Holkot', *Archives internationales d'histoire des sciences* 10 (1957), pp. 227-35.

A. G. G. Thurlow, 'The Bells of Norwich Cathedral', *Norfolk Archaeology,* 29(1946), p. 89.

G. J. Toomer, *Ptolemy's Almagest* (London, 1984).

N. M. Trenholme, 'The English Monastic Boroughs', *University of Missouri Studies,* 2 (1927) p. 31.

A. J. Turner, 'The Anaphoric Clock in the Light of Recent Research', in Menso Folkerts and Richard Lorch, eds, *Sic igitur ad Astra: Studien zur Geschichte der Mathematik und Naturwissenschaften* (Wiesbaden, 2000), pp. 536-47.

Malcolm Vale, *The Princely Court: Medieval Courts and Culture in North-West Europe, 1270-1380* (Oxford, 2001).

Graziella Federici Vescovini, 'Michel Scot et la *Theorica planetarum Gerardi*', *Early Science and Medicine* 1 (1996), pp. 272-82.

Victoria History of the County of Hertford, see William Page.

Vitruvius, *Ten Books on Architecture*, trans. Ingrid D. Rowland, commentary and illustrations by Thomas Noble Howe et al. (Cambridge, 1999).

Linda Ehrsam Voigts, 'The *Declararacions* of Richard of Wallingford: A Case Study of a Middle English Astrological Treatise', in I. Taavitsainen and P. Pahta, eds, *Medical and Scientific Writing in Late Medieval English* (Cambridge, 2004), pp. 195-208.

Henry Wansborough and Anthony Marett-Crosby, eds, *Benedictines in Oxford* (London, 1997). Note especially the chapters by Alban Léotaud, James Campbell and Joan Greatrex.

A. Watson, 'The St Albans Clock', *Antiquarian Horology*, 11 (1979), pp. 372, 576.

John Whetehamstede, *Registra quorundam abbatum monasterii S. Albani, qui saeculo XVmo. floruere*, ed. Henry Thomas Riley, 2 vols (London, 1872-73).

L. F. R. Williams, *History of the Abbey of St Alban* (London, 1917).

Marie-Thérèse Zenner, ed., *Villard's Legacy: Studies in Medieval Technology, Science and Art, In Memory of Jean Gimpel* (Aldershot, 2004).

Index

The notes are indexed only where they include more than bibliographical references. (Page references always precede references to notes.) Modern authors (see the Bibliography) are included here only where their views are discussed. Reference to Arabic writers is by short versions of their Arabic names. For simplicity's sake, those beginning with 'ibn' or the definite article are grouped together: al-Ashraf, etc. Cross-reference from medieval Latin equivalents is given when these were widely used.

Many western names are indexed according to modern, and not medieval, preferences. Surnames were rarely stable and self-sufficient before the fifteenth century. Thus Richard of Wallingford was occasionally Richard of St Albans, but never Wallingford. Robert Grosseteste could be Robert of Lincoln, but also Grosseteste, or even Lincoln. Modern historians, however, have made the second names of many (but not all) leading scholars standard, and we follow this rather unpredictable usage. (Typical examples are Aquinas, Ockham, Pecham, Bradwardine, Heytesbury, Buridan, Oresme.) For some of the famous, and most of the less famous, first names remain more reliable.

When monastic status is indicated without qualification (abbot, monk, prior, cellarer, and so forth) the person may be assumed to belong to St Albans. The years of office of St Albans abbots, and of popes, and the regnal years of kings, are all added.

Abba Mari ben Moses, n. 133
Abbasids (Baghdad), 232, 240
Abbo of Fleury, 287
abbot as mesne-lord, 93
abbot's chamber, 223
Abbreviated Almagest, 288-89, 361
Abelard, 33
Abingdon, 112
—, storming of the abbey, 69
Abraham bar Ḥiyya (Savasorda), 240, 288; n. 136
—, astronomical tables of, n. 174
Abraham Ibn Ezra, 240-41, 287; n. 130
absolute power of God, 274-75, 383
Abū Jaʿfar al-Khāzin, n. 231
Abū Kāmil, 236
Abū Maʿshar (Albumasar), 58, 236, 322, 324, 329-30; n. 205
Abū Naṣr Manṣūr, 335
Abū Nawas, 249
Abū'l Wafāʾ, 335
Abū'l-Salt Umayya, 355
Abulcasim, *see* Ibn al-Ṣamh
Adam de Brome, 48
Adam of Buckfield, 265; n. 153
Adam of Orleton, 70-72
Adelard of Bath, 234, 287, 337; n. 124
Æthelbert, 5
Agas, Ralph, 40
Agricola, Georgius, 21, 121, 196

Aḥmad ibn Yusūf al-Miṣrī, 338; n. 224
Aignel, John (not Sir John), 133-35
Aignel, Sir John, 96, 105
Airy, Sir George, n. 9
Aix-la-Chapelle, 285
al-Andalus, *see* Muslim science
al-Ashraf, sultan of Damascus, 163
al-Battānī (Albategni), 344, 361
al-Bīrūnī, 335
al-Ghazālī, n. 169
al-Khwārizmī, astronomical tables, 234-35, 241, 287, 335
al-Kindī, 236, 264, 323
al-Qabīṣī (Alcabitius), n. 205
al-Raqqa, meridian of, n. 174
al-Rāsi, 236
al-Sijzī, Abū Saʿīd, n. 224
Alban, saint, 6-8, 64, 96; n. 2
Albategni, *see* al-Battānī)
Albert of Saxony, 277, 316; n. 246
Albertus Magnus, 168, 239, 242, 265-66, 281, 383; n. 153
Albigensian crusades, 244
albions, surviving, 373-74
albion, the name, 64
—, *see also* Richard of Wallingford
Albumasar, *see* Abū Maʿshar
Alcabitius, *see* al-Qabīṣī
alchemy, 274

Alcuin, 9
Alexander IV, pope (1254-61), n. 49
Alexander of Hales, 242
Alfonsine canons, 303-4
Alfonsine astronomical tables, 171-72, 216, 288, 303-4, 365
—, for Cambridge, 307
—, for Colchester, 307
—, for Leicester, 307
—, for London, 307
—, for Oxford, 307-8
—, Parisian modifications, 303-4, 306-7
—, for York, 307
Alfonso I king of Aragón (alias Alfonso VII of Castile), 233; n. 123
Alfonso X, king of León and Castile, 131, 156, 171, 245, 246, 303, 356; n. 129
Alfraganus, *see* al-Farghānī
Alfred, king of the West Saxons (871-99), 146
Algazel, *see* al-Ghazālī
al-Ghazālī (Algazel), 261
Alhazen, *see* Ibn al-Haytham
Alice of Hakeneye (Hackney), 105
Alice of Pekesdene, 105
Allah's plan for Aristotle, 249
almanac, planetary, 245, 307-8
Almohad persecutions of Jews, 243
almuri, 364
Alnan, St, 112
Alpetragius (al-Bitruji), 296, 312
altitudes, observing with double accuracy, 352
Ambrose, 314
Amphibalus, saint, 8, 104
Anagni, 81
Andaló di Negro, 296
Andronikos of Kyrrhos, 144; n. 71
angels, 314
Angelus and bell-ringing, 193
animals and plants, Aristotle on, 250
Anne Boleyn, queen of Henry VIII, 4
Anselm, 314-15
Antikythera, geared astronomical mechanism from, 163
Apianus, Petrus, 348, 357, 376-79; nn. 244-45
Apollonius, 236, 297
appulse of Venus with Regulus, n. 217
apsides, line of, 305
Aquinas, Thomas, Saint, 36, 44, 239, 242, 272, 277, 281, 313-14, 383
Aquitaine, 73
Aratus, 323
—, *Phaenomena*, 285-86
Archimedes, 163, 225, 236
Archivo de la Corona de Aragón, water-clock drawing, 149

Aristotelian philosophy, prohibitions on teaching, 35-37, 238, 242, 259, 274-75
Aristotelian science in decline, 280-82
Aristotle, 28-30
—, categories, 278, 383; n. 165
—, cause and necessity, 251-52, 310
—, on chance, 252
—, and Christian ideas of God, 251
—, commentaries on, 235, 237
—, continuum, motion, mechanics, 253-55, 261, 275-78; n. 162
—, cosmology and world order, 32, 253, 256, 261-62, 287, 290, 293, 296-97, 308-10, 312-14, 322, 356
—, *De interpretatione*, 252
—, First Mover (*Primum Mobile*), 202, 255, 309, 315
—, *On Generation and Corruption*, 236, 276, 383
—, geometry, 266-67, 272
—, *On the Heavens*, 236, 256, 284, 313, 383
—, in medieval Jewish thought, 241-42; nn. 132, 135
—, introduction to his natural philosophy, ch. 17 (249f)
—, mathematics, 251
—, medieval astrology and, 328-29
—, *Metaphysics*, 253, 277
—, *Meteorology*, 236, 257-58, 263, 276
—, on Nature, 276
—, on his predecessors, 250
—, optics, 260, 263-64, 266
—, Oxford influenced by, 35-37, 237-39, 259, 263-65, 271, 284, 325, 328; nn. 51, 144
—, Paris influenced by, 36-37, 272, 322; n. 24. *See also* Tempier.
—, 'The Philosopher', 230
—, *Physics*, 55, 55, 236, 238, 252, 276-77
—, *Politics*, 60, 60
—, *Posterior Analytics*, 236, 249, 251, 259, 269
—, potential and actual, 254, 274, 261-62
—, Ptolemy a follower of, 289-90, 309
—, rival medieval traditions, 243
—, scientific method, 30, 30, 260, 269-72; n. 159
—, *Sophistical Refutations*, 259
—, *On the Soul* (*De anima*), 259, 263, 383
—, translations of, 29, 230, 237-38; n. 119
—, vacuum (void), matter and place, 254, 275, 312, 317
arithmetic, expertise in, 352
Arles, 242
armillary sphere, 345, 346-47; n. 77
armour, iron and steel, 20
arts as propaedeutic to theology, 271-72
arts, faculty of, 48-50, 268
Ashenden, John, 324, 326-29
Ashmole, Elias, 173

INDEX

aspects, planetary (astrology), 331
astrarium, 178-79, 182
astrolabe, 64-67, 341, 343, 346
—, conceivably derived from anaphoric clock, 150
—, dial on clock, various conventions, 151-52, 164; nn. 72, 82
—, exploded view of its parts, 65
—, horary instrument, 341; n. 225
—, lines for unequal hours, 370, 372; n. 225
—, projection, 66
—, solid, or spherical, 341
—, techniques borrowed for albion, 363, 370, 372-73
—, uses for, 66-67
astrolabe, see also clock
astrolabe quadrant, 174, 245, 333, 341-43; n. 226
astrology, 58-60, 78, 242, 296, 319-32
—, as a natural science, 320
—, biblical interpretation, 323
—, Byzantine, 322
—, determinism, 320
—, Hellenistic, 321
—, in Oxford, 324-32
—, Indian, 321-22
—, mathematics, 327-28
—, personal, 329
—, physics, 322, 328-29
—, politics, 326-27
astronomical motives for mechanical timekeeping, 161-69
astronomical sphere, self-moving, 157-58
astronomical tables, Hebrew, 243
astronomical tables, procedures for using, n. 189
astronomy, 283-317
—, divisions within, 325-26
—, elementary texts, 292-96
—, and mathematics, 333-34
—, planetary, 62-64
—, and religion, 244
Athens, Tower of the Winds, 144
atomism, nn. 146, 147
—, and atheism, 260
—, and hylomorphism, 252-53
Atwood, Seth, n. 90
Augustine of Hippo, Saint, 261, 265, 268, 313-14, 316-17; n. 148
Augustine of Canterbury, Saint, 9
Augustinian (Austin) friars, 46
Augustinian hermits, 43
automata, water-driven, 148-49, 163-64, 168-69
aux (apogee), 304-5, 331; n. 188
Avempace, see Ibn Bājja
Averroes, see Ibn Rushd
Averroism and Christian theology, 47, 237-38. see also Tempier.
Avicebron (Solomon ben Judah ibn Gabirol), 261

Avicenna, see Ibn Sīnā
Avignon, 110
—, bridge, 86-87
—, clock builders, 192
—, clockmakers, 198
—, papal court, 53, 79-90
—, popes, 81
axiomatics, 270, 272
Azarchel, see Ibn al-Zarqāllūh

Babylonian water-clocks, 146
Bacon, Roger, xxii, 47, 156-63, 168, 225, 246, 258-59, 265, 272-73, 277, 295, 312, 325, 331, 340
Baconthorpe, John, 47, 281, 325-26
Baghdad, 343; n. 226
Bājja, see Ibn Bājja
Baldock, 71
Baldwin IV, king of Jerusalem, 101
Bale, John, 5, 204
Balliol College, Oxford, 34, 43
Balliol, Edward, 73
Balliol, John, 17
Baltic grain, 132
Bannockburn, battle of, 73, 101, 324
Barcelona, 242; n. 174
—, astronomical tables, 243
Barnet Wood, 125
Barnwell, 71
Bartholomew the clockmaker (St Paul's, London), 154
Bartholomew the Englishman (Bartholomaeus Anglicus), 314-15
Bassantin, Jacques, 378-79
Batecombe, William, astronomical tables (1348), 307
bearings, materials for, n. 97
Beatrice (in Dante), 315
Beaumont Palace, Oxford, 40, 43
Beauvais clock, 167
Beckett, Edmund, 1st Baron Grimthorpe, 15-16; n. 9
Bede, the Venerable, 9, 206-7, 231, 284-85, 313-14
Bedford, 221
bell, curfew, 220
—, words for, 193
bell-ringing clocks, 167-69
bell-ringing mechanism as potential escapement, 166, 175-76, 182-94 passim
bellows in the smithy, 18, 21-22
—, water-powered, 122
bells, small, in alarm-ringing devices, 149-50
Belvoir priory, 23
Benedict XII, pope (1334-42), 55
Benedict Spichfat, burgess of St Albans, 129
Benedict, saint, 7-8, 51, 147, 315

Benedictine houses (map), 19
Benedictine provincial chapters, 25, 39, 221-22
Benedictine rule, 75, 94, 102-3, 111, 114, 119, 129, 148
Bernard (writer on alchemy), n. 211
Bernard of Treves, n. 211
Bernard of Verdun, 242
Bianchini, Giovanni, 307
Bible and knowledge, 273
biblical cosmology, 283-84
bibliography, 411-23
Big Ben, n. 9
Billingsley, Henry, n. 114
Black Death, 56, 94, 280, 326
Black Monks, *see* Benedictine rule, etc.
blacksmith, *see* smithy
Blake, Thomas, 327
bloomery, 18-19
Blund, John, 259
Boethius, 29, 230, 268, 284, 337; n. 119
Bolingbroke, Roger, 327
Bonaventure, Saint, 36, 271; n. 157
Boniface VIII, pope (1294-1303), 81, 105
Boreman, Richard, 224
Boroughbridge, battle of, 70
Boulogne, 85
Bourges, length of church, n. 4
—, clock, 152-53
Bovelli, Antonio, plombarius and clockmaker, 198-99
Bradwardine, Thomas, briefly archbishop, 54-56, 201-2, 261, 266-67, 276; nn. 147, 155, 164, 202
—, on God's location and attributes (in *De causa dei*), 316-17
—, on geometry, 266
brass or bronze axle bearings, 196
Bredon, Simon, 296, 324, 333, 335, 357; nn. 155, 169, 220
Britten, F. J., 177
Brouncker, William, viscount, 327
Buridan, John, 242, 274, 277-78, 316; n. 246
Bury St Edmonds, Benedictine monastery, 11, 71
Bury, Richard de, 55, 72, 104, 107-8, 110-11, 113
Butler's Farm, n. 45
Byzantine scholars at Charlemagne's court, 285

Cabbala, 319; n. 123
Caen, abbey of St Etienne at, 12
caesium-133, 146
Cairo, 241
Calais, 85
Calcidius, 285
Calculator, *see* Richard Swineshead
calendar, for a queen, 131-32
—, lunar, 147
caliga (tube), 215

Cambrai clock, 154
Cambridge, 30, 35, 52, 71
Campanus of Novara, 239, 267, 291, 338, 345
—, equatorium treatise, 356-57, 77
candle-clocks, 146, 148
canons, iron, 20
Canterbury, 7
—, abbot of St Augustine's abbey, 108
—, automaton, 169
—, cathedral, 12, 18
—, clock, 154, 167, 197
—, province of, 10, 221
—, St Augustine's abbey at, 11
Canterbury College, Oxford, 38
Canute, king of England (1016-35), 17
Cardano, Girolamo, 335
carillons, 188-89
Carmelite Friary, Oxford, 40, 43
Cartesian geometry, 375
Castle Rising, Norfolk, 73
cata conjuncta, cata disjuncta, 339
Catalan astronomical tables, 243
categories (Aristotelian), n. 165
Catherine of Aragon, queen of Henry VIII, 4
Celestial Hierarchy, see pseudo-Dionysius
cellarer, 98
Chaldaean hours, 147
Channel crossing, 85
Charlemagne, 9, 231-32, 285
Charles V, emperor, 178
Charles IV, king of France, 68, 71
—, death of, 90
Chartres automaton, 169
Chaucer, Geoffrey, 46, 105, 202, 286, 323
—, equatorium treatise, 357; n. 29
—, *Treatise on the Astrolabe*, 67, 229, 375; n. 225
Chauntecleer, 202
chevage, n. 55
Chicago, n. 90
Chichester cathedral, 12
chiming barrel on carillons, 188
Chinese clock, n. 83
chords and sines (trigonometrical), n. 176
Christ in Majesty (St Albans mural), 14
chronometer, 146
Church Fathers, 230, 314
Cistercian order, 10
Cistercian rule, and timekeeping, 149
civil war, *see* Edward II, downfall
Clement IV, pope (1265-68), 273
Clement V, pope (1305-14), 81
Clement VI, pope (1342-52), 243
clepsydra, *see* water-clocks
clergy, secular and regular, 27-28
clerks (university students), 27-28

clockmakers, professional, 141-43, 154-56, 166-69, 194-95
clock, anaphoric, 150-53
—, ancillary crafts, 197-99
—, astrolabe display, 150-53, 200, 204, 208-11
—, complexity implies complexity of motives, 165-66
—, dial, 142
—, early Italian, 160
—, frames and wheels of iron, 21, 194-200, 215-16
—, Grand (Vosges) fragment, 151-52
—, Imser's, 378
—, invention of, and social change, xxi, 143, 190-94
—, mechanical, first putative records, 153-54
—, 'public', 191
—, reliability, 199
—, Roman (anaphoric), 151-52, 202
—, St Albans, high cost, 140-41
—, St Albans manuscript, 167
—, Salzburg (anaphoric) fragment, 151
—, as scientific instrument, 202-5
—, as theatre, 169, 218
—, tower, turret, 191
—, Vitruvian type, 150-53
clok, as word describing a bell, 187
clokchambre (at St Albans), 208
Cluniac rule, 149
Cluny, abbey of, 10
Cobham, Thomas, bishop, 48
Codicote mill, 132
colleges, developing Oxford tradition of, 35
comets, 257
—, as portents, 323-24
computus, 231, 284-85
—, disc or ring, 344
—, teaching at Oxford, 346
confession, right to hear, 46, 105-6; nn. 29, 49
conjuration among the monks of St Albans, 99
Constantine the African, 233
constellations, 293
constructibility, proofs of, 267
contrate spiral wheel, 195-97
Conventuals, 82
coordinate systems, astronomical, 347-49
Copernicus, Nicholas, 62, 225, 282, 289, 298, 305, 307
copyhold, 116
Corbie, 286
corrodies, 23, 79
counting devices for bell-striking mechanisms, 186-89
Courtenay, *see* William Courtenay
Creation, 261
Crokesley (Croxley), manor, 80, 90
Crucifixion, darkening of the Sun at, 283, 294

crystalline, 314-15
Ctesibios, 144
curriculum, university, 25, 28, 251, 259, 265-68, 282
cylinder dial, 344-45; n. 228
Córdoba, emirate of, 231-47, passim

Dalembert, Jean, 383
Damascus, 232
Daniel of Morley, 235-36, 323
Dante, 238, 245, 286-87; n. 74
—, clock description in, 154; n. 74
—, on the spheres, 315
Darwinian ideas, 252
David and Goliath, 77-78
David Bruce, 73
day, sidereal and solar, n. 110
de la Mare, *see* Thomas de la Mare
De sphera (*On the Sphere*), *see* Sacrobosco; Grosseteste; Pecham
decans (astrology), 329-30
Dedekind, Richard Dee, John, nn. 114, 222
deferent circle, 299, 301, 304-5
Deffand, Madame du, 383
degrees of qualities, 278-79
demonstration, *quia* and *propter quid*, 270-71
demonstrative knowledge and theology, n. 171
denominations (mathematics), 338; n. 223
Dent, Edward John, n. 9
Descartes, René, 375-76
Despenser, Hugh (father and son), 70-72
determining in arts, 49
determinism, 323
diagrams of planetary motions, early, n. 172
Dionysius the Areopagite, 201
diopters (sighting pinnules), 347
directorium (astrological instrument), 345
Dispenser's Land, 96
distances, planetary, *see* scale of the planetary system
Dohrn-van Rossum, Gerhard, 191-92
Domesday survey, 17, 117, 119, 124-25
domiciles (astrology), 329-30
Dominicans in Oxford, 43-44, 47, 269
Don Vidal Menahem ben Solomon Me'iri, n. 133
Dondi, Giovanni de', 178-79, 182
Dover, 85
Dragon, Head and Tail of (lunar nodes), 216-17, 368
drawings as aids to mathematical manipulation, 337
Dream of Scipio (Cicero), 285-86
Duns Scotus, John, 53, 242, 272, 274
Dunstable priory, 24
—, clock, 153-54, 167, 197
Durham, Benedictine monastery at, 12
—, clock, 167
Durham College, Oxford, 38, 43, 45, 110

Easter calculation, *see* computus
eccentric motion, 298, 301
eccentric of the epicycle of Mercury (curve on albion), 365
eclipse calculation, use of albion for, 367-71
eclipse theory, Ptolemy's, 216-17, 301
eclipse, miraculous (at the Crucifixion), 284, 294
Edmund of Abingdon, 259
Edward I, king of England (1272-1307), 68, 116; nn. 54, 61
Edward II, king of England (1307-27), 48, 68, 110; n. 36
—, downfall, 69-72
Edward III, king of England (1327-77), 68, 110; nn. 33, 36
—, claim to French throne, 73
—, *coup d'état* (1330), 72-73
—, marriage, 72
Edward IV, king of England (1461-83), 68
Edward the Confessor, king of England (1042/3-66), 130
Edward VI, king of England (1547-53), 224
Edward, prince (future King Edward III), 71-72
Edward, the Black Prince, 68, 132
Egyptian hours, 147
eighth sphere, motion of, 333
Eleanor Cross at St Albans, 115; n. 53
Eleanor, duchess of Gloucester, 327
Eleanor of Castile, queen of Edward I, 68, 131, 197
elections, astrological, 321
elements and compounds, language of, 328
elements, Aristotle on, 250-51
elliptical orbits, 211-12
Ely, Benedictine monastery at, 8
—, clock, 154, 197
————, custodian, 167
empyrean heaven, 314-15
ephemeris, *see* almanac, planetary
epicycle, 299, 301, 304-5
epochs of astronomical tables, 303
equant, 301, 304-5
equation (planetary), 304-6, 351
—, of days (related to our equation of time), 365
—, of the argument, 306, 364
—, of the centre, 306
—, of time, 215
equator of lunar velocity (albion curve), 371
equatorium, 63, 345, 162
—, early, 351-58
—, materials for making, 357
—, mechanical difficulties, 352-54
—, Merton College, 352, 357-58
—, 'mirror of the stars' (1319), 359
escapement (mechanical), ideas for, 155-61, 175-85
—, type of the first, 176-78, 182-85
eternity, 275

Euclid, 236, 244, 263-64, 266, 334, 337-38
Eudoxus, 286, 297, 337, 340; n. 222
Evesham Abbey, 279
exaltations (astrology), 329-30
excommunication, 107
Exeter clock, 154, 154, 167, 197
Exeter College, Oxford, 70
Exeter Hall, Oxford, 43
exhalation theory of the weather, 257
experimental method and experience, 203-4, 229, 260, 272-74
experimenters, 158

faces (astrology), 329-30
famine, 132-33; n. 55
al-Farghānī (Alfraganus), 236, 296, 312
Fermat, Pierre de, 376
feudal tenure, 93-94, 115-18
Fez, 241
Fine, Oronce, 341; n. 227
Finzi, Mordechai, 307
firmament, 313-14
Firmicus Maternus, 322
Firmin of Belleval, n. 213
Florence clock, 160, 192
foliot, n. 85
forged charter, 130
Fortuna, Roman goddess, 207
fortune, wheel of, 141-42
Fourth Crusade, 95
Francis de Marchia, 277
Franciscans in Oxford, 43-47, 260, 269
Franco of Polonia, 348
frankpledge, 125-26
Frederick II, emperor, 163, 246
Frederick of Austria, 83
free tenants, 115-16
friars, defence of (1288), n. 49
Froissart, Jean, 177-78, 184
Fulbert of Chartres, 233
Fulk de Spitalstrate, 117
fulling mills, 122-23; n. 57
functional dependence, new instrumental procedures (albion) expressing idea, 364, 372-73

Gabriel, a St Albans town bell, 220
Galen, 236, 259, 263-64
—, his scale of health, n. 166
Galileo Galilei, 225, 277-78, 279, 282, 338; n. 79
—, jovilabium, 357
Galvano Fiamma, 186-87, 193
gear ratios, mathematics of, 172-73, 204-5
—, bevel, spiroid, hypoid, worm, n. 98
Gebir, *see* Jābir
Genesis, Book of, 261-62, 283, 312, 314
—, commentary on, 37

genethlialogy (casting nativities), 321, 329
Genoa clock, 160
geomancy, 323
geometry, algebraic, 375-77
geometry, *see also* quadrivium
Gerald of Wales, 28
Gerard of Brussels, 280
Gerard of Cremona, 235-37, 296, 335
Gerbert of Aurillac, later Pope Sylvester II (999-1003), 232-33, 287
Germanus, bishop of Auxerre, 8
Gersonides, *see* Levi ben Gershom
Gesta abbatum, 5, 42, 63, 74, 89, 99, 101, 122, 136, 140, 142, 171, 223
Gilbert, William, 79
Giles of Rome, n. 195
Giovanni Colonna, cardinal, 90
glans (acorns or mast), n. 67
Glastonbury, hour-striking in, 194
Gloucester Abbey, 39
Gloucester College, Oxford, 38-45, 47, 340; n. 27
Gloucester Hall, Oxford, 39
Gmunden, John of, 367
goblet, the abbot's, 96
God and the void, n. 25
Godstow, n. 12
Gough, Richard, 221
Gower, John, 105
grain, grinding, 121-22
Granada, 244
Granary, a work by Abbot John Whetehamstede, 203, 340-44
Grant, Edward, 383
graphing of functional relationships, 372-73; nn. 242-45
Gravesend, 85
great conjunctions, astrological doctrine of, 323-24
Great Schism, 81; n. 38
Greek and Hebrew at Oxford, 246
Greeks paid to assist in translation, 246
Gregory I, pope (590-604), 9
Gregory IX, pope (1227-41), 35
Gregory XI, pope (1370-78), 81
grey friars, *see* Franciscans
Grimthorpe, *see* Beckett, Edmund
Grosseteste, Robert, 11, 30-33, 37, 47, 381; nn. 25, 26, 28, 29, 145, 146, 148, 159, 204
—, Aristotelian natural philosophy in, 239, 49, 258-60
—, astrology, 59, 323, 328-29, 331
—, astronomy, 201, 239, 293, 314
—, bishop of Lincoln, 32-33
—, ecclesiastical reformer, 295
—, and the Franciscans, 32
—, handling of Greek, 246
—, library of, 32, 260, 295

—, metaphysics of light, 260-62
—, optics, 263, 265-66
—, Oxford 'master of the schools' or chancellor, 31
—, as scientist, 33, 280
—, theology, 260, 269, 272-73, 313
—, value of mathematics, 340
Guide for the Perplexed, see Maimonides
Guy de Foulkes, cardinal, 273

Halidon Hill, battle of, 73
hallmote courts, 118
haloes, rainbows, and mock suns as optical phenomena, 258, 265
Hamadan, 238
hand-mills, 121-23
Harriot, Thomas, 264
Hatfield Peverel Priory, 23
Haward, Peter, n. 90
Haytham, *see* Ibn al-Haytham
heavens (theology), 312-17
—, concentric, in the Aristotelian cosmos, 255-56, 261-62
Hebrew translation of the Batecombe astronomical tables, 307
helical pinions in Richard of Wallingford's clock, 196-97
Hell, 315
Henry I, king of England (1100-35), 43, 287
Henry II, king of England (1154-89), 18
Henry III, king of England (1216-72), 68, 80, 116, 119
Henry IV, king of England (1399-1413), 68
Henry VI, king of England (1422-61 and 1470-71), 327
Henry VIII, king of England (1509-47), 4, 8, 88, 224
Henry, duke of Lancaster, 68
Henry, earl of Lancaster, 68; n. 33
Henry Arnaut of Zwolle, 307
Henry Bate of Mechelen, 239
Henry of Harclay, 53
Henry of Langenstein, 308; n. 192
Henry of Staunton, 34
Henry Prat, poacher, 116
Hereford, 288
heriot, n. 55
Hermann the Dalmatian, 236
Hermann the Lame, 233
Hermes Trismegistus, 202
Hermetic philosophy, 319
Heron, 172
Hertford Priory, 23
Hertfordshire, disenfranchised boroughs, n. 43
—, sheriff, 60, 127
Hesdin, palace at, 168-69
Heytesbury, William, 279-80

Hezekiah, 283
hierarchies, celestial and political, n. 181
hierarchy, love of, 268
Hijra, 303
Hindu-Arabic numerals, 293
Hipparchus, 291-92, 297, 334, 375
historical periods, astrologically determined, 322
holidays for the monks, 104
Hollar, Wenceslaus, 43-44
horary quadrants, universal character of, n. 225
horologe, *horologium*, ambiguity of the words, 3, 145-46
—, *see also* clock; Richard of Wallingford
horologiarii, *see* clockmakers
horologium astronomicum named as an instrument due to Richard of Wallingford, 340, 345
horoscopes, 62, 131-32, 227, 234-35, 375
Hortus deliciarum, 206
hour glass, 346
hour-striking (technical term), 186-87
hours, canonical, of prayer, 147, 165-66
—, division of the day and night into twelve, 146
—, equal and unequal (seasonal), 146-47, 370; n. 225
—, point from which counted, 147, 192-93
Huberd, brother, 46
Hugh of Eversdone, abbot (1308-26), 13, 51, 69, 77-79, 99, 104, 117, 124-26, 141
Hugh of Langley, 98
Hugh ('Master Hugh', Oxford master), 259
Hugh of St-Victor, 268
Humfrey, duke of Gloucester, 195, 308
hundred (county division), 126
hundred court of the abbot's liberty, 119
Hundred Years War, 73
hunting, 104-5
Hurley (Benedictine cell of Westminster), prior, 77
hylomorphism and atomism, 252-53
hypotheses, investigation of, 274-75

iatromathematics, 321
Ibn Bājja (Avempace), 241, 277
Ibn al-Haytham (Alhazen), 236, 244, 264-66, 268, 312
Ibn Rushd (Averroes), 237-39, 244, 257, 259, 261, 265, 277, 281; nn. 30, 171
Ibn al-Ṣamh, 231, 355
Ibn Sīnā (Avicenna), 236, 238, 261, 265
Ibn Tibbon, Jacob ben Machir (Profatius Judaeus), 244-45; n. 139
Ibn Tibbon, Judah ibn Moses, n. 139
Ibn Tibbon, Moses ibn Samuel, n. 139
Ibn Tibbon, Rabbi Samuel, 241-42
Ibn Yūnus, 306
Ibn al-Zarqāllūh or Ibn al-Zarqellu (Azarchel), 235-36, 244, 335, 345, 355-56, 361, 367

illumination, divine, 272
illustrations, list of, xi-xvi
Ilsley, 112
Immanuel Bonfils, n. 136
impetus, 277-78
Imser, Philip, 205, 378-79
incepting as master, 49
Indian water-clocks, 148
inertia, 277
infinity, Aristotle on, 254
Innocent IV, pope (1243-54), 33
Innocent VI, pope (1404-6), 198
instruments, for common and expert use, 341
—, importance of size, 364
interrogations, astrological, 321, 327
Ireland, computus in, 284
iron, and clockmaking, 21, 194-200, 215-16
—, European trade in, 18
—, grille at St Albans, 14
—, working of, 18-22
irrational numbers, 338
Isaac ben Sid (Ishaq ibn Sid, or Rabi Çag of Toledo), 156, 303; n. 75
Isabella of France, queen of Edward II, 69-74, 110, 132
Isabella, mother of Richard of Wallingford, 17, 23
Isaiah, 283
Isidore of Seville, 325
Italian clockmaker in London, 160
Italian geared planetarium, thirteenth-century, 163-64, 204
Italian hours, 147, 192-93
Italy and the invention of the mechanical escapement, 192

Jābir ibn Aflaḥ (Gebir, astronomer), 324, 326, 335
Jābir ibn Ḥayyān (Gebir, alchemist), n. 219
jacks, jacquemarts, *see* automata
Jacques of Cahors (Pope John XXII), 84
Jacob ben David Bonjorn, n. 136
Jacques Duèse, *see* John XXII, pope
Jerome, 111, 314
Jewish hours, 147
Jewish mysticism, 319
Jewry, 24
Jewry, English, n. 138
—, expulsion of (1290), 244
Jews in Spain, 232-35, 240-45
Jews, persecution of, 242-44; n. 129
Joan I, queen of Navarre, 68
John, king of England (1199-1216), 30, 43, 117
John Ball, 118
John Lutterell, 53, 82
John Neckham, 37
John of Brescia, 245

John de Cella, abbot (1195-1213 or 1214), 13; n. 108
John Damascene, 314
John Dumbleton, 279
John of Garland, 35
John of Gaunt, duke of Lancaster, 68
John of Gmunden, 63
John of Hertford, abbot (1235-60), 102, 149-50
John Howe, poacher, 117
John of Jandun, 83
John of Lignères (or Linières), 304, 357
John of London, 35
John of London (fl. 1290), 331
John Loukyn, subsacrist, 174
John of Maryns, abbot (1302-8), 89, 119
John of Muridene, coroner, 128
John of Murs, 304
John of Northampton, 344
John of Reading, 53
John of Saxony, 304
John of Seville, 236
John of Sulsull, 97
John of Tywynge, 98
John of Walden, abbot's chaplain, 128
John of Wallingford, monk, 207
John of Woderove, 98
John Pecham, *see* Pecham, John
John Smith, poacher, 116
John Taverner, alias Marchal, townsman, 126-27
John the Beadle, poacher, 116
John Whetehamstede, alias Bostok, abbot (1420-40 and 1452-65), 42, 203-4, 208, 340-44, 347, 357
John XXII, pope (1316-34), 59, 81-90, 112
Johnson, Samuel, 333, 381
Joshua, 283
jovilabium, 357
jubilee year, n. 49
Judah ben Moses ha-Cohen, 303
Judah ben Verga, n. 136
Jupiter's satellites, equatorium for, 357
jurisprudence (canon and civil law), 29
justiciars of trailbaston (judges), 127

Kepler, Johannes, 211, 266
Keplerian ellipses (planetary orbits), 355
Killingworth, John, 307-8
Kilwardby, Robert, archbishop, 35-36, 268, 281, 325; n. 156
kinematics, 278-80, 338
—, graphical methods, n. 168
—, Oxford advances in, 37
King's Langley, royal palace, 132
knight service, n. 44
Knights Templar, 101
knowledge, nature and acquisition of, 262
Kyngesburye, Thomas, 5

Kyrfoth, Charles, 346

Lafeu (*All's Well That Ends Well*), 319
Lambourne, Reginald, 328
Landes, David S., 164-65
Lanfranc, 12
Langley, manor, 91
languages, acquisition of, in the fifteenth and sixteenth centuries, 246-47
—, learned for missionary purposes, 246
—, of learning, 229
Languedoc, 244
Lateran councils, 11, 221
latitude of forms, 375; n. 169
latitude, theory of planetary, 300-1
Laurence of Stoke, St Albans monk and clockmaker, 141-43, 167, 169, 197
lay brothers, 23
Le Goff, Jacques, 192
lectures on astronomical calculation, 308
Lee, Sir Richard, 224
Leland, John, 3-5, 8, 101, 173, 204, 207, 213
Lemaître, Georges, 262
Leonardo da Vinci, 184-85, 189-90, 194
Leopold of Austria, 329; n. 213
leprosy, 99-102, 111-12
—, medieval rituals connected with, 101
Levi ben Gershom (or Gerson), 243; nn. 135-36
Lewis of Caerleon, 216, 327
leywrite, n. 55
Limia, 236
limits (astrology), 329-30
Lincoln, bishop of, 112
—, diocese of, 31
Lionel, duke of Clarence, 68
Lisbon astronomical tables, 243
Little Langley mill, 131-32
Littlemore, nuns of, 48
Liuprand of Cremona, 168
Llull, Ramon, 246
locking lever, 185, 187
lodestone, 159
logarithmic slide-rule, spiral, n. 237
Loggan, David, 41, 54
logic, 282
—, syllogistic, 271
Loire, 86
Lombardy, iron crown, 84
London Bridge, 87
—, *see also* tides and tidal theory
London clock at Old St Paul's, 154, 168, 197
London astronomical tables, 287
longitude, planetary, 304-5
Lord of the Year, 329
Lorraine, schools of, 233
Louis X, king of France, 68

lubricants, n. 97
Lucera, 159
Lucretius, 201
Ludwig IV, emperor, 83-84
lunar mansions, 322
lunar nodes, 216-17
lunar theory, *see* Moon
lunar velocity, astrological concern for, 370
Luton mill, 132
Luttrell Psalter, 134
Lydgate, John, 8
Lyons, 86

machina mundana, n. 103
machina mundi, universe as cosmic machine, 161-69, 201-2
Macrobius, 285
magic, 59, 274, 383
Magna Carta, 125
magnetism and perpetual motion, 158-60
Maimonides (Rabbi Moses ben Maimon), 241-42
Malden, Surrey, 34
Malmesbury, abbey, 39, 41
—, monk of, 186
malt, manufacture of, 132
—, mill, 132
Manilius, 227, 322
mannikins, *see* automata
manuscripts, 411
—, Archivo de la Corona de Aragón, Barcelona, MS Ripoll 225, n. 122
—, Biblioteca Ambrosiana, Milan, Codex Atlanticus, 184-85; n. 89
—, Biblioteca Civica, Padua, MS 631, 179
—, Biblioteca Nacional, Madrid, MS 8937 ('Codex Madrid'), 184-85, 189-90; n. 89
—, Bodleian Library, Oxford, MS Ashmole 1796, 173-76, 186, 349-50
———, Oxford, MS Digby 176, 328
—, British Library, Cotton Claudius E.iv, 5-6
———, Cotton Nero D.7, 170
—, Brussels, Bibliothèque Royale, MS 10117-26, 173
—, Gonville and Caius College, Cambridge, MS 116, 173
Margaret Newman, 116
Margaret of France, queen of Edward I, 68
Marie of Luxembourg, queen of France, n. 213
Markyate Hermitage, cell of St Albans, 23
Marrakech, 237
Marriage of Philology and Mercury (Martianus Capella), 285-86
Marseilles, astronomical tables, 288
—, Jewish scholars of, 242
Marsilius of Padua, 83
Martianus Capella, 285-86, 314

Māshāʾallāh (Massahalla), 174
Master G., chaplain of Merton College, 166
mathematics, as instrument, 333-34
—, as the language of nature, 273, 340
—, ordinary language and notations, 336-37
mathesis, n. 1
Matilda, empress, 17-18
Matthew of Aquasparta, n. 157
Matthew Paris, 6, 13, 23, 45
Maudith, John, 56, 58, 174, 333, 336
Maurolico, Francesco, 266
mechanical clock, possible motives for creating, 161-69
mechanical drawing in MS Asmole 1796, 175-76
memory and learning, 57-58
Menaechmus, 376
mendicant friars, *see* Franciscans, Dominicans
Menelaus, 236, 290-91, 297, 334-35, 337; n. 177
—, theorem of, 338-39
merchet, n. 55
mercury clock, 156-57, 163
Mercury, apparent path in the year 2003, 297
—, deferent of, technique adopted for albion, 363-64
—, oval deferent, 355, 363
—, simplified Greek model to account for its motion, 299
Merle (or Morley), William, 328-29
Merlin, 323
Merton College, Oxford, 34-35, 38, 43, 54-55, 63, 307, 324; n. 28
—, clock, 154, 166
—, equatorium, and lost disc of epicycles, 358-59
Merton mean speed theorem, 279
Merton priory, 34
Mertonian kinematics, 279-80
Mesopotamian omens, 320
Michael Bryd, 123
Michael of Cesena, 83
Michael of Mentmore, abbot (1336-49), 111, 142, 221
Michael Scot, 312
Milan, church of San Eustorgio, clock, 154
—, church of San Gottardo, automaton and clock, 160, 169, 187, 191, 193
Milky Way, 257
mill machinery, wooden, 194, 196
mill types, breast-wheel, 120-21
—, overshot, 120-21
—, undershot, 120
mills, 119-36; nn. 56-57
millwrights, 194
miracles, 319, 383
—, and Aristotle on chance, 252
Modena clock, 191
monasteries, dissolution of, 4-6, 224

monastic alarms, 148-49
monastic towns, 94-95
monetary equivalents, n. 102
monitoring of time as a motive for clock design, 165
monopolies, 119
Monte Cassino, 8, 233
Monte Mario Observatory, Rome (owner of an albion), 374
Montpellier, 242
Monza clock, 160
Moon ball, rotating, on Richard of Wallingford's clock, 196, 211, 216-18
—, complexities in Ptolemaic theory, 369-70
—, theories of (Hipparchus and Ptolemy), 292
Moor mill, 132
mortmain, 116
Moscow, hour-striking in, 194
motus (mean and true), 305
Muchelney, Benedictine abbey, 344
multiplication of species, 262, 265
multure, 123
Muslim science in Spain (al-Andalus), 231-47, 354-56
Muslim Spain, astronomy in, 62, 136
mycobacterium leprae, 100
Münster, Sebastian, 378-79; n. 245

nails for Christ's crucifixion, 19
Nāṣir al-Dīn al-Ṭūsī, 335
navicula, 343-44
Neckham, Alexander, 29
Nevers, 86
Newcome, Peter, 75, 173; n. 67
Nicholas IV, pope (1288-92), n. 103
Nicholas of Autrecourt, 274
Nicholas of Flamstede, prior of Hertford, 80, 103, 112-13, 128, 135-36
Nicholas of Lyra, n. 130
Nicholas V, disputed pope (1328-30), 84
Nigel Niger, Æthelwine Swart (Black, Niger), 130
nocturnal (timetelling instrument), 286
Northampton, 91
—, centre of study rivalling Oxford, 28
—, parliament (1328), 91
—, treaty of, 72
Northumbria, n. 205
Norwich, priory clock, 142, 153, 160, 167-68
Notker of St Gall, 287
Nottingham Castle, 73, 80, 109
novices, 24
numerology, 323
Nun Hall, Oxford, n. 28
Nuremberg clock, 182

obedientiaries, 98, 104-5
oblation, 23

occult, 273, 319
occultation of Aldebaran by the Moon, n. 217
Ockham, William, 53, 82, 263, 274, 281, 316
Offa, king of Mercia, 5-6, 8
oneiromancy, divination through dreams, 323
Oporto, cardinal bishop of, 88
optics, 268, 274
optics and theology, 265
optics, theories of, 262-66
Orange, astronomical tables, 243
Order of St Lazarus, 101
ordination as priest, age of, n. 32
Oresme, Nicole, 165, 375; n. 246
—, graphical methods in kinematics, n. 168
organ (*organum*), 167
Oriel College, Oxford, 48
Orleans, 86
Orvieto clock, 160, 191
oscillations, forced, 178-79
Oseney (Oxford), n. 12
Otto, cardinal and papal legate to England, 97, 221
Ottobon, cardinal and papal legate (Ottobono Fieschi, later Pope Hadrian V, 1276), 97, 221
oval contrate wheel, of 219 teeth (Richard of Wallingford's clock, early trial), 215
—, of 331 teeth, 205, 212-14
oval curves on the albion, 364-65
Oxford, 295
—, arts faculty, 50, 271
—, beginnings of the university, 27-30
—, castle, 18
—, church of St Mary the Virgin, 26, 48
—, Congregation House, 26, 48
—, curriculum, 32-37, 47-50, 251, 265-66, 268, 281, 308, 346
—, division into northerners and southerners, 32
—, Dominican scholarship in, 33, 70
—, foreign students, 29
—, lectures and disputations, 49
—, Lincoln diocese, 31
—, maps of, 44-45
—, parliament (1258), 81
—, as a *studium generale*, 34
—, theology the highest faculty, 32, 51-53
—, Thomism, n. 26
—, town and gown quarrels, 30-32
—, university library, 26
—, university life, 27-50

Padua, 281-82
—, clock, 160
—, hour-striking in, 194
papal chancery, 88
Paris and its university, 28-30, 31, 32, 33, 52-53, 268, 272, 295, 308, 316, 325
Paris, churches of, 85-86

Park Manor, court book of, 98, 116-17; n. 55
Park mill, 122, 132-33
parliamentary system and precedence, 74, 107-9
Parma clock, 160, 191
Paul of Caen, 12-13, 15, 23, 139
Paul, apostle, 14, 147
Peasants' Revolt (1381), 119
Pecham, John, archbishop, 34, 36, 47, 258, 265-66, 294-96; nn. 26, 49, 180-83
Pelagius, 201
pendulum escapement, 178
Pere IV, king, 198
perpetual motion, 154-60
Perpignan, 242
—, clock builders, 192
—, clock, building of, 198-99
perspective (*perspectiva*), *see* optics
Peter Lombard, *The Sentences*, 33, 314; n. 19
Peter Nightingale (Peder Nattergal), 343, 357
Peter of Corbara (Pietro Rainalducci, Pope Nicholas V), 84
Peter of Mucheleyo (Muchelney), 344
Peter of St-Omer, 245
Peter, apostle, 14
Peterborough Abbey, 11
Petrarch, 89, 110
Petrus Alfonsi (formerly Moshe Sefardi), 233-34, 240, 287
Peurbach, Georg, 307, 311-12
Philip III, king of France, 68
Philip IV, king of France, 68, 81
Philip V, king of France, 68, 68
Philippa of Hainault, queen of Edward III, 68
Philo, 144
Philobiblon, 110
Philoponus, John, 277
philosophers, 137
Piccolomini, Alessandro, 256
Pierre de Maricourt (Petrus Peregrinus), 158-60, 168, 202, 274
pin-wheels (part of escapement), 179-81
Pisa, 241
Pits, John, 204
planetarium, n. 81
—, early, 163-66
planetary calculation using tables, 302-8
—, models, integrating, 354
—, motions, theories of, 292, 296-301
—, paths, apparent, 297-300
—, spheres (shells), 309-14
Plantagenet, royal house of (genealogical table), 68
Plato, 230, 261, 313, 337
—, on motion, 253-4
—, *Timaeus*, 239, 285
Plato of Tivoli, 240
Pliny, 284-85, 323, 324

—, *Natural History*, 231
poaching, 116, 124
points, a cosmology of (Grosseteste), 260-61
—, creation of a line from, 261
Pontefract, monks of, 70
Pontius Pilate, 136
Porphyry, n. 119
—, *Isagoge*, 230
Posidonius of Apamea, 163, 257
post-mills, 133-35
Poulle, Emmanuel, 205
poverty, Franciscan debate on, 82
Prague and Silesian copies of the Batecombe astronomical tables, 307
prayer cycle, 147, 165-66
Price, Derek de Solla, 164
Proclus, 239
Profatius Judaeus, *see* Ibn Tibbon, Jacob ben Machir
Profatius quadrant, 174, 245, 333, 341-43
prophecy, desire for, 322-23
proportion and proportionality, theories of, 337-38; n. 222
proportionality, medieval terminology of, 338
Provence, 244
—, as staging post for eastern science, 243-44
pseudo-Dionysius, 265, 295, 314-15
Ptolemy, 334, 340, 344, 361; n. 205
—, accepts much of Aristotle's cosmology, 289-90, 309
—, *Almagest*, 58, 62, 217, 235, 211, 216, 225, 273, 287-92, 324, 326, 335, 339, 347
—, *Analemma*, 289
—, armillary sphere, 347-48
—, *Geography*, 289
—, *Handy Tables*, 287, 356
—, 'King', 303
—, life, 289
—, *Optics*, 263-64, 289
—, *Planetary Hypotheses*, 289, 309, 313, 354; n. 193
—, *Planisphere*, 289
—, *Tetrabiblos* (or *Quadripartitum*), 59, 321-22, 325, 327
Purgatory, 315

quadrant, horary (with cursor), 341-43; nn. 225-26
—, 'of Israel', 245
—, new, 174, 245, 333, 341-43
quadratura plumborum (part of escapement), 179
quadrivium (arithmetic, geometry, astronomy, music), 49, 282
querns, 122-23
Quintillian, 111
quodlibetal questions, 49

Rabanus Maurus, 323
Ragusa (Dubrovnik) clock, 160, 191

Ralph of Hengham, 166
Ralph of Sempringham, 33
Ramon Sans, clockmaker, 198
Rashdall, Hastings, 273
rationalism, 272
Ravenna, 287
Raymond of Marseilles, 341
Reading, 30
reality of the astronomers' circles, 308-9
Recorde, Robert, 374
rectangulus, *see* Richard of Wallingford
Redbourn, 96, 104; n. 65
—, bondsmen of, 129-30
—, mill, fire at, 124
—, priory, 23
Rede, William, 63, 307
refraction, laws of, 264
refutation in science and philosophy, 259
Regiomontanus, Johannes, 63-64, 307, 334-35
Reichenau, Benedictine monastery, 233
religious sects, rise and fall of, 323-24
retreats in London and Great Yarmouth, 104
retrograde motion, 298
Rheims, 287
—, clock, 167
Rhône, 86
Richard de Bury, *see* Bury, Richard de
Richard de Mediavilla (Middleton), 274
Richard Fitzralph, 55
Richard I, king of England (1189-99), 43
Richard II, king of England (1377-99), 118, 216, 327
Richard III, king of England (1483-85), 68
Richard of Bordeaux (later King Richard II), 327
Richard of Hetersete, 98
Richard of Ildesle (or Hildesley), 112-14, 135
Richard Kilvington, 54-55
Richard of Paxton, archdeacon, 80
Richard Swineshead, 225, 279
Richard of Tring, confessor, 77, 80, 97
Richard, prior of Tynemouth, 96-97
Rickmansworth, n. 65
Rico y Sinobas, Manuel, 157
Robert of Baldock, 70-71
Robert the Bruce, 73, 101
Robert of Chester, 236
Robert, clockmaker at Norwich, 197
Robert Graystanes, 54
Robert Holcot, 54-55, 326
Robert of Lymbury, 124
Robert the Mason, 12
Robert of Norton, prior, 114
Robert Stikford, monk, 344-45
Robert of York (Perscrutator), 323-24, 328-29
Robertus Anglicus, 147, 154-55, 159, 168
Rocamadour, 86

Rocher des Doms, Avignon, 86
Roger of Hereford, 288, 296
Roger Mortimer, 70-71
Roger of Norton, abbot (1260-91), 117, 123, 131
Roger of Stoke, clockmaker (layman?), 141-43, 167, 197
Roger of Wendover, 6
Rolle, Richard, 50
Roman astronomical texts, 284-85
— astronomical traditions, 230-31
— calendars, 231
— surveyors, 375
— tiles at St Albans, 13
Rome, 287, 295; n. 183
Rome, A., 347
Rome, P., 347
rood-screen as site of Dunstable clock, 153
Royal Society, president of, 327
Russell, Bertrand, 245
Richard of Wallingford, abbey finances, building and repairs, 139-41, 223
—, adoption by William of Kirkeby, 22-25
—, agricultural experiment, 133; n. 67
—, albion, and treatise on, 60-64, 162, 171, 211, 216, 245, 303, 309, 327, 340, 345, 351-79
————, eclipse calculation, 367-71
————, eclipse instruments, n. 242
————, general arrangement of scales, 360
————, spiral scales, 361, 371, 373
—, astrology, 77
—, beginning of infirmity, 90
—, builds gatehouse, 140
—, *Calendar* for a queen, 131-32
—, *Canons* on John Maudith's tables, 56-57, 336
—, censure of the obedientiaries, 98-99
—, character, 222-23
—, clock, 136, 145-218 passim
————, escapement, 175-85
————, treatise, 172-75
————, eclipse mechanism, 217
—, computus, 231
—, coordinate transformation, methods for, 348-49, 366-67; n. 241
—, dating of his abbacy, n. 40
—, *De sectore*, 57, 340
—, death, 73, 221
—, depicted with clock, 170
—, design for St Albans clock, 171-200
—, early life and education, 17-25
—, election as abbot, 77-79
—, *Exafrenon*, 58-59, 78, 131, 327-32
————, in English translation, n. 21
————, audience for, 331-32
—, fosters school education, 140
—, gives books to de Bury, 110-11
—, intellectual isolation, 382

—, journey to Avignon, 79-90
—, as lawyer, 96
—, leprosy, 99-103, 111-13, 136, 171
—, ordained deacon and priest, 51
—, oval contrate wheel, of 219 teeth (early trial), 215
———, of 331 teeth, 212-14, 364
—, Oxford study and teaching, 47-50
—, portraits (imaginary), 171-72
—, proceeds bachelor of theology, 52
—, *Quadripartitum*, 57-58, 60, 171,173, 290, 333-40
—, on the reality of theoretical entities, 309-10
—, rectangulus, and treatise on, 60-61, 203, 340, 345-51
—, *Sentences* commentary, 280, 382
—, tomb, 15, 221
—, uses a universal astrolabe projection, 64, 366-67
—, version of Benedictine rule, 97, 221-22
—, visitation of his monastery, 97-99
—, visitations of St Albans cells, 104
—, vows taken, as Benedictine monk, 51
Rushd, *see* Ibn Rushd (Averroes)

sacristan and sub-sacristan, 98
Sacrobosco, John of, 154, 201, 284; n. 178
—, treatise on the quadrant, 341-42
St Alban Hall, Oxford, 48; n. 28
St Alban, Robert de, n. 28
St Albans, monastery, 7-16
—, abbey church (plan), 76
—, abbey debts, 79, 139
—, abbey gatehouse, 15, 140; n. 8
—, abbey mills, 123-24, 131-32
—, abbot, feudal obligations, 95, 115-19
———, London house, 109
—, burgesses in parliament, 95
—, clock, 145-218 passim, 171-200, 331
———, dial convention, 151-52
———, escapement, 175-85, passim
———, striking mechanism, 185-90
———, tower, 14; n. 94
———, display for positions of Sun and Moon, 209-18
———, long time of beat, 210
———, position of, 208
———, possible planetary display, 204-5
———, wheel of fortune on, 206-8
—, collapse of part of church (1323), 16, 139
—, commune, attempt to establish, 117-18
—, concessions exacted by townsmen, 69
—, dependent priories (*see also* Belvoir; Hatfield; Hertford; Redbourn; Tynemouth; Wallingford; Wymondham), 23
—, dimensions of the church, 13
—, at Dissolution, 224

—, gatehouse as town school, n. 8
—, hospital of St Julian, 101; n. 46
—, ironwork in presbytery, 195
—, library, 111; n. 51
—, market, 115
—, mills and hand-mills, 119-36
—, nave, length of, 14
—, Norman period of building at, 12-13
—, paintings in the church, 13-14
—, parliament (1295), 95
—, pilgrimage to, 14
—, prosperity of, 117
—, retreats in London and Great Yarmouth, 104
—, Richard of Wallingford's clock depicted, 170
—, riots, 117-18, 124-25, 132
—, ritual of installation of new abbot, 149-50
—, shrines to St Alban and St Amphibalus, 15-16, 104; n. 47
—, townsmen's seal and liberties, 129
—, visit of Queen Isabella, 74
St Albans School, n. 8
St Basil, n. 148
St Bernard's College, Oxford, 38
St Bénézet, 87
St Frideswide's, Oxford, n. 12
Saint-Germain-en-Laye museum, 152
St German's hospital, 133, 140
St John of Jerusalem, prior of, 108
St John's College, Oxford, 40, 43
St Mary's College, Oxford, 38
St Paul, 201
St Paul's cathedral, London, 13, 119
—, automaton, 169
St Peter's parish, St Albans, 142
St Scholastica's Day, 1355 (town and gown riots in Oxford), n. 14
St Swithun's College, Oxford, 38
Saladin, 241
Salamanca, astronomical tables, 243
Salerno, medical school of, 233
Salisbury clock, 154
Salzburg clock, nn. 72-73
San Fernando, father of Alfonso X, 303
Sandon windmill, foundations for, 134
Sandwich, 85
Santa Maria de Ripoll, Benedictine staging post for science, 232-33
—, monastic alarm, 148-49, 232
Sasanian Iran, 322
Saturn, Jupiter and Mars, conjunctions of, *see* great conjunctions
Savasorda, *see* Abraham bar Ḥiyya
Savile, Sir Henry, 225
sawmills, 122
scale of planetary system, 311-12
Schöner, Johannes, 307, 368-69

science, as hostile to religion, 243
—, changing meaning of, 229-30
—, as religious obligation, 242
—, and Talmudic learning, 244
—, links with theology, 269, 281, 312-17, 383; n. 171; *see also* Tempier, Etienne
sciences, division of, 267-68
scientific progress supposed hindered by theology, 381-82
scientific system, overarching, 258, 269
Scots, treaty with France, 73
Scott, Sir George Gilbert, n. 9
scripture rarely used for scientific ends, 383
seasons, 294
sector figure, 339
seisin, 96
Selden, John, 166
semicirculus (part of escapement), 179-82
Sentences, by Peter Lombard, 37, 51-53, 271, 278-79
Seneca, 136
sensation in the eye, place of, 263-64
Servius, 324
Seville, 237
Sforza, Ludovico, duke of Milan, 184
Shakespeare, William, 319
shawms ('mules'), 149
shooting stars, 257
shuttling motion, Leonardo's mechanism for, 190
signs (zodiacal) of 30 and 60 degrees, 307; n. 190
Silas, 147
Simon the leper, 99
Simon Tunsted, *see* Tunsted, Simon
Smith, John (d. 1652), 201
smithy at Wallingford, 18-22
smithy, noise and smoke from, 22
social control, the clock as, 219-20
social unrest (1327), 69
Solomon ben Adrat, n. 133
sophisms, 280
Soto, Domingo, 280
Southcote, John, 166
Southwell, Thomas, 327
space-filling figures, 267
Speaker of parliament, 108
species, sensible, 263
spheres, celestial, *see* heavens, concentric; planetary spheres
spheres, number of, 315
spherical astronomy, 284, 345, 347, 351
spiral tracking for hour-striking, 188-89
spirals, uses for, n. 237
spiritualities and temporalities, n. 47
Spirituals, 82-83
spiroid gears, 197
spring, n. 91

spring-powered clocks, 145
Stacy, John, 327
Stankfield mill, 132
Stapledon, Walter, bishop of Exeter, 329
star catalogue, 365
star polygons, 267
stars, said to 'incline but not compel', 320
—, located on a sphere, 292
steel, production of, 19-20
Stephen, king, 18
Stoic philosophy, 290
Stoke, place name, 142-43
Stoke, *see also* Laurence of Stoke; Roger of Stoke
Stonehenge, 334
Strasbourg clock, 167
striking barrel on the St Albans clock, 188
strob (part or whole of escapement), 180, 182
Sun marker on clock dials, 152-53
Sun, Greek model for apparent motion, 298
sundials, 344-45
Super cathedram (papal bull), n. 49
Sussex ironwork, 15
Sutton, bishop of Lincoln, 33-34
Swarte-smeked smethes, n. 11
Swein (or Swegn) Forkbeard, king of Denmark, and of England (1013-14), 17
Swineshead, Richard, *see* Richard Swineshead
swords, steel, 20
syphilis, 101

Tables of the Prince (Savasorda), 288
tables, astronomical, 302-8
Tabulae resolutae, 307
tallage, 129; n. 55
Tarascon astronomical tables, 243
taxation, n. 55
Technisches Museum, Vienna, 205
teleology, 252
Tempier, Etienne, bishop of Paris, and his prohibitions of 1277, 36-37, 39, 47, 238, 242, 268, 274, 281, 317; nn. 22-24
Terence, 111
terms (astrology), 329-30
Thābit ibn Qurra, 291, 294, 338-39; n. 224
Thales, 60
Thames, river, 20, 45, 85
Thames valley area, map of, 19
The Sentences, *see* Sentences
Themon Judaeus, n. 246
Theodoric of Freiburg, 266
Theorica planetarum (*Theory of the Planets*, various authors), 296, 312, 356; nn. 169, 184
Third Crusade, 134
Thomas Aquinas, *see* Aquinas, Thomas
Thomas Bradwardine, *see* Bradwardine, Thomas

Thomas of Leighton (Leighton Buzzard), blacksmith, 197-98
Thomas de la Mare, abbot (1349-96), 38, 42, 130, 141
Thomas of Boningdone, monk, 102
Thomas Walsingham, 5-6, 28, 52, 78,
Thomas, earl of Lancaster, 55, 68, 70; n. 33
tides and tidal theory, 3, 59, 206-7, 320, 331
Time Museum, Rockford, Illinois, 200, 209; n. 90
timekeeping, non-mechanical, 161-62
tithes, 98, 129
Toledan astronomical tables, 235, 288, 303-4, 339, 365
Toledo, 235, 236, 323
Tolpade mill at Cashio, 133
Toreham mill, 123
torquetum, 348-50; n. 229
Toulouse, astronomical tables, 288
—, university of, 35
Tower of London, 216, 327
Tower of the Winds, Athens, n. 71
townsmen's revenge (1381), 135-36
trailbaston, n. 61
translation (especially in Muslim Spain, Provence, Constantinople, and Sicily), 235-40, 244-45; n. 187
—, technique of double, 245
Trieste clock, 160
trigonometrical identities, 335
trigonometry, 56-57, 243, 290-92
—, definitions of, 334
—, histories of, 334
—, Indian, 334, 336
Trinity College, Oxford, 38
trip-hammer, 21
triplicities (astrology), 329-30
triquetrum, n. 229
Triton, 144
trivium (grammar, rhetoric, dialectic or logic), 48-49, 282
tuberculosis, 101
tubes, in axles of mechanical devices, 164, 189, 205, 215-16
Tunsted, Simon, provincial minister of the Franciscans, 63, 350, 365, 367, 374
Turkey, 245
two truths, doctrine of, n. 171
Tyburn, 73
Tycho Brahe, 282, 314
Tynemouth, astronomical table for, 365
—, prior, 77,96
—, priory, 23
—, journeys to, 96
Tyttenhanger, 224

Udall, Nicholas, 224

universal astrolabe (*saphea Arzachelis*, Azarchel's plate), 245, 345, 366-67
universities, continental, 28
university chairs of Greek, Hebrew, Arabic, and Chaldaic, 246
University College, Oxford, 34, 43
Urania, muse of astronomy, 341
urban society and the mechanical clock, 191-92
Uzbekhistan, 238

Valenciennes clock, 191
variable velocity drive for the Sun's position (Richard of Wallingford's clock), 211-16
Varro, Marcus Terentius, 268; n. 71
Venice, little ship of, 343-44
Ver, river, 12, 104, 121
verge and foliot, 176-79
Verulamium, 7, 12
Vicenza clock, 160
Vick, Henri de, 178
Vienna, 308
Vienne, council of (1311), 246
view of frankpledge (court), 126
Villard de Honnecourt, 155-56, 169
villeinage, 94, 115-18
virgaters, 116
Virgil, 111, 324
Virgin Mary, 14, 77, 85, 169, 315
Virginia and its Dissenters, n. 67
Visconti, Azzo, duke of Milan, 193
Visigothic rulers, 240
vision, the psychology of, 266
—, theories of the nature of, 263
Viterbo, 295; n. 183
Vitruvius, 151, 344
void, *see* Aristotle

Walcher of Malvern, 233-34, 287
Wallingford, 17-25, 72
—, Benedictine priory, 22-25
—, castle, 18, 20
—, Holy Trinity church, 23
—, and Oxfordshire, n. 10
—, Richard of, *see* Richard of Wallingford
Walsingham, Thomas, *see* Thomas Walsingham
Walter of Amundesham, abbot's marshall, 127, 129
Walter Burley, 55, 281
Walter Chatton, 53
Walter de Merton, bishop of Rochester, 34, 48
Walter of Odington, 279
Walter of Stapledon, 70
Walters, John, 326
Wat Tyler's rebellion, 118
water power, 20-21
water-clock, 144, 146-53
—, depicted in French moralised Bible, 149

Watford poachers, 124
Watling Street, 11-12
wayfarer's dial, n. 228
weather, explaining and forecasting, 58-59, 78, 257, 324, 328-29, 331
weights, lead for clock, 198
Westminster Abbey, 11
—, abbot of, 108
Westminster Hall clock, 166
Wheathampstead (or Whetehamstede) mill, 123
wheel of fortune, 206-8
Whetehamstede, *see* John Whetehamstede
Wigginton windmill, near Tring, 133
William I (William the Conqueror), king of England (1066-87), 17, 220
William of Auvergne, 157-58, 242
William de Brock, 39
William Courtenay, archbishop of Canterbury, 38
William Cumnor, abbot of Abingdon, 113
William, count of Hainault, 71
William Heyron, prior of Wallingford, 77, 79
William Heytesbury, *see* Heyetesbury, William
William of Kirkeby, prior of Wallingford (adoptive father to Richard of Wallingford), 22-25
William of Moerbeke, later archbishop of Corinth, 238-39, 295
William Montagu (or Montacute), 72-73, 109-10
William of Nedham, sub-prior, 102
William of Ockham, *see* Ockham, William
William of Pikewell, clockmaker in London, 154
William Rede, *see* Rede, William
William, father of Richard of Wallingford, 17-23
William Rishanger, 6

William of Trumpyngtone, abbot (1214-35), 104
William Walsham, monk, 169
William of Winslow, kitchener, 92, 102
Winchelsea, 85
Winchester, 241
—, astronomical tables, 287
—, cathedral, 12-13
windmills, 119, 133-34
winds, n. 205
Windsor Castle clock, 160
Witelo, 239, 265-66, 295; n. 182
Wolsey, Thomas, cardinal, 23, 224
women and the smithy, 18-19
Wood, Anthony, n. 28
Woodstock, royal residence, 19, 221
wool, embargo on export, 73
Worcester College, Oxford, 39
world as sponge in an infinite sea, 317
Wroxeter, 12
Wycliffe, the Lollards, and Grosseteste, n. 18
Wymondham Priory, 23, 77
—, prior, 77

Yazdijird, Persian era, 303
York Minster, 12
York, monastery at, 12
—, province of, 11, 221

Zacut (or Zacuto), Abraham, n. 136
Zarqāllūh, *see* Ibn al-Zarqāllūh
Zeno, 278
Zeno's paradoxes, 253-54